International Conference on Modern Technologies and Materials of New Generations (MTMNG-2017)

IOP Conference Series: Materials Science and Engineering
Volume 286

Tomsk, Russia
9 – 13 October 2017

ISBN: 978-1-5108-5501-4
ISSN: 1757-8981

Printed from e-media with permission by:

Curran Associates, Inc.
57 Morehouse Lane
Red Hook, NY 12571

Some format issues inherent in the e-media version may also appear in this print version.

Printed by Curran Associates, Inc. (2018)

For permission requests, please contact the Institute of Physics
at the address below.

Institute of Physics
Dirac House, Temple Back
Bristol BS1 6BE UK

Phone: 44 1 17 929 7481
Fax: 44 1 17 920 0979

techtracking@iop.org

Additional copies of this publication are available from:

Curran Associates, Inc.
57 Morehouse Lane
Red Hook, NY 12571 USA
Phone: 845-758-0400
Fax: 845-758-2633
Email: curran@proceedings.com
Web: www.proceedings.com

Table of contents

Volume 286

The International Conference "Modern Technologies and Materials of New Generations"

9–13 October 2017, Tomsk Polytechnic University, Tomsk, Russia

Accepted papers received: 14 December 2017
Published online: 2 January 2018

Preface

Papers

IOP Conf. Series: Materials Science and Engineering **286** (2017) 011001 doi:10.1088/1757-899X/286/1/011001

Preface

The International Conference "Modern technologies and materials of new generations"" (MTMNG-2017) was held at Tomsk Polytechnic University, October 9-13, 2017, Tomsk, Russia. The conference is aimed at attracting young researchers to the current problems of the development of materials based on new ideas about the nature of the solid body behavior under external influence. A complex approach to the description of a material behavior under external influences contributes to understanding the laws of multilevel nonlinear processes in hierarchical condensed systems. This approach is one of the promising directions in the development of new materials and technologies for their processing.

The conference proceedings include the accepted papers discussing problems and methods for creating new materials used for a variety of purposes: structural, instrumental, materials for electronics, medicine, construction materials, surface hardening and protective coatings.

The mains topics of the conference were as follows:

- *Problems of strength, ductility and fatigue life of advanced structural materials*
- *Surface hardening and protective coatings*
- *Additive technologies for obtaining new materials*
- *Silicate and refractory nonmetallic materials from natural and technical raw materials*
- *Photonic Materials and technologies*

The Organizing Committee would like to thank Institute of Strength Physics and Materials Science SB RAS (Tomsk, Russia), Russian Material Science Center (Tomsk, Russia) and National Research Tomsk Polytechnic University (Tomsk, Russia) for their participation in the conference organization and helpful peer review of the papers.

The Organizing Committee of MTMNG-2017

Prof. Sergey Panin

Institute of Strength Physics and Materials Science SB RAS, Tomsk, Russia

Tomsk Polytechnic University, Tomsk, Russia

IOP Conf. Series: Materials Science and Engineering **286** (2017) 011002 doi:10.1088/1757-899X/286/1/011002

Peer review statement

All papers published in this volume of *IOP Conference Series: Materials Science and Engineering* have been peer reviewed through processes administered by the proceedings Editors. Reviews were conducted by expert referees to the professional and scientific standards expected of a proceedings journal published by IOP Publishing.

IOP Conf. Series: Materials Science and Engineering **286** (2017) 012001 doi:10.1088/1757-899X/286/1/012001

Computer simulation of low-temperature composites sintering processes for additive technologies

A O Tovpinets, V N Leytsin and M A Dmitrieva

Immanuel Kant Baltic Federal University, Kaliningrad, 236016, Russia

E-mail: tovpinets_a@mail.ru

Abstract. This is impact research of mixture raw components characteristics on the low-temperature composites structure formation during the sintering process. The obtained results showed that the structure determination of initial compacts obtained after thermal destruction of the polymer binder lets quantify the concentrations of main components and the refractory crystalline product of thermal destruction. Accounting for the distribution of thermal destruction refractory product allows us to refine the forecast of thermal stresses in the matrix of sintered composite. The presented results can be considered as a basis for optimization of initial compositions of multilayer low-temperature composites obtained by additive technologies.

1. Introduction

A relevant task is to research technological methods of sintering of multilayered low-temperature ceramic. Additive manufacturing technologies using these materials are distinguished by layer-by-layer material growth and subsequent sintering. The heterogeneity of particle sizes and the density of their packing during the formation of initial compacts are technological parameters of layers packing. And the formation of interlayer interfaces should be taken into account. The under study model structure of the initial compact should reflect the heterogeneity of concentrations and sizes of original powder components and pores in the consolidated layers and interlayer interfaces.

The computer modeling approach is developed to study the determining factors of low-temperature composites structure formation with considering the hierarchy of rheological processes in powder bodies and modification of hierarchically organized structures. The physical mesomechanic approach was taken in this research. It is combining the ideology and tools of solid state physics and mediums micromechanic with a structure. [1]. Local processes and materials characteristics are simulated based on the classical results.

The initial compact preparation regimes, low-temperature ceramics sintering, and the effect of structural-mechanical characteristics on residual strains in the sintered material are examined.

The initial compact of low-temperature ceramics is formed layer by layer from a mixture of components with certain porosity. Material matrix is forming during the composite sintering by the low-melting mixture raw components in a temperature lower than its melting point. The effective characteristics of the sintered ceramics are determined by the macroscopic structure of the initial compact, components and pores distribution and concentrations inhomogeneity of the phase and fractional initial components composition.

2. Methods of modelling

IOP Conf. Series: Materials Science and Engineering **286** (2017) 012001 doi:10.1088/1757-899X/286/1/012001

The raw polydisperse mixture material of the initial compact refractory elements is presented by a set of particle sizes and volume fractions of each set in a mixture of refractory components. Coarse and microheterogeneous particles of the initial powder refractory components are different in sizes and referred as "large" particles because of ability to form the interacting particles power frame. Ultra-microheterogeneous refractory particles are attached to the dispersion phase forming and modifying its characteristics. The prediction of forming the refractory particles framework on macro-, meso-, or micro levels determines possibility to forecast the minimal sintered body porosity in accordance with the conditions for achieving mechanical contact of the fraction particles. It depends on the type of discrete distribution. The forming possibility of refractory components mixture frame at any structural level decreases with a local refractory components proportion decrease (and with increasing low-melting component proportion). The forming possibility of refractory components frame at any indenture level makes a key contribution to the pore structure of the sintered sample formation [2].

Gas-phase decomposition products and a crystalline residue are forming during the binder's thermal destruction. The initial porosity of the green compact is increasing after removing the gas-phase.

The crystalline residue of thermal destruction is an extra refractory phase in the sintered samples matrix. This leads to the appearance of thermal residual stresses in the matrix caused by the difference in the thermal expansion coefficients of the disperse compositions and the matrix.

The following expression can be used from the position of thermo elasticity to estimate the residual thermal stresses in the sintered ceramic matrix:

$$\sigma_T = \int_{T_0}^{T_p} K_{glass}(T) \left[\frac{\beta_{glass}(T) C_{glass} - \beta_{ceramics}(T) C_{ceramics}}{C_{glass} + C_{ceramics}} + \frac{\beta_{glass}(T) C_{glass} - \beta_{SiO2}(T) C_{SiO2}}{C_{glass} + C_{SiO2}} \right] dT, \quad (1)$$

where K_{glass}(T) is the temperature dependence of the elastic bulk modulus of glass; β_{glass}(T), $\beta_{ceramics}$(T), β_{SiO2}(T) are the temperature coefficient dependences of thermal matrix and dispersed refractory components expansion; C_{glass}, $C_{ceramics}$, C_{SiO2} are the volume concentrations of the matrix and dispersed refractory components; T_p is the temperature point of transition to an elastic stage, T_0 is the initial temperature.

Calculation of residual thermal stresses should take into account the stage of viscoplastic stress relaxation when the sintered ceramics are cooled to laboratory temperature. In [3], it is proposed to introduce the temperature transition point of the material into the elastic state T_p in order to estimate the residual thermal stresses. The assumption is made of an abrupt transition to an elastic state, with allowance for which it can be assumed that in the temperature range from the sintering temperature to the temperature point of the transition, the temperature stresses relax to a negligible level. It is assumed that the transition to an elastic state stops the relaxation of residual stresses, and the evaluation of thermal stresses is possible from the position of thermoelasticity. For borosilicate glass we can use the temperature value Tp = 900 K, indicated in Figure 1 as the point of transition to the elastic state.

The type of the residual stresses (tensile or compressive) is determined by numerator signs in integrand in the temperature range T_p - T_0. It is assumed that for temperature values $T \geq T_p$ the values of temperature stresses $\sigma_T \approx 0$.

3. Results and discussion

Numeric evaluations were made using the component's material characteristics in [4-7].

The sintered ceramic sample model structure forming conditions were examined. The initial compact is formed by three successively applied layers. Two compositions of the initial mixture, differing in the mass content of components are considered to study the factors determining thermal stresses in the matrix of sintered composite.

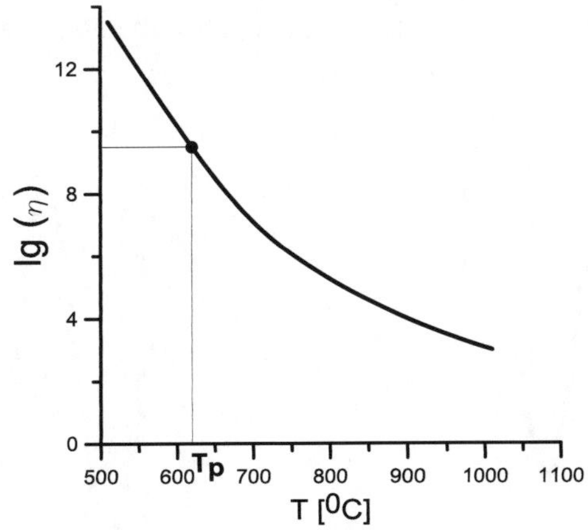

Figure 1. Temperature dependence of the viscosity of borosilicate glass.

Figure 2 shows the distributions of the volume concentrations of mixture components by the thickness of the model sintered sample at different mass compositions of the components in the initial compact:
a) αAl_2O_3 – 40 mass.%, borosilicate glass – 40 mass.%, polyorganosiloxane (the binder) – 20 mass.%,
b) αAl_2O_3 – 30 mass.%, borosilicate glass – 50 mass.%, polyorganosiloxane (the binder) – 20 mass.%.
The porosity and the standard deviation of the components volume concentrations in the initial compact are equal (=10%) in both cases.

Areas bounded by dashed lines represent the interlayer interfaces of the simulated layers. They are characterized by high content of refractory component αAl_2O_3 and low content of sintered composite matrix which is borosilicate glass. A - B intervals are the external boundaries of the simulated periodicity cell, which is developing in each phase of additive initial compact preparing process. The model periodicity cell consolidates over layer and the interlayer interface.

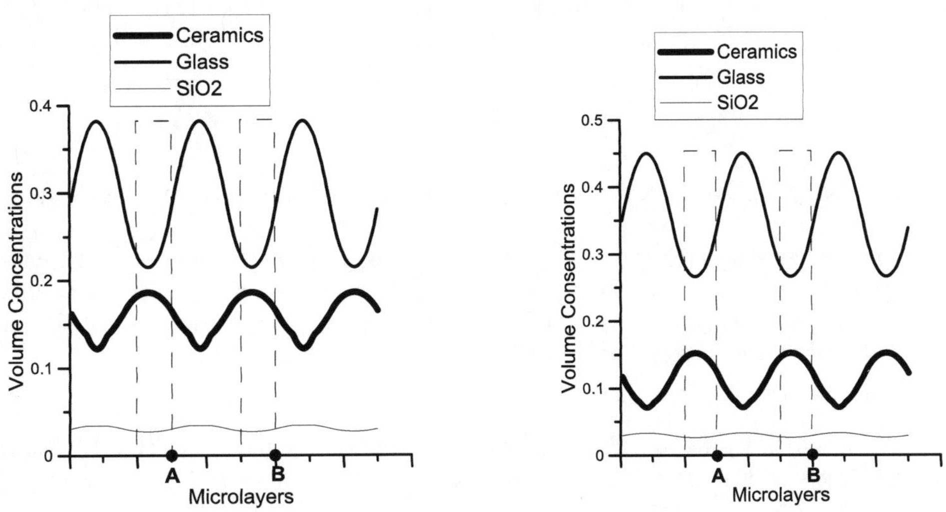

a) b)

Figure 2. Distribution of volume concentrations of sintered ceramics components by the thickness of the modeled layer.

The prediction of thermal stresses in the matrix of simulated samples is made using the relation (1) at the next step of computational experiments.

IOP Conf. Series: Materials Science and Engineering **286** (2017) 012001 doi:10.1088/1757-899X/286/1/012001

Figure 3 shows the residual thermal stress distributions for compositions (a) and (b) by the thickness of the model layer.

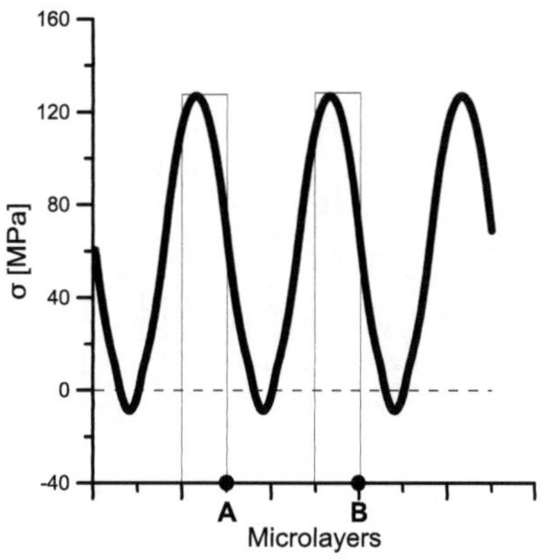

a)

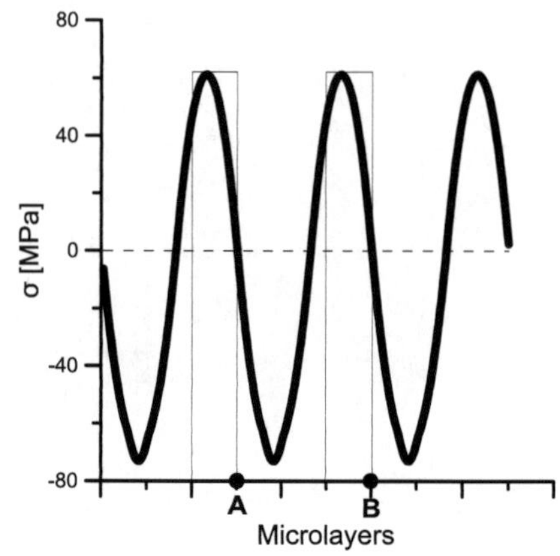

b)

Figure 3. Distribution of residual thermal stresses in the matrix of sintered ceramics by the thickness of the modeled layer. Mass content of components in the initial compact: a) αAl_2O_3 – 40 mass. %, borosilicate glass – 40 mass.%, polyorganosiloxane (the binder) – 20 mass.%, b) αAl2O3 – 30 mass.%, borosilicate glass – 50 mass.%, polyorganosiloxane (the binder) – 20 mass.%.

The local tensile stresses are observed in interlayer interfaces (areas bounded by thin lines). The presence of significant tensile stresses exceeding the strength limit can cause defects and cracks in model composite matrix. As follows from the results shown in Figure 2, it is possible to change residual stresses ranges and to accomplish the strength criteria by varying the mass fraction balance. The tensile stresses for composition (b) in the interlayer interface are 2 times less than for the composition in (a). In addition, the significant local compressive thermal stresses in the ceramics sintered from composition (b) are predicted in the deposited (additive) layer, which prevent the development of defects.

4. Conclusion

The computer simulation of binder thermal destruction, which is the initial stage of composite sintering process, makes it possible to obtain a prediction of the sintered samples structure and residual thermal stresses.

Analysis of the results shows the possible appearance of both tensile and compressive quantities of residual stresses in the matrix of sintered ceramics. This can lead to destructive or hardening effects in the resulting products. Changing of value and residual stresses sign in the sintered material is possible by optimizing the concentration composition of initial compact components.

References

[1] Leitsin V N, Dmitrieva M A and Kolmakova T V 2008 Governing Factors of Physical and Chemical Behavior of Reactive Powder Materials *Powder Metallurgy Research Trends* (New York: Nova Science Publishers)

[2] Leitsin V N, Dmitrieva M A, Tovpinets A O, Ivonin I V and Ponomarev S V 2016 Analysis of the low temperature ceramics structure with consideration for polydispersity of initial refractory components *Advanced materials with hierarchical structure for new technologies and reliable structures*: *Proceedings of the International Conference on Advanced Materials with Hierarchical Structure for New Technologies and Reliable Structures* (Tomsk) pp 146-147

[3] Chernyshev G N, Popov A L, Kozintsev V M and Ponomarev I I 1996 Residual stresses in deformable solids *Nauka* (Moscow) p 240
[4] Imanaka Y 2005 Multilayered low temperature cofired ceramics (LTCC) technology *Springer science business media* p 205
[5] Saito K 1981 Use of Organic Materials for Ceramic-Modeling Process-Binder, Deffloculant, Plasticizer, Lubricant, Solvent, Protective Colloid *J. of Adhesion Society of Japan* **17** **3** 104-113
[6] Kamehara N, Imanaka Y, and Niwa K 1987 Multilayer Ceramic Circuit Board with Copper Conductor (Denshi Tokyo) **26** 143- 148
[7] Tummala R.R. et al. 1992 High performance glass-ceramicl copper multilayer substrate with thin-film redistribution *IBM J. Res. Develop.* **36** **5** 889-904

IOP Conf. Series: Materials Science and Engineering **286** (2017) 012002 doi:10.1088/1757-899X/286/1/012002

Increasing durability of bearings by the application of technological operations on the basis of cyclic elastic deformation of the rings

A V Korolev, A A Korolev, A F Balaev, A S Yakovishin and A A Korolev

Yuri Gagarin State Technical University of Saratov, 77 Politechnicheskaya str. Saratov, 410054, Russia

E-mail: yakovishinas@gmail.com

Abstract. The appearance and growth of fatigue microcracks during the relaxation of residual stresses under the action of high-cycle loads is one of the main reasons of decreasing of durability of products. To reduce residual stress of the bearing rings, there was developed a technology on the basis of the cyclic elastic deformation instead of using energy-consuming thermal tempering. The technology consists of continuous running of the rings between three rollers under the load. A key feature of the technology is the controllability of cyclic loading of parts. The conducted study of an experimental batch of rings proved the reduction of residual stresses by more than 2 times in comparison with the average value of residual stresses after thermal tempering.

1. Introduction

Durability is one of the most important performance indicators of the quality of products which are made of structural materials. Therefore, provision of the required durability of structural materials under high-cycle loading is extremely relevant. As is known [1], up to 94% of the causes of machine and structure failures are related to the appearance and propagation of cracks. The majority of the cracks, leading to equipment failure, are caused by fatigue damage in parts and components that are subject to high-cycle dynamic loads. For example, among the failures of aircraft engines up to 60% have mechanical causes, 80% of which are associated with the accumulation of fatigue damage and propagation of fatigue cracks under vibration [1].

Rolling bearings are products, widely used in various types of mechanical engineering that are susceptible to high-cycle loading during operation. Therefore, to ensure the required durability of bearings, one have to meet particularly high requirements. The emergence and propagation of cracks in the bearing rings and reduction of their service life are the result of residual stresses arising in technological processes of forming and heat treatment of the part. Residual stresses are concentrated at the grain boundaries of the material, and this accumulation of residual stresses leads to the appearance of microcracks [2]. Subsequent cyclic loading of the bearing rings in the course of their operation leads to accumulation of internal energy and microcrack propagation.

To reduce residual stresses in bearing rings, they are subjected to thermal tempering, which is traditionally carried out via heating and slow cooling before the final operations of the technological process. Thermal methods of geometry stabilization, having low performance and relatively high energy consumption, do not allow complete relaxation of residual stresses. Therefore, under the guidance of A.V. Korolev there was developed the technology of residual stress relaxation in the bearing rings on the basis of cyclic elastic deformation of products by their centerless running [3].

2. Technology description

The technology of residual stress relaxation consists in a continuous running of the rings between three rotating rolls under the action of external load, which ensures elastic deformation of the ring in the contact area with the rolls. A gradual accumulation of internal energy in the zones of residual stresses occurs under the action of cyclic elastic strain of the rings during running. During this accumulation of internal energy of the dislocation formations, there occurs shear plastic strain of the micrograins of the material, which is followed by the relaxation of residual stresses.

The controllability of cyclic elastic strain is a key feature of the technology. On the one hand, the magnitude of the energy input to the elastic strain per cycle should be sufficient for movement, integration and subsequent collapse of intergranular dislocations, accompanied by the relaxation of internal stresses. On the other hand, when a large amount of energy imparted per loading cycle, there occur the collapse of a critical number of dislocations with the emergence of cracks in the region of localization and the propagation of existing microcracks. As was noted above, the last phenomenon is the cause of the reduction of product life.

To implement the conditions of the rings loading, the authors developed a model [4] for the calculation of rational technological modes of processing:

$$\tau = \frac{c_P}{c_o} \cdot \frac{\sigma_m^2 \cdot K_P}{\sigma_{mo}^2 \cdot n} \cdot \frac{\int_{\frac{\alpha}{2}}^{\theta_P} J_M(\theta)^2 d\theta + \frac{4k_\sigma^2 \cdot h^6}{W^2} \int_{\theta_P}^{0} J_P(\theta)^2 d\theta}{\int_{\frac{\alpha}{2}}^{0} J_M(\theta)^2 d\theta} \quad (1)$$

where is the processing time; $J_M(\theta)$ and $J_P(\theta)$ are the introduced notations (they are not shown due to their inconvenience) [14]; σ_m is the maximum stress of elastic deformation of the part, determined without taking into account plastic deformation; σ_{mo} is the maximum stress occurring in the material of the ring during processing; c_o is the number of contact areas between rolls and the part; θ is the central angle of the cross-section; θ_P is the angle at which the zone of elastic strain transits to the plastic one; α is the angular pitch of the points of application of external load P (c_o); K_P is the ratio of energy absorption; the value cP should be taken depending on the error of the geometric shape of the ring: if second harmonic prevails in the ring, then $c_P = 2$; if there prevails third harmonic, then $c_P = 3$; h is the distance from the central axis of the cross section to the extreme point of the cross section; W is the moment of resistance to bending; n is the frequency of the ring rotation.

3. Experiment description

To assess the effectiveness of this technology, the authors in cooperation with the bearing enterprise, JSC "EPK Samara", conducted experimental research. Forty five outer rings of the bearings 7000806 (GOST 8338-75) were selected as the object of research. The rings were numbered and divided into five batches of 9 rings each:

1) the first batch of the rings was used as a control one and was not subjected to treatment;

2) the second batch was subjected to tempering in a furnace at the temperature of t=200°C for 2.5 hours with cooling in open air;

3) the third batch of the rings was subjected to centerless running (CR-1) in the following mode: the frequency of rotation of the ring is n=240 rpm, the processing time is T=5 min, the technological deformation of the rings is Δ=0.016 mm;

4) the fourth batch was subjected to centerless running (CR-2) in the following mode: the frequency of rotation of the ring is n=240 rpm, the processing time is T=3 min, the technological deformation of the rings is Δ=0.030 mm;

5) the fifth batch of the rings was subjected to centerless running (CR-3) in the mode: the frequency of rotation of the ring is n=350 rpm, the processing time is T=3 min, the technological deformation of the rings is Δ=0.016 mm.

To determine the residual stress, each ring was cut in the center plane. The magnitude of residual stresses was calculated by the magnitude of the ring deformation after cutting, which was defined as the difference of the diameter of the ring in the plane perpendicular to the cutting plane, before and after cutting. According to the results of measuring each batch of the rings, there was carried out calculation of the mathematical expectation (ME) and standard deviation (SD) of the values of deformation and residual stress of the rings, which is shown in Figure 1 and Figure 2.

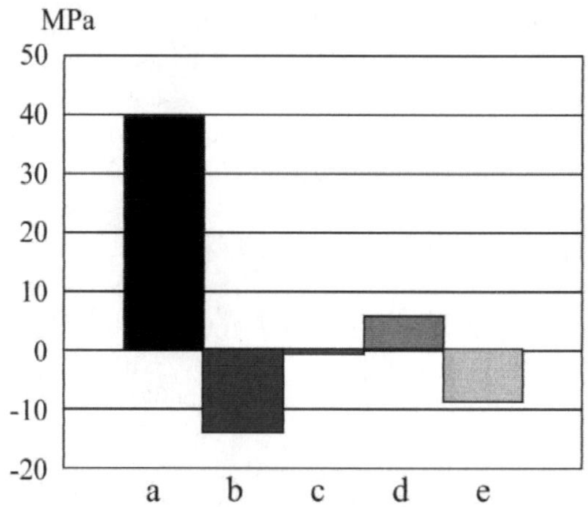

Figure 1. The mathematical expectation of the magnitude of residual stresses after 5 different types of processing: a – without treatment; b – after thermal tempering; c, d, e – after centerless running when 1, 2 and 3 are the processing modes, respectively.

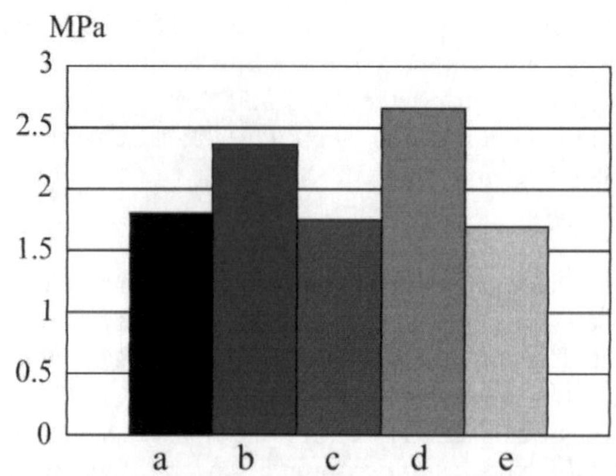

Figure 2. The standard deviation of the magnitude of residual stresses after 5 different types of processing: a – without treatment; b – after thermal tempering; c, d, e – after centerless running when 1, 2 and 3 are the processing modes, respectively.

The diagram in Figure 1 shows that significant tensile stresses dominate without relaxation of residual stresses (c). The reason for this can be explained by mechanical operations of edge cutting and abrasive machining of the ring, which are accompanied by localized heating of the surface layers. This is due to the high rate of plastic strain and friction between the tool and the material of a part. The subsequent rapid cooling of the part leads to the appearance of significant tensile residual stresses in its surface layer.

Thermal tempering, used for relaxation of residual stresses after machining, consists in slow heating of the entire volume of the part and its subsequent slow cooling in the open air. As the velocity of the heat generation of the part surface is above the core, and then after thermal equilibrium of the surface layers is reached, the core will continue to cool for some time. With further cooling, the deformation of the surface layers stops and the core continues to shrink, which leads to the occurrence of compressive residual stresses. This phenomenon is confirmed by the section b of the diagram in Figure1. It should be noted that the relaxation of residual stresses, while using the thermal method, has a diffusion mechanism and takes place by climbing of edge dislocations, and that is accompanied by significant energy consumption.

During cyclic elastic strain, the stress relaxation occurs due to the slip of dislocations caused by plastic strain of the micrograins of the material under the action of energy of elastic deformation. Thus less energy is required in comparison with the previous method. Energy of elastic strain, input with each cycle, due to viscoelasticity of the material, is absorbed until the equilibrium energy is reached. This state is characterized by the absence of residual stresses. Further communication of energy to elastic strain leads to the accumulation of internal energy and the growth of compressive residual stresses.

Comparison of the processing modes CR-1 and CR-2 shows that the number of cycles during the regime CR-2 is noticeably smaller at the same frequency of loading. Despite the large magnitude of elastic strain, the total amount of energy reported to the part at the CR-2 mode is less than at the CR-1, and that is insufficient for complete relaxation of residual stresses, as shown by section d in Figure 1.

Having compared the processing modes CR-1 and CR-3, it can be seen that for the same amount of elastic strain and a slightly smaller total number of cycles in the mode CR-3, the material of the part has compressive residual stresses (section e in Figure 1). This characterizes the transition of the equilibrium state and the subsequent strengthening of the part. This is because the frequency of elastic strain at the mode CR-3 is 1.45 times higher than the frequency at CR-1; and with the increase of frequency of elastic strain, part of the energy absorbed per cycle increases.

4. Conclusions

Durability of bearings depends on the stability of geometric shape throughout their operation. The presence of residual stresses is one of the main reasons for changing the geometry of the bearing rings. A gradual relaxation of residual stresses takes place during operation and this leads to the change of the rings geometry. During the relaxation of compressive residual stresses, there occurs the expansion of the volume of rings, which creates additional interference and increase in the contact area of surfaces of the rolling elements. During the relaxation of tensile residual stresses, the volume of the rings contracts and that creates a gap and wobbling of the bearing while in operation. In each of the two variants there occur some changes of the required operating conditions, which reduce the service life of the rolling bearings. The above mentioned results show that the stabilization technology based on cyclic elastic deformation of the rings by running has the advantage over thermal tempering , enabling us to achieve significant reduction of residual stresses, thereby increasing the durability of bearings.

References

[1] Bolotin V V, Shipkov A A 2001 Prediction of the growth of fatigue cracks taking environmental factors into account *Journal of Applied Mathematics and Mechanics* **65(6)** 1001-1010

[2] Wang M, Fei Q and Zhang P 2016 A Modified Fatigue Damage Model for High-Cycle Fatigue Life Prediction *Advances in Materials Science and Engineering* **2016** 2193684

[3] Kocanda D, Jasztal M 2012 Probabilistic predicting the fatigue crack growth under variable amplitude loading *International Journal of Fatigue* **39** 68-74

[4] Yazdanipour M, Pourgol-Mohammad M, Choupani N, Yazdani M 2013 Probabilistic Fatigue Crack Growth Analysis for Life Prediction of Automotive Components *ASME 2013 International Mechanical Engineering Congress and Exposition* **15** 0110002

[5] Perera R, Marin R, Ruiz A 2013 Static–dynamic multi-scale structural damage identification in a multi-objective framework *Journal of Sound and Vibration* **332(6)** 1484-1500

[6] Kumar S S, Balaji V N, Prrithvi P M, 2014 Influence of Residual Stress on Stress Intensity Factor Estimation of Multiple Cracks in a Dissimilar Welded Joint *Procedia Engineering* **86** 234-241

[7] Song X M, Wang G Z, Xuan F Z, Tu S T 2015 Investigation of residual stress effects on creep crack initiation and growth using local out-of-plane compression *Engineering Fracture Mechanics* **149** 45-57

[8] Xu M, Xu J, Lu H, Chen J, Chen J, Wei X 2015 Fractal and probability analysis of creep crack growth behavior in 2.25Cr–1.6W steel incorporating residual stresses *Applied Surface Science* **359** 73-81

[9] Shin J, McMahon C J 1984 Mechanisms of stress relief cracking in a ferritic steel *Acta Metallurgica* **32(9)** 1535-1552

[10] Yu Z-S, Zhang J-X, Wang H-Z, Zhou R-C, Yuan Y 2017 Mechanism of Stress Relief Cracking in a Granular Bainitic Steel *Acta Metallurgica Sinica (English Letters)* **30 (2)** 156–163

[11] Sandoval-Pineda J M, Urriolagoitia-Sosa G, Urriolagoitia-Calderón G, Hernández-Gómez L H, García-Lira J, Beltrán-Fernández J A and Rodríguez-Martínez R 2009 Numerical and experimental evaluation of the residual stress relaxation and the influence zone due to application of the crack compliance method *Journal of Physics: Conference Series* **181** 012069

[12] Korolev A V, Yakovishin A S, Korolev A A, Davidenko O Yu, Iznairov B M, Vasin A N, Balaev A F, Mukhina E V, Mazina A A, Sidorenko A D, Savran S A, Konovalov V V 2017 Geometric stabilization of bearing races by centerless rolling *Russian Engineering Research* **37(1)** 79–81

[13] Korolev A V, Korolev A A, Balaev A F, Savran S A and Yakovishin A S 2016 Experimental study of residual stresses relaxation in ring details during multicyclic loading *IOP Conference Series: Materials Science and Engineering* **156** 012011

[14] Balaev A F, Zakharov O V, Korolev A V, Korolev A A and Kochetkov A V 2015 Modeling the mechanism of stress relaxation of ring parts at high-cycle loading *Proceedings of 2015 International Conference on Mechanical Engineering, Automation and Control Systems, MEACS 2015* 1–4

IOP Conf. Series: Materials Science and Engineering **286** (2017) 012003 doi:10.1088/1757-899X/286/1/012003

Chemiluminescence and reactivity of the composites based on blends of polypropylene and polyamide

N V Vorontsov[1,2], A A Popov[1,2] and A L Margolin[2]

[1]Plekhanov Russian University of Economics, 36 Stremyanny per., Moscow, 117997, Russia

[2]Emanuel Institute of Biochemical Physics, Russian Academy of Sciences, 4 Kosygina str., Moscow, 119334, Russia

E-mail: nikolayvorontsov1993@gmail.com

Abstract. The effect of the composition of blends based on isotactic polypropylene (PP) and aliphatic polyamide 6/66-4 (PA) on the rate of photo-oxidation of their mixtures in air at room temperature has been studied. The decay of photoinduced chemiluminescence was studied to determine the kinetics of peroxyl radical termination in composites and the rate constants of this process depending on the composition of the mixtures. In the presence of PA, the rate of photo-oxidation of mixtures is much higher than the rates of photo-oxidation of separately taken components, PP and PA. Thus, the kinetics of photo-oxidation of mixtures differs from the simple sum of photo-oxidation kinetics of PP and PA, which should be expected in the absence of chemical and physical interaction of the components of the mixture. A decrease in the rate constants due to PA additives indicates a decrease in the mobility of molecules in the composites and explains the observed increase in photo-oxidation of mixtures.

1. Introduction

Blending has been recognized as the most versatile, economic method to produce materials able to satisfy complex requirements for application characteristics of products [1]. Blending may improve product performance by producing materials having a full set of the desired properties at lowest cost and improving specific properties such as toughening, heat-resistance, modulus, reactivity etc. [2, 3]. Chemiluminescence (CL) of polyolefins arises in the reaction of peroxy macroradical termination and is a convenient way to test the molecular mobility, which determines the rate and direction of polymer reactions. It is important that in this method the molecular mobility is tested in a place where the reaction centers are located [4]. Thus, free radicals act as molecular probes and inform about the changes of molecular movement of main chains of the polymers [5]. In addition to CL and electron spin resonance (ESR) methods, molecular mobility of blends can be tested by nuclear magnetic resonance (NMR). The molecular motion of poly(epichlorohydrin) (PECH) was studied by ^{13}C NMR and showed different behavior in pure PECH and blend of PECH/PMMA. The line shape and temperature dependence of T_1 and T_2 in PECH indicate that the molecular motion of PECH is inhibited by surrounding PMMA chains and that the decrease in molecular motion depends on miscibility [6]. The radical decay in PMMA-PEO blends has been studied in the paper [7]. A slower decrease in the concentration of free radicals in a sample of a PMMA/PEO blend containing 50 vol% PEO compared to samples with 10, 30 and 70 vol% PEO is observed. The mixing for a blend PMMA-

IOP Conf. Series: Materials Science and Engineering **286** (2017) 012003 doi:10.1088/1757-899X/286/1/012003

PEO has been shown to affect molecular motions as a result of specific interactions between dissimilar PMMA and PEO chains [7, 8].

This work represents the study of chemiluminescence and reactivity of composites based on isotactic polypropylene (PP), which is the second most produced polymer, and polyamide 6/66 (PA). Creation of new polymeric materials with enhanced reactivity, which can improve the degradation by the action of photo-oxidation and other environmental factors, is a promising area of research projects aimed at reducing environmental pollution by polymer waste. The problem is solved effectively by creating polymer blends [3]. CL analysis is sensitive to detect the very first stage of oxidation of polymers, and does not require long periods of time to obtain the information [9].

2. Experimental

Isotactic PP (Kaplen 01030, TU 2211-015-00203521–99) was obtained from OAO Gaspromneft'-Moskovskiy NPZ. The melt flow index (MFI at 230°C and 2.16 kg load) was 3,0 g/10 min. PP's melting and crystallization temperatures were 163°C and 111°C, respectively as determined by differential scanning calorimetry (DSC). Aliphatic polyamide (PA 6/66-4, OST 2224-438-02099342-93) was obtained from OOO Anid. This PA is the polycondensation product of caprolactam and the salt of hexamethylenediamine and adipic acid with the viscosity number of the polyamide in metacresol equals to 134 cm^3/g, moisture content no more 2,6 %, melting temperature 142°C, crystallization temperature 92°C as determined by DSC.

Compositions were prepared using a laboratory rotor mixer of Brabender type. Total sample weight was made 20 grams. Blending was carried out at a temperature of 190°C under an argon atmosphere. At the rotor rotation speed of 15 rpm, pelletized polyamide and pelletized polypropylene were added together into the mixing chamber, then rotation speed was increased to 30 rpm and blending continued for 5 minutes. After that, the blends were removed from the mixing chamber and cooled down at the room temperature. Composite materials of the following composition were objects of this study: PP with addition of 5, 10, 20, 30, 40 and 50 wt % of PA.

Film samples were obtained from crushed blends using a laboratory hydraulic press equipped with an electronic module of plate heating at a temperature of 190^0 C and a pressure of 60 atm on a cellophane substrate. Crushed samples of the blends were evenly distributed on the substrate. Pressing time was equal to 2 minutes. After that, the obtained samples were quickly cooled down in air at temperature of approximately 24°C. The obtained samples were films of 110 ± 20 μm thickness.

The films of PP, PA and PP-PA blends were irradiated by light of UV lamp OKUF-5M with a wavelength of 254 nm in air at room temperature. After a short-term irradiation of the films the light was switched off, the films were put in CL camera and the CL intensity was measured [10]. The kinetics of CL decay in the dark was measured one minute after irradiation.

3. Results and discussion

The low temperature photo-oxidation of polymers is a complex process consisting of several main stages: photoinitiation, termination, propagation (see Scheme 1). According to common scheme of low temperature polymer oxidation CL arises in reaction of peroxy radical termination. Free radicals are produced by UV light in the first stage of oxidation. In the second stage they transform into peroxy radicals via their reaction with oxygen, then free radicals take part in reaction of oxidation chain propagation. At low radical concentrations as in the case of measurements by sensitive CL method the rate of second order termination is negligibly low compared to linear chain termination. In the case of the first order termination CL intensity J_{CL} is proportional to the rate of radical termination and quantum yield ϕ of CL equation (1). So the rate of photo-oxidation W is proportional to CL intensity according to equation (2).

$$J_{CL}=\phi k[ROO] \qquad (1)$$

$$W=k_p[RH][ROO\cdot] \sim J_{CL} \qquad (2)$$

IOP Conf. Series: Materials Science and Engineering **286** (2017) 012003 doi:10.1088/1757-899X/286/1/012003

In equations (1, 2) k is the rate constant of linear radical termination, *[ROO·]* is a free peroxy radical concentration, k_p is the rate constant of oxidation chain propagation. As can be seen from equation 2, CL intensity in the moment of light off is proportional to photo-oxidation rate.

Scheme 1. Reactions of polymer photo-oxidation

RH (or ROOH) + $h\nu \rightarrow$ R	photoinitiation of chain oxidation
R· + $O_2 \rightarrow$ ROO	propagation of oxidation chain
ROO· + RH $\rightarrow$ ROOH + R·	propagation of oxidation chain
ROO· $\rightarrow$ products + $h\nu$	linear chain termination and CL
ROO· + ROO· $\rightarrow$ ROOR + O_2+ $h\nu$	second order chain termination and CL

Figure 1 shows the example of dependence of CL on the duration of UV irradiation for PA. As can be seen the intensity of CL rapidly reaches a stationary level, corresponding to the stationary concentration of radicals during photo-oxidation.

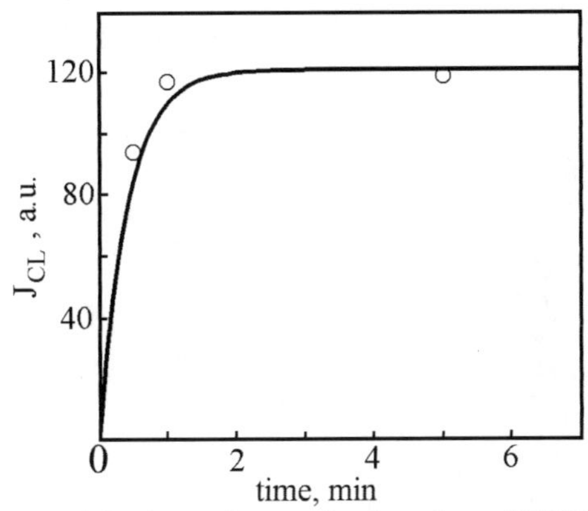

Figure 1. The dependence of CL intensity on the duration of UV irradiation for pure PA.

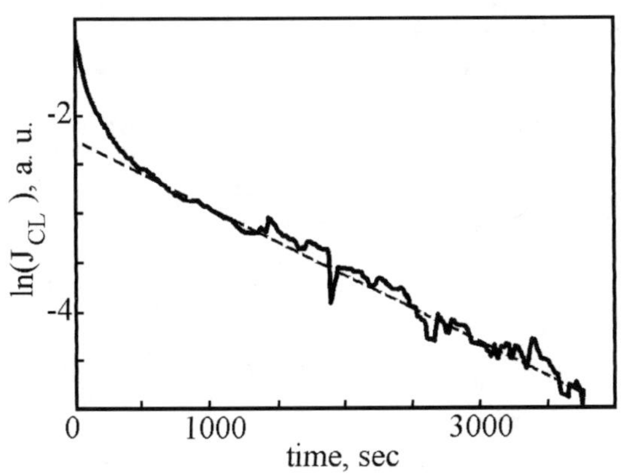

Figure 2. CL decay of pure PP in the air at room temperature after 1 min UV-irradiation.

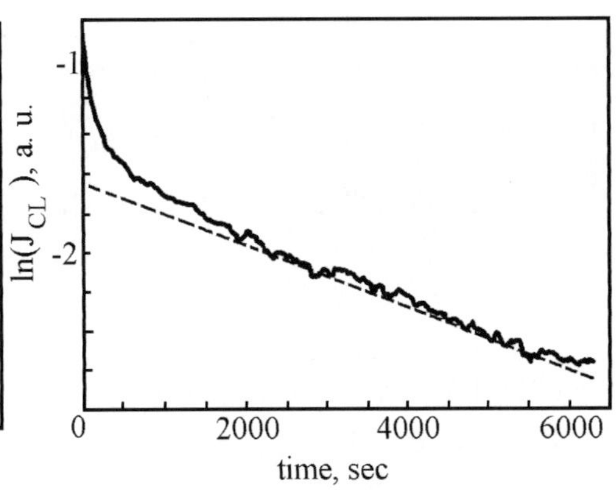

Figure 3. CL decay of pure PA in the air at room temperature after 1 min UV-irradiation.

Cl decay of pure PP and pure PA after irradiation are shown in figures 2 and 3 respectively. After relaxation, the radical decay of the first order is established with the rate constant, the value of which is equal to the slope of the trend line (dashed line in figures 2 and 3). It was studied how the composition of PP-PA mixtures affects the rate of composite photo-oxidation and the rate constant of peroxy macroradicals linear termination. The data on CL decay in two samples of PP-PA composites are presented in figures 4 and 5.

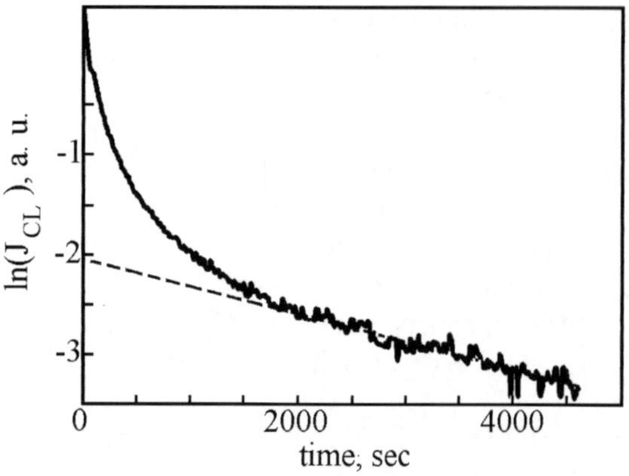

Figure 4. CL decay of blend with 90 wt.% PP and 10 wt.% PA after UV-irradiation in the air at room temperature.

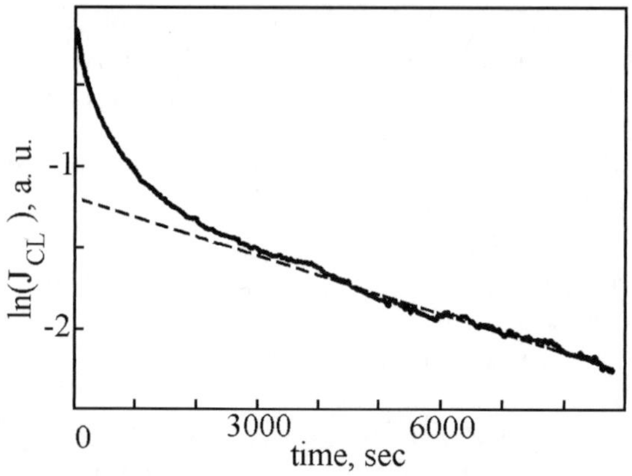

Figure 5. CL decay of blend with 70 wt.% PP and 30 wt.% PA after UV-irradiation in the air at room temperature.

The results of measurements of both initial CL intensities of composites and rate constant of radical decay are presented in figures 6 and 7 respectively.

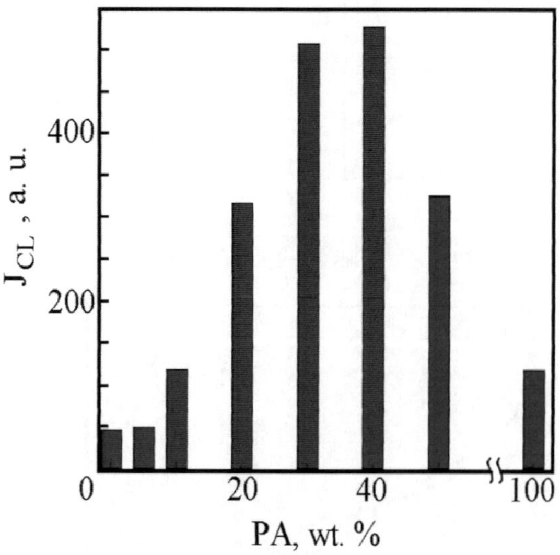

Figure 6. The dependence of initial CL intensity of PP-PA blends on PA content.

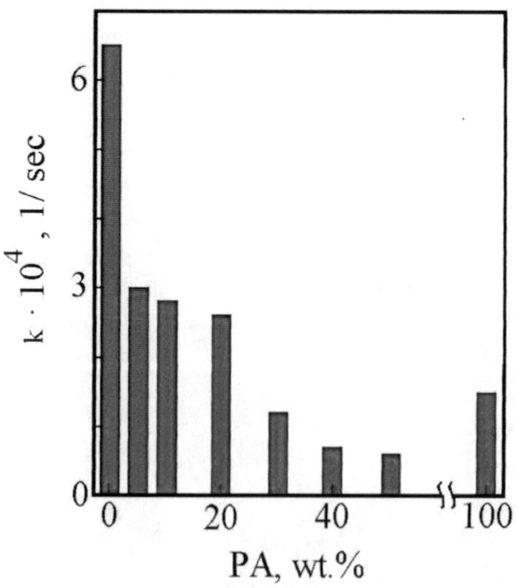

Figure 7. The dependence of rate constant for radical decay of PP-PA blends on PA content.

As can be seen from figure 6, the PA additives significantly increase the rate of photo-oxidation of composites. In the presence of PA, the rate of photo-oxidation of mixtures is much higher than the rates of photo-oxidation of separately taken components, PP and PA. Thus, the kinetics of photo-oxidation of mixtures differs from the simple sum of photo-oxidation kinetics of PP and PA, which should be expected in the absence of chemical and physical interaction of the components of the mixture. The highest photo-oxidation rate is observed at 40-50% PA in the PP-PA mixture.

The dependence of rate constant for radical decay in PP-PA blends on PA content is presented in figure 7. The rate constant of radical decay decreases with increasing PA content and reaches its lowest value at 40-50% PA. From the comparison of figures 6 and 7 one can see some relation between photo-oxidation rate and rate constant of radical decay for different values of PA content.

IOP Conf. Series: Materials Science and Engineering **286** (2017) 012003 doi:10.1088/1757-899X/286/1/012003

Thus, the photo-oxidation rate increases with decrease in the rate constant of radical decay. The rate constant of radical decay is a measure of molecular mobility. So decrease of rate constant indicates a decrease in the molecular mobility of composites. Thus, increased both photo-oxidation rate and radical concentration is due to lowered molecular mobility in blends. Immiscible polymers in mixtures are known to affect the structure of each other due to specific interactions between polymer chains. For instance changes in polymer structure are observed in blends of PP and polyethylene [11]. Other examples of lowering the molecular mobility in polymer mixtures are given in [6, 8].

4. Conclusion

In this paper, the rate constants for the peroxy radical decay in PP-PA mixtures were measured from the kinetics of photoinduced CL decay . These rate constants were compared with the data on the initial intensity of CL which is proportional to the concentration of peroxy radicals and consequently to the rate of photo-oxidation of the PP-PA blends. The photo-oxidation rate is shown to increase with decrease in the rate constant of radical decay. The free radical decay of irradiated polymers is controlled by segmental movements of macromolecular chains. The rate constant of free radical decay provides information about the molecular dynamics of the examined polymer system. It is assumed that increase in both the rate of photo-oxidation and the concentration of radicals during photo-oxidation (which are proportional to the initial CL intensity) is due to decrease in the mobility of polymer chains of PP-PA mixtures.

Acknowledgement

The materials under investigation were studied with the equipment of the common use centers at Emanuel Institute of Biochemical Physics of Russian Academy of Sciences and Plekhanov Russian University of Economics. This study was financed by a grant from the Plekhanov Russian University of Economics.

References

[1] Utracki L 2014 *Polymer Blends Handbook* ed L Utracki and C Wilkie (New-York: Springer Reference) p VIII

[2] Chandrasekharakurup S, Shanks R A and S Thomas 2014 *Nanostructured Polymer Blends* ed S Thomas and R Shanks et al (Oxford: Elsevier) p 1

[3] Krištofič M and Ujhelyiová A 2012 Compatibilisation of PP/PA Blends *Fibres & Textiles in Eastern Europe* **20** 93

[4] Nedorezova P, Chapurina A, Klyamkina A, Aladyshev A, Popov A, Shibryaeva L, Monakhova T, Margolin A 2011 *Polymer Science, Series* B **53** 448

[5] Szocs F, Klimova M, Chodak I and Chorvath I 1996 *European Polymer Journal* **32** 401

[6] Yamanaka A, Kaji A and Murano M 1998 *Polymer Journal* **30** 210

[7] Miklesova K and Szocs F 1992 *European Polymer Journal* **28** 553

[8] Sonnier R and Taguet A 2012 *Functional Polymer Blends: Synthesis, Properties, and Performance* ed V Mittal (New-York: CRC Press) p 261

[9] Corrales T, Catalina F, Peinado C, Allen N, Fontan E 2002 *Journal of Photochemistry and Photobiology A: Chemistry* **147** *213*

[10] Margolin A, Shlyapintokh V 1999 *Polym. Degrad. Stab.* **66** 279

[11] Mastalygina E, Shatalova O, Kolesnikova N, Popova A and Krivandin A 2016 *Inorganic Materials: Applied Research* **7** 58

Investigation of chlorinated modifier content influence on the physical-mechanical properties and vulcanizing characteristics of rubber and rubber mixture

K V Sukhareva[1,2], I A Mikhailov[1,2], Yu O Andriasyan[2] and A A Popov[1,2]

[1]Plekhanov Russian University of Economics, 36 Stremyanny Lane, Moscow, 117997, Russia
[2]Emanuel Institute of Biochemical Physics, Russian Academy of Sciences, 4 Kosygina str., Moscow, 119334, Russia

E-mail: aspirantras@mail.ru

Abstract. Present paper is devoted to the novel solution for the chlorination technology of polyolefin and diene rubbers – mechanochemical halide modification, as a more efficient way to obtain halogenated elastomers with a wide range of halogen content (from 3 to 7 %). The vulcanizing characteristics of elastomeric compounds based on the chlorinated rubber compounds as well as the production conditions of vulcanization process have been studied by the methods of dynamic mechanical rheometry. The performed investigation revealed the influence of chlorine content in different types of rubber compounds on rheological and physical-mechanical properties. New halogen-containing polyolefins rubbers produced by this technology proved themselves in the conditions of the rubber production.

1. Introduction

Today demand for elastomeric materials that have a complex set of specific properties ensuring their operation in the extreme conditions is constantly increasing. In this regard, the necessity to develop novel production technologies for a variety of industrial sectors is growing. Chemical modification of produced elastomers becomes the main trend in the field of producing elastomeric materials with a new complex of properties. One of the most important directions of elastomers modification is halide modification [1]. It is possible to obtain elastomeric materials and composites with a wide range of new specific properties on the basis of halogenated rubbers [2, 3] such as high adhesion, advanced resistance to fire, heat, petrol, oil, ozone, aggressive media and microorganisms, non-flammability, high strength, good gas impermeability, etc. A new method was developed for the halide-containing elastomer production, which is an alternative to the existing method (polymer solution halogenation), namely, the method of mechanochemical halide modification that sufficiently simplifies and cheapens the process of halide-containing rubber production. The developed method is based on the effect of initiation of radical polymer cleavage [4, 5] and participation of produced radicals in the bimolecular reaction with radical scavengers [6–9]; it allows us to halogenate polymers without halogen gases in solution and in the solid phase.

The research is aimed to determine influence of chlorine contain in rubbers on the physical and mechanical properties of rubber compounds based on different chlorinated rubber obtained through thermomechanochemical halide modification [10,11].

2. Materials and Methods

In this work, styrene-butadiene rubbers SBR-30RP and SKMS- 30RP, isobutylene isoprene rubber IIR 1675 and ethylene propylene rubber EPDM-E60 were used as a basic raw material. Chlorine-containing saturated paraffin, which have the length of the carbon chain C30 and contain ~70% of chlorine (wt) was used as modificator. On the base of this rubbers were obtained rubber compounds with a total chlorine content from 5% to 15% by weight.

To determine presence and content of chlorine in modified elastomeric samples, the Schoniger flask combustion method was used.

To determine the bound chlorine in the composition of the rubber macromolecules, the samples of the chlorinated rubber were subjected to extraction to remove the unreacted modifier. Extraction was carried out in Soxhlet's apparatus by adding acetone during 20 hours. Rubber compounds based on SBR, SKMS, EPDM and IIR were prepared to the standard formulation according to GOST 54554-2011.

The research of the vulcanization kinetics of rubber compounds with different halogen content was carried out in accordance with GOST 12535–84 (Rubber compounds. The method for determining the cure characteristics with vulcametric) through dynamic mechanical rheometry (DMR) (RPA 2000 Monsanto).

Physical and mechanical properties of the vulcanizates were determined by tensile testing machine (DEVOTRANS, Turkey) according to GOST 270–75–75 (Rubber. The method of determining the elastic properties of tensile strength.). The hardness was estimated according to GOST 263–75 (Rubber. Determination of Shore A hardness.) Rebound resilience tests of specimens were conducted on a Schob instrument according to GOST 27110-86.

3. Results and Discussions

The first stage of the work is devoted to the study of vulcanization characteristics of elastomeric compositions based on non-chlorinated rubbers (SBR-30RP, CSKMS-30RP, IIR, EPDM-E60) and chlorinated rubbers (CSBR-30RP, CSKMS-30RP, CIIR and CEPDM) containing different amount on combined chlorine. The number in brackets after the rubber compound (CSBR-30RP (5)) indicates the amount of modificator (mass parts (wt.) in terms of 100 grams of rubber). The research results of viscosity and vulcanization characteristics with the help of a dynamic mechanical rheological tester are given in Table 1.

Vulcanized characteristics shows, that mechanochemical modification of SBR-30RP by chlorine-containing organic compound refers to slight reduces the vulcanization rate (2.70 %/min for SBR-30RP and 2.56 %/min for CSBR-30RP) and to increase optimum time of vulcanization (42 min for SBR-30RP and 46 min for CSBR-30RP). This can be explained by an increase of an unsaturation level resulting from dehydrochlorination reaction, which proceeds by the vulcanization temperature (150°). The same dependence was observed for rubbers CSKMS-30RP. Increase in chlorine content of CEPDM rubbers doesn't have significant effect on vulcanization characteristics.

The next stage of our investigations was to study physical-mechanical properties of rubber compounds based on chlorine-containing rubbers, obtained by halide mechanochemical modification and properties of serial rubber mixtures. In the course of the experimental studies, the following parameters were determined: conventional strength at 200% elongation (M200 MPa), conventional strength at 300% elongation, (M300 MPa), conventional strength at 500% elongation (M500 MPa), tear resistance (бp, MPa), tensile strain of the material before it breaks, %, elongation set after destruction during 1 min, %, Shore hardness number, c.u. and rebound elasticity, % (Table 2).

We can see from the table 2, that properties of all chlorine containing rubber compounds have higher values that the serial rubber mixture. The results of carried out investigations show, that adding of chlorine in macromolecular structure refers to some increase of physical-mechanical properties of CIIR,

CSBR-30RP and CSKMS-30RP and to some decrease CEPDM-E60. The strength properties of CSBR-30RP decrease (from 13.9 MPa to 11.4 MPa) apparently due to the increase in the rigidity of the polymer chain with an increase in the chlorine content. Hardness and elasticity of CEPDM-E60 rubbers fast does not change, but strength decreases (from 17.2 MPa to 8.1 MPa) which may be due to an increase in the irregular structure. For CSKMS-30RP the strength and elasticity changes are insignificant.

Table 1. Rheological and vulcametrical parameters of the rubber mixtures.

Rubber compound based on:	Amount of combined chlorine (% wt.)	Vulcanized characteristics						
		M_{st}[a] (dN·m)	M_{min}[b] (dN·m)	M_{max}[c] (dN·m)	M_{opt}[d] (dN·m)	t_s[e] (min)	t_c[f] (min)	V_c[g] (%/min)
SBR-30RP	0	5.5	5.0	19	17.6	5	42	2.7
CSBR-30RP (5)[h]	2.6	5.0	4.0	19	17.5	10	50	2.5
CSBR-30RP (10)[i]	5.0	5.8	5.0	22	20.3	7	46	2.6
CSBR-30RP (15)[k]	7.5	5.8	5.0	21	19.4	7	46	2.6
SKMS-30 RP	0	6.8	6.0	31	28.5	6	47	2.4
CSKMS-30 RP (5)	1.8	5.5	4.8	23	21.2	7	54	2.1
CSKMS-30 RP (10)	3.5	5.8	5.0	26	23.9	7	50	2.3
CSKMS-30 RP (15)	5.4	5.8	5.0	26	23.9	6	45	2.5
BR	0	10.8	9.8	31	28.5	2	40	2.6
CBR (5)	2.7	11.3	10.5	32	29.9	2.5	41	2.6
CBR (10)	5.2	12.5	10.5	34	31.9	2	38	2.8
CBR (15)	7.5	15.2	13.0	27	25.4	2	30	3.6
EPDM-E60	0	11.0	9.0	57	52.2	2.5	34	3.2
CEPDM-E60 (5)	2.1	13.0	10.0	60	55.0	2.5	30	3.6
CEPDM-E60 (10)	4.0	14.5	11.5	63	57.9	1	28	3.7
CEPDM-E60 (15)	5.5	13.0	10.2	58	53.2	1	33	3.1

[a] M_{st} - starting torque value
[b] M_{min} - minimum torque value
[c] M_{max} - maximum torque value
[d] M_{opt} - optimal torque value
[e] t_c - optimum time of vulcanization
[f] t_s - compound scorch time
[g] V_c - vulcanization velocity
[h] (5) - amount of the introduced chlor-modificator 5 wt% (mass parts (wt) in terms of 100 grams of rubber).
[I] (10) - amount of the introduced chlor-modificator 10 wt% (mass parts (wt) in terms of 100 grams of rubber)
[k] (15) - amount of the introduced chlor-modificator 15 wt% (mass parts (wt) in terms of 100 grams of rubber)

Table 2. Physical-mechanical properties of rubbers based on chlorine-containing rubbers.

Rubber compound based on:	Amount of combined chlorine (% wt.)	Physical-mechanical properties							
		M_{200} (MPa) (Δ± 0.5 MPa)	M_{300} (MPa) (Δ± 0.5 MPa)	M_{500} (MPa) (Δ± 0.5 MPa)	σ_b, (MPa) (Δ± 0.5 MPa)	Tensile strain (%) (Δ±0.5 %)	Elongation set after destruction (%) (Δ±0.5 %)	Shore hardness number (c.u.) (Δ±2 c.u.)	Rebound elasticity (%) (Δ±2 %)
SBR-30RP	0	0.5	1.2	4.4	10.6	720	14	43	37
CSBR-30RP (5)	2.6	1.8	2.7	9.0	13.9	720	14	43	37
CSBR-30RP (10)	5.0	1.8	4.4	11.0	11.9	525	6	52	29
CSBR-30RP (15)	7.5	0.9	2.3	6.6	11.4	680	14	58	23
SKMS-30RP	0	2.5	5.2	-	8.8	350	6	60	32
CSKMS-30RP (5)	1.8	1.3	3.5	9.9	11.7	550	10	54	31
CSKMS-30RP (10)	3.5	1.3	2.9	7.9	12.2	600	12	58	26
CSKMS-30RP (15)	5.4	1.4	4.0	10.3	13.2	580	11	57	25
BR	0	2.7	5.4	13.7	16.3	565	28	62	14
CBR (5)	2.7	1.7	4.0	10.4	15.1	590	40	66	17
CBR (10)	5.2	2.1	4.0	10.2	14.3	650	47	77	12
CBR (15)	7.5	1.3	2.6	6.3	13.5	740	53	76	13
EPDM-E60	0	7.9	15.1	-	16.4	330	4	72	32
EPDM-E60 (5)	2.1	6.1	11.2	-	17.2	475	24	81	27
EPDM-E60 (10)	4.0	9.5	-	-	10.3	225	8	82	25
EPDM-E60 (15)	5.5	8.1	-	-	8.1	200	10	80	25

4. Conclusion

So, a new chlorine-containing rubbers obtained by method of mechanical-chemical halide modification were studied. Some structural characteristics of rubbers, curing kinetics of rubber compounds and vulcanized rubbers physical-mechanical properties, were studied. Thus, it can be seen from the studies carried out that the introduction of chlorine into the structure of the rubber macromolecules somewhat reduces the rate of vulcanization. Also, when the modifier is added to 5 wt. for CSBR-30RP and EPDM-E60, some increase in strength is observed, but when more modifier is introduced, the strength decreases, which is apparently due to the increase in the proportion of unreacted modifier that acts as a plasticizer. In the case of modification of SKMS-30RP, this effect is not observed until the introduction of 15 parts by weight of the modifier. However, when the BR is modified, the strength decreases, since the BR has a highly ordered structure and is broken during mechanical processing, as does the plasticizing effect of the unreacted modifier.

Acknowledgments
This study was financed by a grant from the Plekhanov Russian University of Economics. The authors express an acknowledgement to the shared use centers "New materials and technologies" at the Emanuel

Institute of Biochemical Physics of Russian Academy of Sciences and "New composite materials and technologies" at the Plekhanov Russian University of Economics for provided equipment.

References

[1] Sukhareva K V, Mikhailov I A, Andriasyan Yu O, Popov A A 2016 Thermomechanochemische Modifikation des Butylkautschuks in Anwesenheit von chlorhaltingen Reagenzien *Gummi. Fasern. Kunststoffe* **6** 374-376

[2] Halliwell S 2002 Polymers in Building and Construction (Smithers Rapra Publishing: Shawbury) p 136

[3] Schulz D N and Patil A O Chemical Modification of Synthetic Elastomers 2000 *Handbook of Elastomers, Second Edition* ed A K Bhowmick and H Stephen (Marcel Dekker Inc.: Basel) pp 107–165

[4] Kauzmann W and Eyring H J 1940 The viscous flow of large molecules. *Journal of the American Chemical Society* **62** (11) 3113-3125

[5] Pike M and Watson W F 1952 Mastication of rubber I. Mechanism of plasticizing by cold mastication *Journal of Polymer Science* **9** 229–251

[6] Watson W F Mechanico-chemical reactions of polymers 1959 *Macromolecular Chemistry and Physics* **34** 240–252

[7] Ayrey G, Moore C G, Watson W F Mastication 1956 Part III Chemical verification of the mechanical degradation mechanism of cold mastication *Journal of Polymer Science* **19** 1–15

[8] Popov A A, Zaikov G E 1987 Effect of stresses on polymer oxidation. Quantitative aspects. Kinetics. *Journal of Macromolecular Science: Reviews in Macromolecular Chemistry and Physics* **27** 379-457

[9] Popov A A, Zaikov G E, Semeriov N N 1992 Kinetics of Chemical Reactions of Stressed Polymers. *International Journal of Polymeric Materials* **17** 143-149

[10] Mikhaylov I A, Sukhareva K V, Andriasyan Yu O, Popov A A, Vorontsov N V 2016 Mechanochemical modification of natural rubber *AIP Conf. Proc.* **1783** 1-4

IOP Conf. Series: Materials Science and Engineering **286** (2017) 012005 doi:10.1088/1757-899X/286/1/012005

Investigation of surface halide modification of nitrile butadiene rubber

K V Sukhareva[1,2], I A Mikhailov[1,2], Yu O Andriasyan[2], E E Mastalygina[1,2] and A A Popov[1,2]

[1]Plekhanov Russian University of Economics, 36 Stremyanny lane, Moscow, 117997, Russia
[2]Emanuel Institute of Biochemical Physics, Russian Academy of Sciences, 4 Kosygina str., Moscow, 119334, Russia

E-mail: aspirantras@mail.ru

Abstract. The investigation is devoted to the novel technology of surface halide modification of rubber samples based on nitrile butadiene rubber (NBR). 1,1,2-trifluoro-1,2,2-trichlorethane was used as halide modifier. The developed technology is characterized by production stages reduction to one by means of treating the rubber compound with a halide modifier. The surface halide modification of compounds based on nitrile butadiene rubber (NBR) was determined to result in increase of resistance to thermal oxidation and aggressive media. The conducted research revealed the influence of modification time on chemical resistance and physical-mechanical properties of rubbers under investigation.

1. Introduction

Nowadays obtaining of halogen-containing polymers including synthesis and halide modification is one of intensively developing direction in rubber industry. Due to halide modification of polymers having technological flexibility and effectiveness in production, elastomer materials and composites are managed to obtain with a wide complex of required specific properties: high adhesion, incombustibility, high strength, gas permeability, resistance to fire and heat, corrosive media (oil, gasoline, ozone) and microorganisms, etc. [1-4].

Among the various methods of polymers modification, the surface modification is most promising in the area of polymers practical application. Different methods have been developed to obtain the required surface properties of polymer composites. The most-used method of polymer solution halogenation comprising a heterogeneous reaction of gaseous F_2 or Cl_2 and their mixtures with a polymer surface [5]. One of the main disadvantage of this method is high aggressiveness of initial fluorinating and chlorinating reagents. The method is also characterized by a high labor input, undesirable presence of toxic by-products and complex technical design of production process.

The proposed novel technology of surface mechanochemical modification is one of the new approach of scientific direction “Mechanochemical halide modification of elastomers” [6, 7]. The developed technology is characterized by production stages reduction by treating the rubber compound with a halide modifier.

2. Materials and Methods

In this work, nitrile butadiene rubber (NBR) supplied by SIBUR Holding was used as a basic raw material. Rubber compounds based on NBR were prepared according to the requirements of GOST 54554-2011. 1,1,2-trifluoro-1,2,2-trichlorethane was used as a halide modifier [8]. The modification of rubber was carried out by treating the surface with a modifying compound for different periods of time (24h, 72h, 144h).

Modified rubber compounds based on NBR were oxidized by a temperature of 150°C at an oxygen pressure of 300 mmHg. The kinetics of oxygen absorption was studied using a manometric unit with the absorption of volatile reaction products by solid KOH according to GOST ISO 188-2013. As a result, kinetic curves of oxygen absorption for samples with different modification time (24h, 72h, 144h) and initial NBR rubber compound sample without modification were obtained.

Rubber resistance to aggressive media was determined by a change in mass of samples at relaxed state in accordance with GOST 9.030-74. As aggressive media a set of different compounds were used including industrial mineral oil, "Nefras" gasoline, sulfuric acid, hydrochloric acid, and nitric acid.

Physical and mechanical properties of the vulcanizates were determined by tensile testing machine (DEVOTRANS, Turkey) [9] according to GOST 270–75 (Rubber. The method of determining the elastic properties of tensile strength). The hardness was estimated according to GOST 263–75 (Rubber. Determination of shore a hardness.) Rebound resilience tests of specimens were conducted by a Schob instrument according to GOST 27110-86.

3. Results and Discussions

For the purpose of oxidative stability study, the kinetics of thermal oxidation of untreated and modified rubbers depending on modification time was analysed. The oxygen absorption by the samples was fixed within certain time intervals followed by plotting kinetic curves (Figure 1). According to study results, the influence of modification time on the oxidation resistance of modified rubbers was determined.

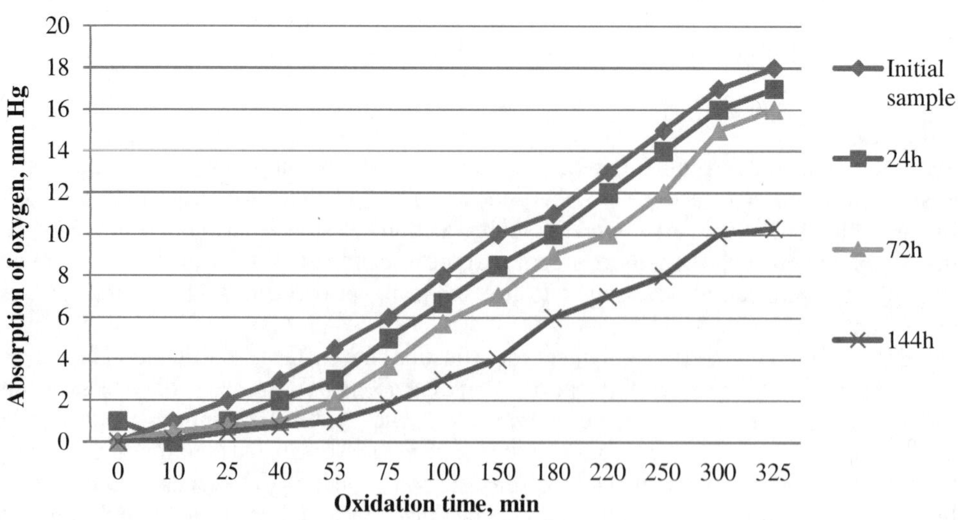

Figure 1. Kinetic curves of oxidation of rubbers based on NBR obtained by mechanochemical surface halide modification.

Figure 1 shows the surface halide modification within 24h÷72h has no significant effect on the resistance to thermal oxidation compared to the oxidation resistance of initial untreated sample. In case of surface halide modification for 144h the resistance to thermal oxidation was increased by a factor of 1.5.

The influence of modification time on the resistance of modified rubbers to different aggressive media was investigated with the application of standard test solutions. Results are presented in Table 1.

Table 1. Resistance of modified NBR-rubbers to aggressive media depending on modification time.

Modification time	Swelling degree (%) (Δ±15%)								
	Initial sample			48h			72h		
Time of exposure, h	24	48	168	24	48	168	24	48	168
Industrial oil	2.7	2.7	2.7	-1.4	-4.1	-4.1	-3.8	-6.0	-6.0
Gasoline	2.7	4.0	5.3	1.3	2.6	2.6	-3.7	-3.7	-3.7
Sulfuric acid	139.0	251.0	251.0	118.0	160.0	170.0	100.0	150.0	152.0
Hydrochloric acid	8.0	9.5	16.0	0.0	0.0	5.9	-3.6	-9.0	-12.0
Nitric acid	48.0	-	-	29.3	32.0	37.0	20.5	26.0	34.0

According to results illustrated in Table 1, the surface haloid modification leads to reduction of swelling in industrial oil and gasoline, as well as in sulfuric and hydrochloric acids. After modification for 3 days in nitric acid swelling of rubber samples completely stopped revealing that the samples have acquired the increased acid resistance [10]. It was suggested that the samples of modified NBR during more than 3 days would also be resistant to acids and oils.

Physical-mechanical properties of modified compounds based on NBR rubbers obtained by surface halide mechanochemical modification and serial rubber mixtures were determined by tensile testing machine. In the course of the experimental studies, the following parameters were determined: tensile strength (σ_b, MPa), relative elongation at break (ε_b, %), elongation set after destruction during 1 min (%), Shore hardness number (c.u.), and rebound elasticity (%) (Table 2).

Table 2. Physical-mechanical properties of initial sample of NBR rubber and rubbers based on modified NBR.

		σ_b (MPa) (Δ±0.5 MPa)	ε_b (%) (Δ±10 %)	Elongation set after destruction (%) (Δ±0.5 %)	Rebound elasticity (%) (Δ±2 %)	Shore hardness number (c.u.) (Δ±2 c.u.)
Initial sample of NBR rubber		9.7	590	8	32	62
Rubbers based on NBR modified during:	24h	12.3	480	8	31	63
	72h	13.1	440	7	31	63
	144h	13.6	460	7	31	64

As shown in Table 2, the properties of all the modified NBR rubbers have higher values than the initial untreated sample of NBR rubber. Summing up the investigation results, the modification induce some increase of such physical-mechanical properties as tear resistance (from 9.7 MPa to 13.6 MPa) and to some decrease of tensile strain (from 590% to 460%). Hardness and elasticity of modified NBR rubbers remained virtually unchanged [11].

4. Conclusion

The optimized technique for the surface modification of rubber products of various types and form using 1,1,2-trifluoro-1,2,2-trichlorethane as a modifier was developed. The increase in duration of halogenation of samples under investigation apparently leads to a proportional increase in degree of samples modification (fluorination and chlorination). The presence of halogen atoms (fluorine, chlorine) in the surface layer of rubber sample are responsible for a complex of enhanced properties, such as chemical resistance to oxidants and aggressive solutions, high tensile strength without a loss of continuity of the halogenated surface layer, and a number of other characteristics. In addition, an increase in the resistance to aggressive media of surface-modified rubbers could be related to a decrease of reaction centres on the rubber surface.

Acknowledgments

This study was financed by a grant from the Plekhanov Russian University of Economics. The authors express an acknowledgement to the shared use centers “New materials and technologies” at the Emanuel Institute of Biochemical Physics of Russian Academy of Sciences and “New composite materials and technologies” at the Plekhanov Russian University of Economics for provided equipment.

References

[1] Halliwell S 2002 Polymers in Building and Construction (Smithers Rapra Publishing: Shawbury) p 136
[2] Schulz D N and Patil A O Chemical Modification of Synthetic Elastomers 2000 *Handbook of Elastomers, Second Edition* ed A K Bhowmick and H Stephen (Marcel Dekker Inc.: Basel) pp 107–165
[3] Dontsov A A and Lozovik G A 1979 Chlorinated polymers (Khimia: Moscow) p 232
[4] Ronkin G M 1979 Halogenated Polymers. The Main Tendencies in Production and Application. Reviews on the Most Valuable Scientific and Technical Industry (NIITEKhim: Moscow) p 82
[5] Nazarov V G 2008 The surface modification of polymers (Moscow: MSUPA) p 474
[6] Sukhareva K V, Mikhailov I A, Andriasyan Yu O, Popov A A 2016 Thermomechanochemische Modifikation des Butylkautschuks in Anwesenheit von chlorhaltingen Reagenzien *Gummi. Fasern. Kunststoffe* **6** 374-376
[7] Mikhaylov I A, Sukhareva K V, Andriasyan Yu O, Popov A A, Vorontsov N V 2016 Mechanochemical modification of natural rubber *AIP Conf. Proc.* **1783** 1-4
[8] Sukhareva K, Mastalygina E, Mikhailov I, Andriasyan Y, Popov A 2017 Novel technology of butyl rubber chlorination and investigation of chlorinated modifier content influence on vulcanizing characteristics of pure-gum compound *AIP Conf. Proc.* **1800** 1-4
[9] Mastalygina E E, Varyan I A, Kolesnikova N N, Karpova S G, Popov A A 2016 Effect of chemical composition and dimensional parameters of natural filler on structure formation and physical-chemical properties of polyethylene composites *AIP Conf. Proc.* **1783** 1-4
[10] Mohan R, Raja S, Saraswathy G, Das B N 2015 Surface modification of tpr sole: an approach to improve slip resistance on quarry and ceramic tiles *Rubber Chemistry and Technology* **88:1** 163-175
[11] Vallat M F, Radabutra S, Thanawan , Bresson B, Chollakup R, Rungsanthien K, Suwanruji P, Klanarong S, Smitthipong W, Mougin K 2014 Influence of Non-Rubber Components on NR Surface Modification by Chlorination *Advanced Materials Research* **844** 369-372

IOP Conf. Series: Materials Science and Engineering **286** (2017) 012006 doi:10.1088/1757-899X/286/1/012006

Fine Structure Study of the Plasma Coatings B_4C-Ni-P

E E Kornienko[1], **V A Bezrukova**[1], **V I Kuz'min**[2], **V S Lozhkin**[1] **and M K Tutunkova**[1]

[1] Department of Material Science in Mechanical Engineering, Novosibirsk State Technical University, Prospekt K. Marksa, 20, Novosibirsk, 630073, Russia
[2] Laboratory of Physics of Plasma Arc and Laser Processes, Institute of Theoretical and Applied Mechanics SB RAS, Institutskaya str., 4/1, Novosibirsk, 630090, Russia

E-mail: vikabezrukova1995@mail.ru

Abstract. The article considers structure of coatings formed of the B_4C-Ni-P powder. The coatings were deposited using air-plasma spraying with the unit for annular injection of powder. The pipes from steel 20 (0.2 % C) were used as a substrate. The structure and phase composition of the coatings were studied by optical microscopy, scanning electron microscopy, transmission electron microscopy and X-ray diffractometry. It is shown that high-density composite coatings consisting of boron carbide particles distributed in the nickel boride metal matrix are formed using air-plasma spraying. The areas with round inclusions characterized by the increased amount of nickel, phosphorus and boron are located around the boron carbide particles. Boron oxides and nickel oxides are also present in the coatings. Thin interlayers with amorphous-crystalline structure are formed around the boron carbide particles. The thickness of these interlayers does not exceed 1 μm. The metal matrix material represents areas with nanocrystalline structure and columnar crystals.

1. Introduction

The topical problem of modern material science is the development of wear resistant materials. One of the most promising materials is the boron carbide (B_4C). Interest in this material is due to a complex of its unique properties: high microhardness (30...48 GPa) and wear resistance, low density (2.51...2.55 g/cm^3), as well as high corrosion resistance [1-3]. Usually thin coatings of boron carbide are deposited on the already finished parts. The main methods of B_4C coatings deposition are considered to be the spraying techniques [3, 5-11].

Unfortunately, some difficulties connected with B_4C properties arise both at the deposition of coatings from boron carbide and at the formation of 3D materials. Low thermal conductivity of boron carbide does not allow to heat the powder particles fully and the high hardness does not allow plastic deformation on the substrate. The solution to this problem is addition of a binder: various metals or chemical compounds possessing higher ductility and thermal conductivity. Although the structure and properties of coatings formed of the blends of B_4C powders and various metals are well studied [4, 9, 11-15], there are few articles describing structural changes in the B_4C-Ni-P coatings [16]. In particular, there is no research relating to the formation features of these coatings using the air-plasma spraying with the unit for annular injection of powder. The purpose of this research is the study of the B_4C-Ni-P coatings fine structure formed using the air-plasma spraying with the unit for annular injection of powder.

IOP Conf. Series: Materials Science and Engineering **286** (2017) 012006 doi:10.1088/1757-899X/286/1/012006

2. Experimental

In this work the research objects were coatings formed from B_4C powder clad by Ni-P. The chemical composition of powder is 70 wt. % B_4C, 25.5 wt. % Ni, 4.5 wt. % P. The powder particles had irregular form.

The air-plasma spraying was performed at the Khristianovich Institute of Theoretical and Applied Mechanics of the SB RAS using PNK-50 plasmatron with the unit for annular injection of powder. The powder was deposited on the pipes from the low-carbon steel 20 (0.2 % C) 25 mm in diameter with 3 mm wall. The spraying regimes were: arc current is 180 A, voltage is 180 V. The spraying distance was 170 mm. The air was used as plasma-forming gas and carrier gas. The blend of air and propane-butane was used as shielding gas.

Structural investigations were conducted using the Carl Zeiss Axio Observer Alm optical microscope and Carl Zeiss EVO50 XVP scanning electron microscope with EDS X-Act microanalyser. The samples for the structural investigations were cut from pipes with coatings, pressed into the polymer matrix and then prepared by the standard method: mechanical grinding and polishing. We used 10 ml of HCl, 0.1 ml of HNO3 and 10 g of FeCl3 solution for etching of the coatings surface. More in-depth structure research was carried out using the Tecnai G2 FEI transmission electron microscope. Workpieces of 3 mm in diameter cut from the middle of the coatings were used as samples. They were mechanically thinned to the thickness of 90...100 μm. Gatan Dimple Grinder was used to obtain spherical dimples. The final operation was the processing of foil by an Ar-beam on Gatan PIPS until receiving a hole in the dimple. The phase composition was studied using ARL X'TRA X-ray difractometer. The microhardness was estimated by Wolpert Group 402MVD microhardness tester at 100 N load.

3. Results and discussion

Figure 1a shows the cross section of plasma coating. The obtained coating thickness is 1200 μm. Figure 1b demonstrates that the coating material consists of dark particles with arbitrary form (area 1) equally spaced in a metal matrix (area 2). The etching of surface allows identifying areas with a eutectic structure (area 3) located predominantly around dark particles.

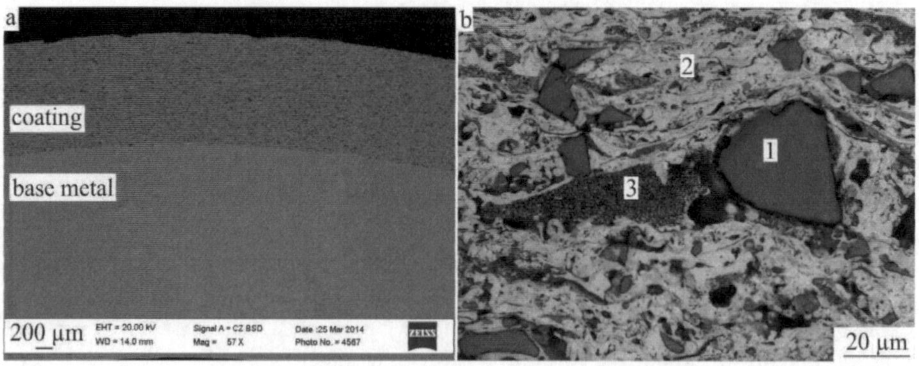

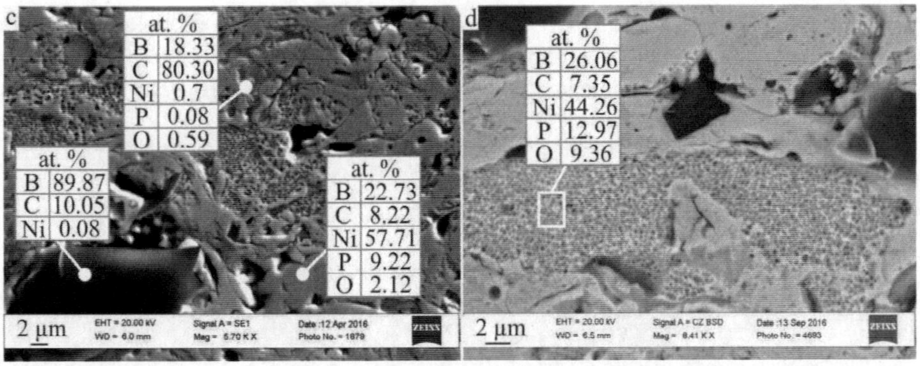

Figure 1. The structure and local chemical composition of coating: a) compound "base metal-coating"; b-d) coatings structure.
a, c, d – scanning electron microscopy; b - optical microscopy.
1 - boron carbide particles; 2 - metal matrix; 3 - eutectic areas.

According to X-ray phase analysis data (Figure 2) as well as micro-X-ray spectral analysis data (Figure 1c) the dark inclusions are the B_4C particles that unmelted in the plasma jet. A metal matrix consisting of various nickel borides (Ni_4B_3, Ni_3B, NiB) is formed as a result of the interaction of Ni and B. High rates of mixing and solidification of the matrix material contribute to the nonuniform distribution of these elements in the coating, as evidenced by the results of micro-X-ray spectral analysis (Figure 1c). The areas with the eutectic structure are characterized by increased content of Ni, P and B. The scanning electron microscopy revealed that eutectic areas constitute round inclusions up to 500 nm located in the metal matrix (Figure 1d). X-ray phase analysis showed that reflections of boron oxide (B_2O_3) and nickel (NiO) are also observed in coatings.

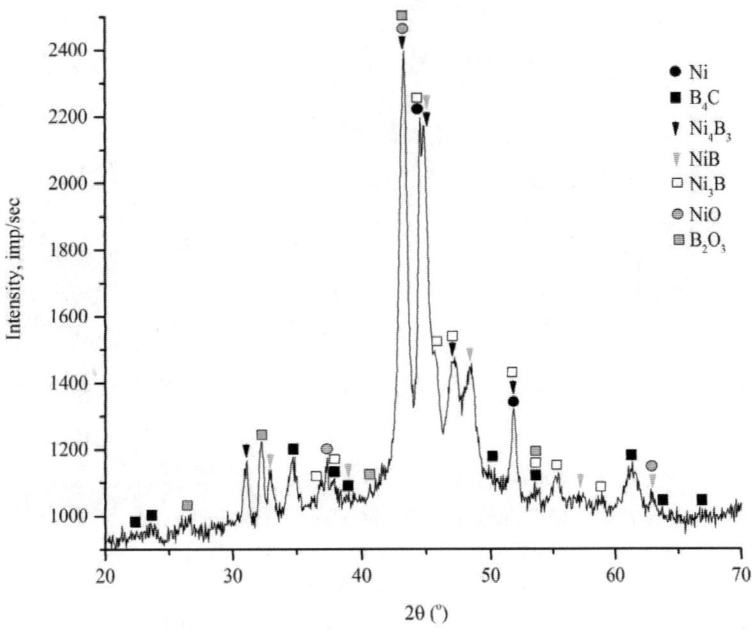

Figure 2. XRD patterns of the coatings.

Studies of the coatings fine structure were carried out using the method of the transmission electron microscopy. Figure 3 shows the coating material. It is clearly seen that there is an interlayer about 1000 nm between the boron carbide particles and the metallic matrix (Figure 3a). The presence of the interlayers is the evidence of high level cohesion between the ceramic particles and the metallic matrix. Figure 3b shows more detailed picture of this interlayer. It is seen to be characterized by an amorphous-crystalline structure, as evidenced by the microdiffraction pattern. The metal matrix is the areas with nanocrystalline structure (Figure 3c) and the columnar structure (Figure 3d). The packets of columnar crystals formation up to 20 nm in width are associated with high degrees' deformation of the sprayed particles upon the concussion with the solidified material. The analysis of the metal matrix allows revealing that areas with a nanocrystalline structure and columnar crystals are randomly distributed within the volume of the coatings.

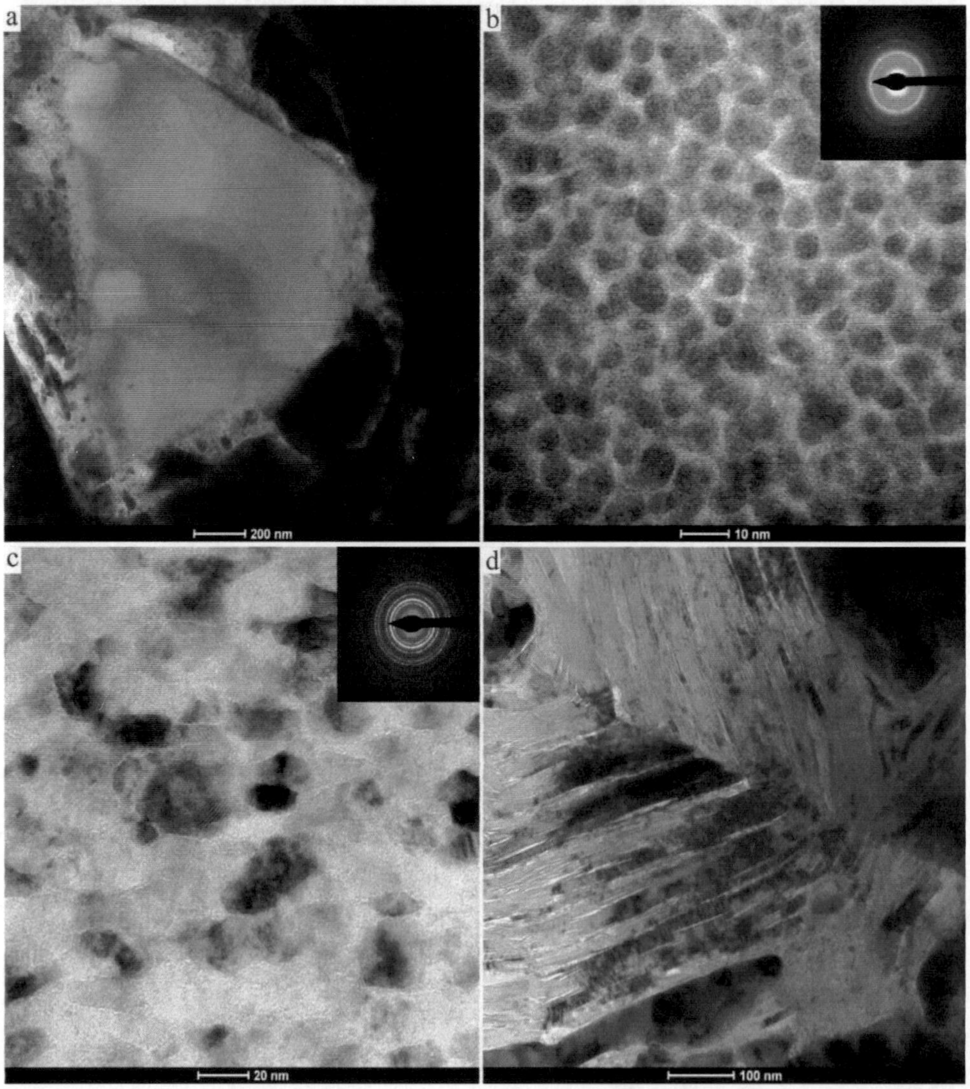

Figure 3. The boron carbide particle and the interlayer between the ceramic particle and the metallic matrix (a); the interlayer structure and microdiffraction pattern (b); the nanocrystalline structure of the metal matrix and the microdiffraction pattern (c); columnar structure of the metal matrix (d).

4. Conclusions

1. The B4C-Ni-P dense composite coatings can be formed using air-plasma spraying.

IOP Conf. Series: Materials Science and Engineering **286** (2017) 012006 doi:10.1088/1757-899X/286/1/012006

2. The optical microscopy and scanning electron microscopy showed that the plasma coating material consists of the boron carbide particles equally spaced in the metal matrix. The metal matrix material is nickel borides. The areas with round inclusions were formed around the particles. These areas are characterized by increased content of nickel, phosphorus and boron. There are boron oxides and nickel oxides in the coating.
3. The transmission electron microscopy showed that thin interlayers with amorphous-crystalline structure up to 1 μm in thickness are formed between the boron carbide particles and the metal matrix in the process of plasma spraying. The metal matrix is the areas with a nanocrystalline structure and columnar crystals.

Reference

[1] Risovany V, Zakharov A, Klochkov E and Guseva T 2011 *Bor v Yadernoi Tekhnike* [Boron in Nuclear Engineering] (Dimitrovgrad: «GNTs NIIAR») p 668
[2] Alekseev A et al. 1986 *Svoystva, Poluchenie i Primenenie Tugoplavkikh Soedineniy* [Properties, Production and Using of Refractory Compounds] (Moscow: Metallurgiya Publ.) p 928
[3] Zhu H, Niu Y, Lin C, Huang L, Ji H and Zheng X 2012 Fabrication and Tribological Evaluation of Vacuum Plasma-Sprayed B_4C *Coating Journal of Thermal Spray Technology* **21** 1216-23
[4] Sarikaya O, Anik S, Aslanlar S, Okumus S and Celik E 2007 Al-Si/B_4C Composite Coatings on Al-Si Substrate by Plasma Spray Technique *Materials and Design* **28** 2443–49
[5] Feng C, Guipont V, Jeandin M, Amsellem O, Pauchet F, Saenger R, Bucher S and Iacob C 2012 B_4C/Ni Composite Coatings Prepared by Cold Spray of Blended or CVD-Coated Powders *Journal of Thermal Spray Technology* **21** 561-570
[6] Cao Y, Huang C, Liu W, Zhang W and Du L 2014 Effects of Boron Carbide Content on the Microstructure and Properties of Atmospheric Plasma-Sprayed NiCoCrAlY/Al_2O_3-B_4C Composite Coatings *Journal of Thermal Spray Technology*. **23** 716-724
[7] Zeng Y, Lee S and Ding C 2002 Study on Plasma Sprayed Boron Carbide Coating *Journal of Thermal Spray Technology* **11** 129-133
[8] Zeng Y, Feng J and Ding C 2000 Microstructure and Properties of Plasma Spraying Boron Carbide Coating *Journal of Materials Science and Technology* **16** 63-66
[9] Moradi M, Moazeni M and Salimijazi H 2014 Microstructural Characterization and Failure Mechanism of Vacuum Plasma Sprayed Ti-6Al-4V/B_4C Composite *Vacuum* **107** 34-40
[10] Salimijazi H, Coyle T, Mostaghimi J and Leblanc L 2005 Microstructure of Vacuum Plasma-Sprayed Boron Carbide *Journal of Thermal Spray Technology* **14** 362-368
[11] Zhu H, Niu Y, Lin C, Huang L, Ji H and Zheng X 2012 Microstructures and Tribological Properties of Vacuum Plasma Sprayed B_4C-Ni Composite Coatings *Ceramics International* **39** 101-110
[12] Rafiei M, Salehi M, Shamanian M and Motallebzadeh A 2014 Preparation and Oxidation Behavior of B_4C-Ni and B_4C-TiB_2-TiC-Ni Composite Coatings Produced by an HVOF Process *Ceramics International* **40** 13599-609
[13] Mao Z, Ma J, Wang J and Sun B 2009 Comparison of the Coatings Deposited Using Ti and B_4C Powder by Reactive Plasma Spraying in Air and Low Pressure *Journal of Materials Science* **44** 3265-72
[14] Lin C, Kong M and Zhu H 2016 Tribological Behavior of Vacuum Plasma Sprayed B4C-Mo Composite Coating *Journal of Inorganic Materials* **31** 100-106
[15] Rafiei M, Salehi M and Shamanian M 2014 Formation Mechanism of B_4C-TiB_2-TiC Ceramic Composite Produced by Mechanical Alloying of Ti–B_4C Powders *Advanced Powder Technology* **25** 1754-60
[16] Ebrahimian-Hoaaeinabadi M, Azari-Dorcheh K and Moomir-Vaghefi S 2006 Wear Behavior of Electroless Ni-P-B_4C Composite Coatings *Wear* **260** 123-127

IOP Conf. Series: Materials Science and Engineering **286** (2017) 012007 doi:10.1088/1757-899X/286/1/012007

Study of the structure and chemical composition of the protective coating of a fist stage gas turbine blade after regenerative heat treatment

D I Davidov, N V Kazantseva, N I Vinogradova and I V Ezhov

M.N. Miheev Institute of Metal Physics of Ural Branch of Russian Academy of Sciences, 18, S. Kovalevskaya Street, Ekaterinburg, 620137, Russia

E-mail: davidov@imp.uran.ru

Abstract. Investigation of the structure and chemical composition of the protective coating of the first stage IN738 gas turbine blade after standard regenerative heat treatment was done. It was found the degradation of microstructure and chemical composition of both the blade feather and its protective coating. Redistribution of the chemical elements decreasing the corrosion resistance was observed inside the protective coating. Cracks on the boundary between the blade feather and the protective coating were found by scanning electron microscopy. The carbide transformation and sigma phase were found in the structure of the blade feather. Based upon the structural and chemical composition studies, it is concluded that the standard regenerative heat treatment of the IN738 operative gas turbine blade does not provide full structure regeneration.

1. Introduction

Heat-resistant nickel alloys are well-known materials for the manufacture of gas turbine blades operating at high temperatures and aggressive environments. Gas turbine blades made from heat-resistant nickel superalloys work in the aggressive corrosive gas environment; hot corrosion is one of the main problems of these materials. Special alloying and application of protective coatings are usually used to improve the corrosion properties of these materials.

The Inconel 738C (IN738) super alloy, developed in 1968, is one of the important nickel based super alloys. This alloy shows the improved creep, hot corrosion, and oxidation resistance and is used in land-based gas turbines [1]. The working temperature of the first stage INC738 gas turbine blade is about of 1100°C [2]. The IN738 super alloy contains refractory elements such as Mo, W, Ta, Cr and Co to prevent local hot corrosion. This super alloy resists oxidation to 1093 °C and shows the improved sulfidation resistance up to 982 °C [1].

In addition to alloying of refractory elements, the different corrosion resistant coatings are used for nickel gas turbine blades. Aluminum coatings have the ability to form a protective oxide film of Al_2O_3 allowing the material to work at temperatures above 900 °C [3]. Yttrium can be added into the aluminide coating for increasing the resistance of the INC738 gas turbine blade against oxidation and hot corrosion [4]. The most effective hot corrosion resistance alloy element is chromium. The presence of chromium in the alloy reduces the amount of aluminum needed for the formation of a protective Al_2O_3 oxide [5]. The chromium combines with oxygen to form Cr_2O_3, which prevents dissolution of protective oxide [2]. MCrAlY coating in which M = Ni, Co or Fe, is successfully used for IN738 superalloy to increase the corrosion protection [5-6]. Co-36.5Ni-17.5Cr-8Al-0.5Y protective coating is applied by low pressure plasma spraying [6] or electrolytic deposition on hot sections of gas turbine [5]. Both processes are followed by a homogenization heat treatment at 1040-1120 °C [5].

During the exploitation gas turbine blade is in stressed thermo-mechanical conditions. Degradation process occurs in both the internal structure of the blade and protective coating of the blade. Structure recovery under heat treatment is important process determining service lifetime of the gas turbine blade.

In this article we present the investigation of the structure and the chemical composition of the protective coating of the operating IN738 gas turbine blade after standard regenerative heat treatments.

2. Experimental

Standard regimes of the regenerative annealing were described in [7]. Service-exposed IN738 blade of the first stage of a gas turbine after a standard two steps regenerative annealing (1121 °C for 2 hr, air cool then 843 °C for 24 hr, air cool) was used for investigation. The annealing of the blade after standard exploitation time was done in accordance to the service plan. The study was carried out using an optical microscope Micromed MET and a scanning electron microscope JSM 6490 with the energy dispersive and wave micro-analyzer Oxford Inca. Small pieces were cut from the different parts of the blade feather for study. The MCrAlY coating was applied by vacuum plasma spray process.

3. Results and discussion

The inner part of the blade feather, which is bordering with the cooling system, is presented in Figure 1a. Thickness of the protective coating in this part of the blade is about of 50 μm. The thickness of the protective coating in the outer part of the blade feather is larger than that in inner part (Figure 1b).This part of the blade works at hot temperature and is more stressed.

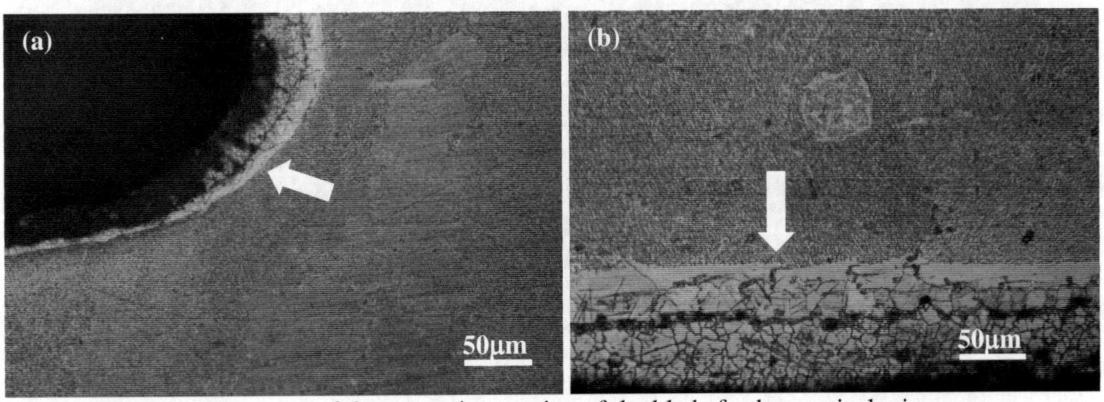

Figure 1. The areas of the protective coating of the blade feather, optical microscopy: a- inner part; b - outer part.

Structure of the blade feather is shown in Figure 2. Carbides are observed inside the grains and their boundaries. Intragranular carbides have a well-defined globular shape which is typical for primary MC carbide. Two types of the carbides are observed at the grain boundaries: the primary MC carbide with the globular shape and secondary $M_{23}C_6$ carbide with an elongated shape (Figure 2a). It is known that $M_{23}C_6$ carbides in nickel super alloys are mainly deposited from the matrix during heat treatment and service (at 760–980 °C) [8]. $M_{23}C_6$ carbide may consist of chemical elements such as Cr, W and Mo. This type of carbide is formed due to super-saturation of carbon in the matrix and degeneration of the MC type carbide [8]. The precipitations of the irregular shape distributed along the grain boundaries and inside grains are associated with the σ-phase (Figure 2b). This phase is enriched with nickel, tantalum, titanium, niobium, and molybdenum.

Needle-shaped precipitates, presumably borides, were found at the boundary between the coating and the outer part of the blade feather (Figure 3a). The straight determination of the chemical composition of these precipitates is difficult due to the limitations associated with the size of the electron

IOP Conf. Series: Materials Science and Engineering **286** (2017) 012007 doi:10.1088/1757-899X/286/1/012007

microscope probe. Attention should be paid that the morphology of the hardening intermetallic γ' -phase near the boundary with the protective coating is changed (Figure 3b). Intermetallic γ'-phase precipitates have elongated shape and are bigger in size than the γ' cuboids of the blade feather (Figure 3b).

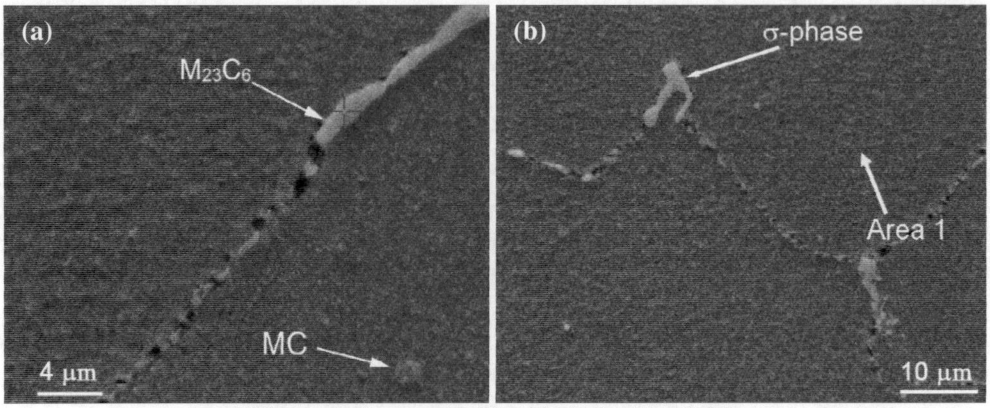

Figure 2. Structure of the blade feather, SEM.

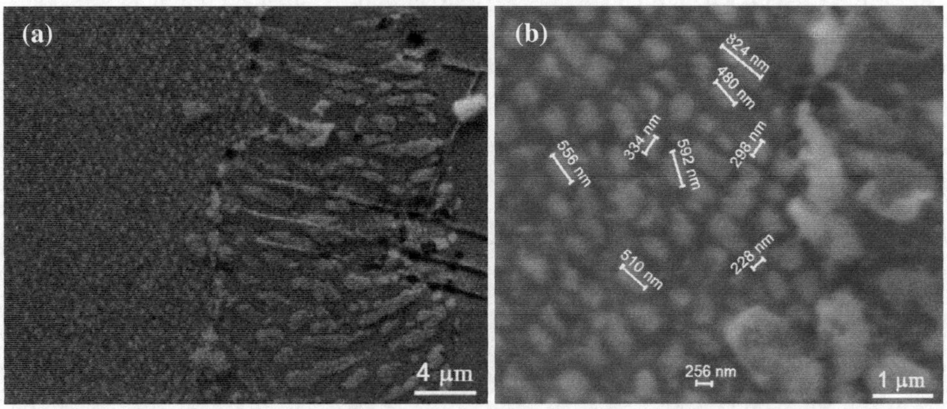

Figure 3. Microstructure of the γ'-phase near the protective coating of the outer part of the blade feather.

The EDS analysis supports these observations. Secondary carbide $M_{23}C_6$ has a high content of nickel, chromium, and cobalt. EDS analysis of the primary MC carbide shows a high content of titanium, niobium, and tantalum; chromium is practical absent. According to the literature data, hardening of the grain boundaries in heat-resistant nickel superalloys is achieved by MC type carbides based on Nb, Ti, W. To ensure a high heat resistance, carbides should have the globular shape, size of about 1 µm or less. MC carbides should be uniformly distributed along the grain boundaries without formation of a continuous grid. The probability of the formation of topologically close-packed (TCP) phases (σ, µ, Laves phases), as well as of M_6C or $M_{23}C_6$ carbides, leading to softening of the alloy, should be minimized [7]. The type of carbide transformations and TCP phases depends on alloy doping and operating temperature. According to the literature, carbides $M_{23}C_6$ and σ-phase may form in the IN738 super alloy after the exploitation for a long time at high temperature [1, 9]. The appearance of σ phase as well as $M_{23}C_6$ carbides points to the degradation of the alloy composition and leads to embrittlement.

Figure 4 presents the microstructure of the protective coating of the inner part of the blade feather. The temperature and stress distributions in the different parts of the blade feather were presented in [10]. According to [10], the inner part of the blade feather is also at high temperature but the bending stress in this part is lower than that in outer part.

IOP Conf. Series: Materials Science and Engineering **286** (2017) 012007 doi:10.1088/1757-899X/286/1/012007

In our blade the protective coating of the inner part of the blade feather was applied on the edge zone near the cooling system. According to EDS analysis, the chemical composition of this coating is complicated by the diffusion of the elements from the blade feather (Table 1).

Chemical composition of the protective coating of the outer part of blade feather has a high percentage of oxygen, aluminum, nickel, chromium, and yttrium. Numerous longitudinal cracks running along the coating are found by scanning electron microscopy (Figure 5b). The oxygen and aluminum content in coating of this part of the blade feather is much lower than that in the inner part. Yttrium, chromium, cobalt and nickel are also present in the composition of the coating. Redistribution of the concentrations of the chemical elements inside the protective coating is observed (Table 1). The EDS results of the chemical composition of the blade feather are given in Table 2.

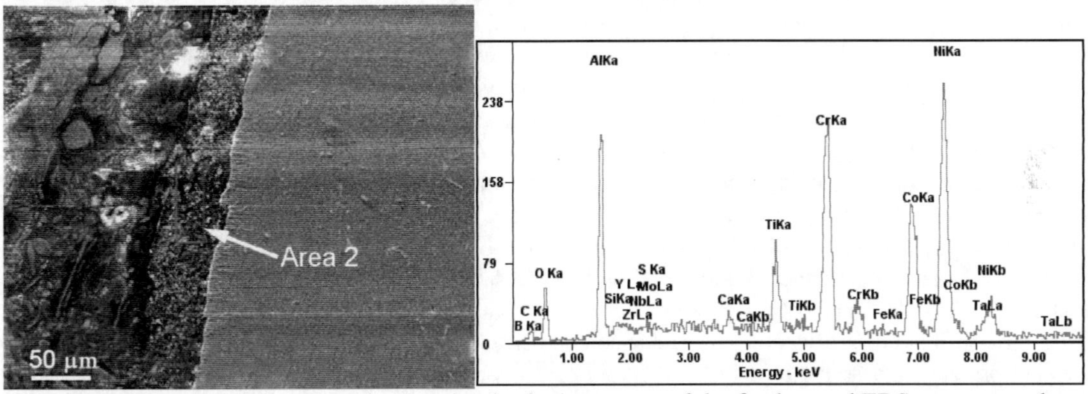

Figure 4. Structure of the protective coating in the inner part of the feather and EDS spectrum taken from the Area 2.

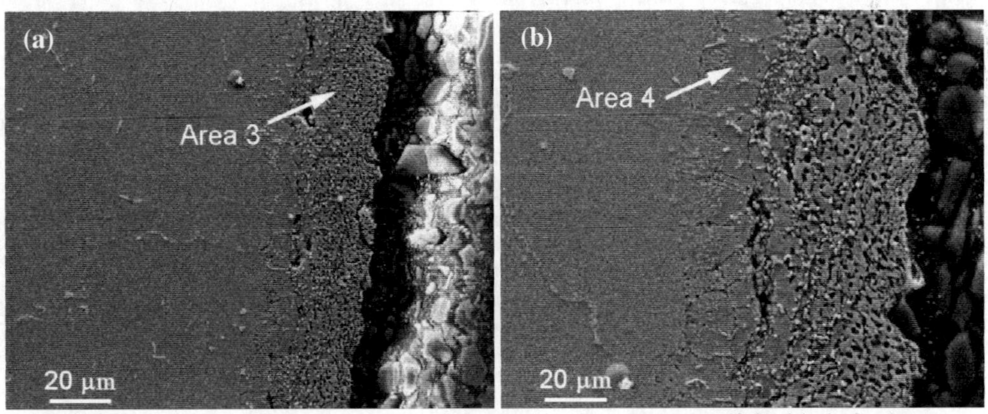

Figure 5. Structure of the protective coating in the outer part of the blade feather.

Table 1. Chemical composition of the protective coating in the different areas, EDS analysis from marked areas in Figures 4-5.

Area	Ni	Al	Co	Cr	Y	Si	Ta	O	C	Fe	Zr	Nb	Mo	Ti
2	29.93	18.66	14.03	14.03	1.03	0.74	2.67	6.69	5.96	0.66	-	-	1.48	4.12
3	29.55	2.01	34.07	24.65	1.08	0.43	1.38	1.36	2.45	0.46	0.34	0.41	1.08	0.73
4	53,58	2.51	9.07	14.95	1.14	1.68	3.23	0.79	0.59	0.69	1.09	1.88	5.21	3.59

Table 2. Chemical composition of Inconel 738C, wt.% (marked Area 1 in Figure 2b).

Ni	Cr	Co	Mo	Ta	Al	Ti	C	Zr	Fe	Nb
balance	15.23	8.73	3.84	5.02	2.18	3.64	1.4	0.89	0.68	1.19

It can be seen from the Tables 1-2 that the degradation of coating occurs not only due to oxidation. The area of the protective coating, which is close to the boundary with the feather (Area 4 in Figure 5b), has the chemical composition which is close to that of the blade feather (Area 1 in Figure 2). In this case, we can say about the diffusion redistribution of the chemical elements between blade feather and protective coating. The process of redistribution the chemical elements may be explained by the Gorsky effect which was observed previously in the gas turbine blade after long service time in [11]. This effect deals with ascending diffusion under elastic-plastic conditions. Under stress, the atoms with small diameter move into the compressed regions and atoms with bigger diameters move into the stretched regions. It is known that the different parts of the working gas turbine blade are under different stress-temperature conditions [10]. The upper edge of the blade feather is subjected to the greatest stresses and temperatures in comparison with that of the inner part of the blade feather which is bordering with the cooling system. In this study we investigate the upper edge of the blade feather and part near the cooling system. Comparison between Table 1 and 2 shows that the oxygen concentration in the coating of the inner part of the blade feather is increased as high as 6.69% and cobalt and chromium contents are reduced to 14.03%. The coating of the outer part of the blade feather has less oxygen content but aluminum content is significantly reduced. Such decrease in content of aluminum testifies that coating does not protect the base metal.

4. Conclusion

The following conclusions can be done from this study:

1. The cracks on the boundary between the feather blade and protective coating are observed. This fact indicates a low adhesion and the possibility of peeling off the coating. Redistribution of the chemical elements decreasing the corrosion resistance is observed in the protective coating.
2. It is found the degradation of microstructure and chemical composition of both blade feather and protective coating. The degradation process is caused by stress, high temperature, hot corrosion, and oxidation. Service induced MC to $M_{23}C_6$ carbide transformation and formation of TCP σ-phase indicate the redistribution of the chemical elements inside the material of the blade feather.
3. Based upon the structural and chemical composition studies, it is concluded that the standard regenerative heat treatment of the IN738 operative gas turbine blade does not effect on carbides transformation, TCP σ-phase dissolution, and the protective coating, and thus do not guarantee the full recovery of the IN738 gas turbine blade.

Acknowledgments

The research was supported by the Russian Science Foundation, grant No 15-12-00001.

References

[1] Jonšta P, Konečná K, Heide R and Gabčová M 2011 *Z Metall-2011* **5.18 -20** 1–8
[2] Rani S, Agrawal A K and Rastogi V 2017 *Case Studies in Engineering Failure Analysis* **8** 1–10
[3] Zielinska M, Sieniawski J, Yavorska M and Motyka M 2011 *Archives of Metallurgy and Materials* **56 (1)** 193–197
[4] Tu D C, Lin C C, Liao S J and Chou *J C 1986 J. of Vacuum Science & Technology A: Vacuum, Surfaces, and Films* **4** 2601–2608
[5] Saremi M and Bahraini M 2003 *Trans. Inst. Met. Fin.* **81(1)** 24-27
[6] Dahl K V, Hald J and Horsewell A 2006 *Defect and Diffusion Forum* **258-260** 73-7

[7] Chabina E V, Filonova E V, Lomberg B S and Bacragze M M 2012 *All materials* (in Russian) **6** 1-11
[8] Kolagar A M, Tabrizi N, Cheraghzadeh M and Shahriari M S 2017 *Case Studies in Engineering Failure Analysis* **8** 61–68
[9] Koul A K, Immarigeon J-P, Castillo R, Lowden P and Liburdi J 1988 *Superalloys* 755-764
[10] Tapse S, Deshmukh A, Bhosale R, Gadhave S and Naidu M J 2017 *International Conference on Ideas, Impact and Innovation in Mechanical Engineering (ICIIME 2017)* **5(6)** 444-452
[11] Kazantseva N V, Korolev A V, Davidov D I, Stepanova N N, Vinogradova N I and Rigmant M B 2013 *Metallography* **4** 18-24.

IOP Conf. Series: Materials Science and Engineering **286** (2017) 012008 doi:10.1088/1757-899X/286/1/012008

Analysis of the ceramic layer microstructure influence on plasma spray thermal barrier coating performance

V I Bogdanovich and M G Giorbelidze

Samara National Research University, 34, Moskovskoe Highway, Samara, 443086, Russian Federation

E-mail: bogdanovich@ssau.ru; m.giorbelidze@ssau.ru

Abstract. This paper outlines the results of analysis and describes the structure of the thermal protection coatings formed by atomic ion stream deposition in vacuum, and plasma thermal spraying method. Crystallite structure features are considered along with the crystallite dimensions, spatial orientation, and position of the boundaries between separate crystallites. Discontinuity, volume, and morphology of the pores has been evaluated. Experimental studies have been accomplished using various fractions of the powder-like material $ZrO_2 - 8\%Y_2O_3$. The influence of the coating microstructure on the coating performance has been analyzed, such as adhesive strength, thermal stability, and thermal conductivity.

1. Introduction

Improvement of efficiency, reliability, and service life are the topical areas of the modern engine building industry. Finding solutions to these problems is provided by various design and engineering means including new materials and special coatings. Protective coatings have become popular in production of modern gas-turbine engines with high operational temperature ratings that are far beyond the potential of the heat-resistant structural materials used for the hot path parts, such as rotary blades and nozzle blades of the turbine, combustion liners, and combustion chamber [1–7]. Currently, from the viewpoint of the structural material thermal protection, the most efficient system is made of heat-resistant bonding metal-ceramic coating Me–Cr–Al–Y and external ceramic coating $ZrO_2 - 8\%Y_2O_3$ [4, 5, 7–10]. Apart from the materials, protection capacity depends on the method and operating practices used to form the coating that predetermine its structural features and performance [10].

2. Materials used in modern thermal protection coatings

It is known [1-3,6] that the first stage gas-turbine engines blades operate in the most severe operating conditions in comparison to other components and practically determine the engine service life. Failure of cooled blades made of heat-resistant superalloys is caused by high-temperature gas corrosion that results in further intergranular corrosion and fatigue cracking under high-temperature and mechanic activation. Attempts to use oxide coatings to protect such alloys revealed a number of problems, the major one is thermal endurance. Plausible solution to this problem has been found in flattening the gradient of thermal expansion coefficient per system thickness by making use of intermediate damping layers, and resulted in creation of a coating material Me–Cr–Al–Y (where Me stands for Ni, Co, Fe or their combinations). This coating was termed metal-ceramic, and for a rather long time period was used as the main thermal barrier coating (TBC) for components with relatively low operation temperature.

To date, more than three million blades have been manufactured with such coatings and are currently being operated [1].

The material is based on heat-resistant alloy components (Ni–Cr), which minimize changes in thermal expansion coefficient, while additional components (A1–Y) are meant to produce high temperature stable oxides on the surface at the end of recrystallization annealing. Processes of diffusion, formation and reduction of yttria-stabilized aluminium oxide still continue against its erosive wear from surface during operation. Thus, this system gradually changes the value of thermal expansion coefficient from values close to the ones of the heat-resistant alloy to values close to ones of the oxide. Metal-ceramic coating service life is over when Al and Y concentration in the material goes under certain critical value below which overall oxide covering is not produced anymore, and the rest components of the material undergo intensive high-temperature corrosion. So, service life is actually specified by the thickness of the coating, as aluminum concentration increase above 13% and yttrium above 0.6% proved ineffective. However, metal-ceramic thickness increase above certain values results in reduction of endurance limit within the coating – metal base system. There exist special manufacturing methods that allow partially balancing this effect, but development of new generation engines with extreme temperatures in combustion chamber required new systems of multi-layer thermal barrier coatings.

Currently, as thermal protection coatings, the systems of metal-ceramic and ceramic layers are used; due to the gradient changing of the properties, these systems are able to provide optimal combination of the adhesive, corrosive, and thermal stability of such coating [4]. As for the ceramic layer, the most popular material is ZrO_2, stabilized 8%Y_2O_3 [7]. ZrO_2 stands out among refractory oxides because of its high melting temperature, low thermal conductivity, relatively high (in comparison to other oxides) thermal expansion coefficient, better chemical stability, structural stability, strength, and resistance to cracks, while being resistant to corrosion. When heated, it is rather stable and loses insignificant amount of oxygen even when melted [4].

Apart from the materials, protection capacity depends on the method and operating practices applied to form such coating, they determine its structural features and performance [10].

3. Dependency of the coating microstructure and properties, and method of application

Notable structure of the coating formed by plasma thermal spraying method is layered and consists of disk-shaped crystallites located along the surface of the structural material, horizontal and vertical borders between the crystallite and micro voids (pores) in the area of the crystallites vertical borders (Figure 1). Thickness of each disk-shaped crystallite is 2 – 10 µm and 10–20 times smaller than its typical horizontal dimension. Vertical borders of the crystallites are limited from both sides by the monolithic material of the other disk-shaped crystallites. The coating has no end-to-end pores, and the percentage of the isolated porosity varies by the spraying modes within 2 – 12 % [8]. Such coatings made from the suitable materials guarantee efficient protection of the structural materials against corrosion, including the high-temperature gas corrosion. Coating adhesive and cohesive compression strength and shear strength is significantly higher than the tensile strength. However, experience has shown [1] that the coating adhesive strength for peeling 25 – 45 MPa and cohesive strength for tension, which is only 0.4 – 0.8 of the monolithic material strength, provide their performance in all known scenarios of complex heavy-duty applications because of their operation under the conditions of external compressive or shearing mechanical forces. One of the main features of this plasma thermal spraying coating structures is its high damping capacity to the alternating mechanical and temperature stresses, as well as the ability to localize the fatigue damage and microcracks inside the crystallite's seed, preventing from growth to the structural material of the substrate.

IOP Conf. Series: Materials Science and Engineering **286** (2017) 012008 doi:10.1088/1757-899X/286/1/012008

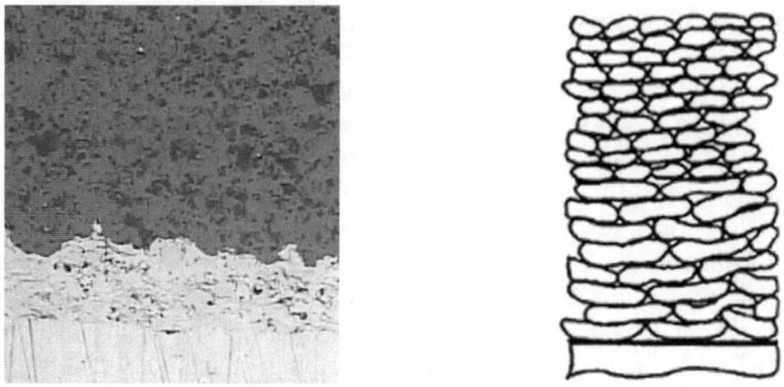

a b

Figure 1. Plasma thermal spraying coating structure (a) and crystallites positioning in its structure (b)

This cluster-like structure of the plasma coatings where the pores are oriented orthogonally to the thermal stream is radically different from the structure of the coatings formed by deposition of the atomic and ionic streams in vacuum (thermal, electron-beam, ionic-plasma deposition) or atomic streams in the galvanic & chemical methods. In this case, the microstructure is column-like. Each separate column of the structure consists of some crystallites with smaller diameter typically oriented normally to the surface of the substrate material (Figure 2). Borders between the columns and crystallites begin on the surface of the coating and end on the surface of the substrate, while the cohesive strength between the crystallites is much greater than the cohesive strength between the columns of the structure [7, 10].

Availability of these vertical borders results in forming of channels for penetration of oxygen ions, harmful impurities, and other chemical elements into the protected structural materials. The vertical borders of columns having thickness more than 5 – 10 μm are the reason of decreasing of endurance strength of the coated product in comparison with the uncoated product. These high-strength borders between the crystallites often contain the emissions of various phases and do not limit the growth of fatigue cracks, but only in some scenarios slow their growth down. Columns and crystallites are relatively non-elastic in the normal direction, and have unsatisfactory toughness. Since the columns are oriented in such a way where the crystallites bond with one end on the small surface area in comparison to their volume, erosion problems may occur [7, 10].

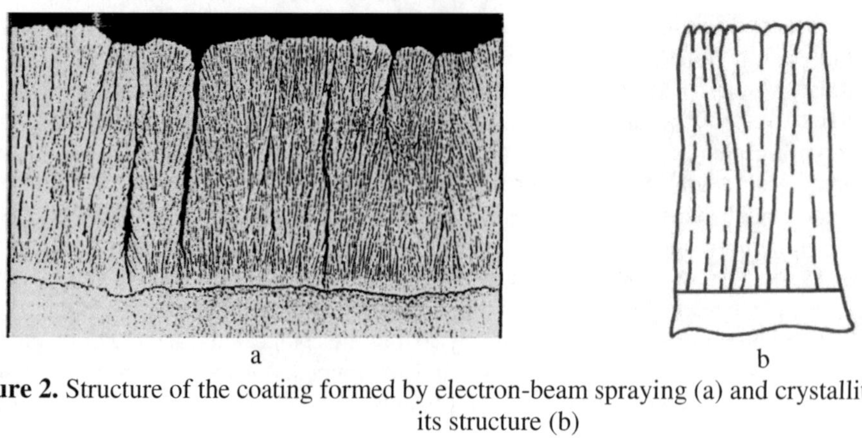

a b

Figure 2. Structure of the coating formed by electron-beam spraying (a) and crystallites positioning in its structure (b)

Structure of the thermal spraying coating also significantly differs from the structure of monolithic structural materials consisting of sizable crystallites with strong bonds between each other through common borders. Structure similar to the layered disk-shaped structure of the plasma coatings is often formed in the near-surface layer of the product to improve the fatigue endurance limit by processing the

IOP Conf. Series: Materials Science and Engineering **286** (2017) 012008 doi:10.1088/1757-899X/286/1/012008

surface with surface plastic deformation methods. However, the possibilities of such methods in creation of disk-shaped crystallites in monolithic materials are a lot scantier than the same of the plasma coating deposition method from powder-like materials.

Often, in the process of development of plasma thermal spraying coating, there is a viewpoint about the need to improve the plasma coatings quality by means of the densification and creation of non-porous structure. According to our experience [10], such structure stays efficient only when used as not thick layer on the typical structure of the plasma coating, since the non-porous monolithic structure results in significant reduction of its service life under alternating mechanical and thermal loads.

Physical nature of layered structure formation from the disk-shaped crystallites when forming the plasma thermal spraying coatings is associated with the step-by-step deposition of powder particles accelerated and heated in the plasma, which – when hitting the surface and with further plastic deformation – becomes disk-shaped [8, 9]. This mechanism of the formation of coating contributes to creation of relatively big pores parallel to the surface of the substrate structural material. Experiments have confirmed [11] that such morphology and positioning of the pores in the plasma gas and thermal coatings provides significantly less thermal conductivity unlike the coatings formed by deposition from atomic streams (Figure 3). Therefore, we can say that the plasma thermal spraying method is the best for protection of surfaces of the gas-turbine engine parts exposed to high temperature loads from the viewpoint of the control over the structure and properties in each separate layer of the coating.

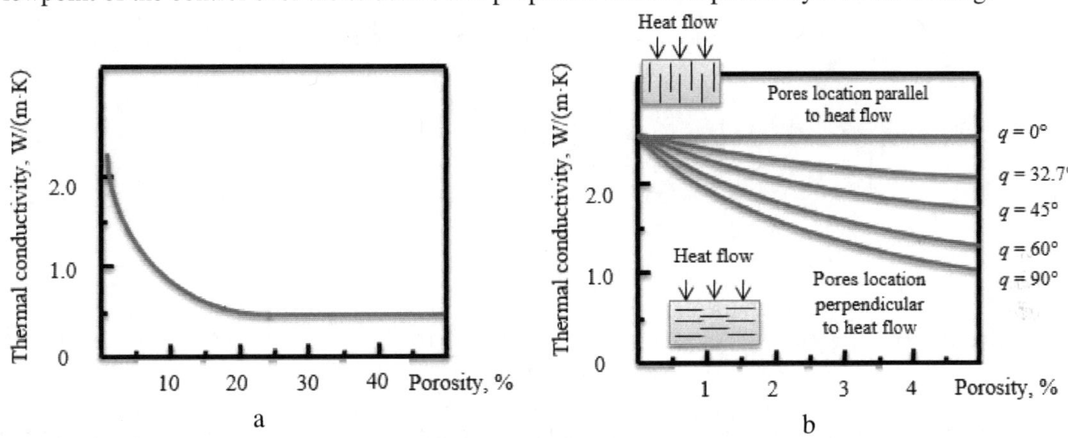

Figure 3. Influence of the pores volume and morphology on the thermal conductivity of coating made of $ZrO_2 – 8\%Y_2O_3$: *a* – experimental data, *b* – calculated data [11].

4. Analysis of powder-like materials granulation influence on the plasma thermal spraying coating microstructure and thermal endurance

Currently, thermal barrier coatings on the hot gas path of the gas-turbine engines are produced in the form of the metal-ceramic layers 100 – 150 μm thick from Me–Cr–Al–Y powder and external layer of ceramic coating 150 – 350 μm thick from $ZrO_2 – 8\%Y_2O_3$ [1–3].

Tests have been accomplished using the fractions: 20 – 40 μm, 40 – 60 μm, and 60 – 80 μm of powder material $ZrO_2 – 8\%Y_2O_3$. The use of fine fraction 20 – 40 μm has reduced the amount of open porosity almost by three times, and reduced the amount of total porosity determined on metallographic specimens to 4 % (Figure 4, Figure 5).

Analysis of the strength of the plasma thermal barrier coatings sprayed from the powders with various granulation, with spraying distance 40 – 80 mm, have demonstrated that no matter what modes are used, the peeling strength is about 28 – 35 MPa that provides durability of the coating.

IOP Conf. Series: Materials Science and Engineering **286** (2017) 012008 doi:10.1088/1757-899X/286/1/012008

a b c

Figure 4. Microstructure of plasma thermal protection coatings formed with the following fractions of the powder material $ZrO_2 - 8\%Y_2O_3$: a – granulation 20 – 40 μm; b – granulation 40 – 60 μm; c – granulation 60 – 80 μm

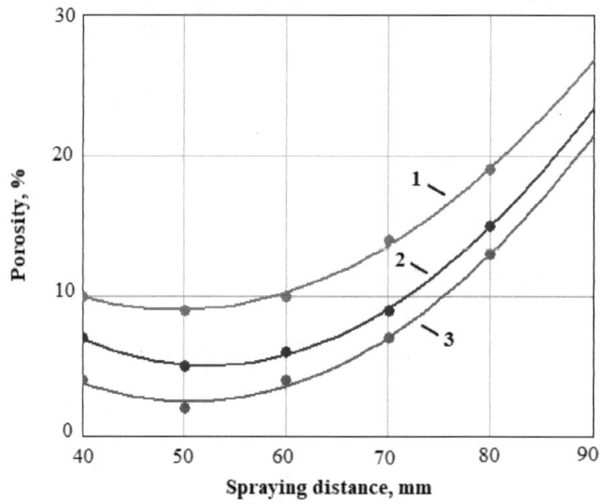

Figure 5. Dependence between the plasma thermal barrier coatings porosity and spraying distance: 1 – granulation 60 – 80 μm; 2 – powder granulation 40 – 60 μm; 3 – powder granulation 20 – 40 μm; — approximated curve; • – experimental data.

Thermal fatigue tests have demonstrated that the specimens with the coating formed using the powder material granulation 40 – 60 μm has the best properties. This is associated with greater damping capacity of the layers formed from the powder particles of this fraction. Increase of the coating porosity to 10 – 12 % results in deterioration of the coating protective properties, and consequently to reduction of thermal fatigue properties (Figure 6).

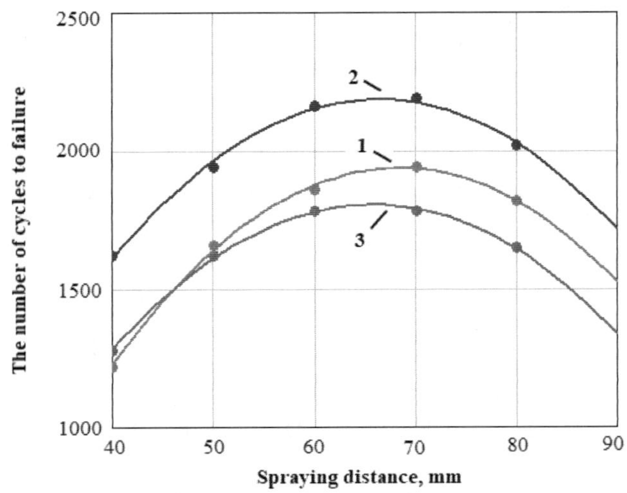

Figure 6. Dependence between the number of cycles to failure of plasma thermal barrier coatings and spraying distance: 1 – granulation 60 – 80 μm; 2 – powder granulation 40 – 60 μm; 3 – powder granulation 20 – 40 μm; — approximated curve; • – experimental data.

5. Results and Conclusions

Apart from materials used, protective capacity of the coating in significant extent depends on the method and operating practices that largely predetermine structural features and performance of the coating [10]. To protect the parts of the modern gas-turbine engines exposed to heavy thermal loads, the optimal way is to use the plasma thermal spraying method from the viewpoint of the possibility to control the structure and properties of each separate layer of the coating, and from the viewpoint of providing significantly less thermal conductivity coefficient in comparison to the atomic and ionic flows deposition in vacuum.

Experimental studies have been accomplished using the various fractions of the powder-like material $ZrO_2 - 8\%Y_2O_3$. Results of the metallographic control has demonstrated that the use of fine fraction 20 – 40 μm has reduced the amount of open porosity almost by three times, and reduced the amount of total porosity to 4 %. It has been established that the increase of the coating density within the entire system leads to the better adhesion of the coating with the substrate, but leads to worse thermal endurance and thermal endurance strength because of damping properties deterioration of thermal protection coating layers. Therefore, some compromise is required in the structure of the system to improve its resistance against the high-temperature gas corrosion and increase of resistance to multi-cycle alternating thermal and mechanical loads.

Acknowledgments

The reported study was funded by the Ministry of Education and Science of the Russian Federation according to the research project No.11.2744.2017.

References

[1] Barvinok V A 2005 *Plasma in Technology: Reliability and Resource* (Moscow: Nauka i technologii Publishers) p 456

[2] Eliseev Yu S, Boytsov A G and Krymov V V 2003 *Technology of aviation gas turbine engines production* (Moscow: Mashinostroyeniye) p 510

[3] Eliseev Yu S, Krymov V V, Kolesnikov S A and Vasilyev Yu N 2007 *Non-metallic composite materials in structural elements and the manufacture of aviation gas turbine engines* (Moscow: Bauman University Publishers) p 368

IOP Conf. Series: Materials Science and Engineering **286** (2017) 012008 doi:10.1088/1757-899X/286/1/012008

[4] Bobrov G V, Ilin A A and Spektor V S 2014 *Theory and technology of inorganic coatings formation* (Moscow: Alfa-M) p 928
[5] Puzryakov A F 2003 *Theoretical bases of plasma spraying technology* (Moscow: Bauman University Publishers) 458
[6] Kablov E N and Muboyadzhyan S A 2012 *Metals* **1** 5-13.
[7] Tamarin Yu A and Kachanov E B 2008 *New technological processes and reliability of gas turbine engines* **7** 144-158
[8] Dokukina I A 2012 *VESTNIK of Samara University. Aerospace and Mechanical Engineering* **5** 49-54
[9] Bogdanovich V I and Giorbelidze M G 2017 *Key Engineering Materials* **743** 118-123
[10] Bogdanovich V I and Giorbelidze M G 2016 *IOP Conference Series: Materials Science and Engineering* **156(1)** 012016
[11] Hasselman D.P.H. 1978 *Journal of Composite Materials* **12** 403-407

Biological degradation of gas-filled composite materials on the base of polyethylene

E A Grigoreva[1,4], N N Kolesnikova[1], A A Popov[1, 2] and A A Olkhov[1, 2, 3]

[1] Emanuel Institute of Biochemical Physics of Russian Academy of Sciences, Kosygina st. 4, Moscow, 119334, Russia,
[2] Plekhanov Russian University of Economics, Stremyanny per. 36, Moscow, 117997, Russia
[3] Semenov Institute of Chemical Physics of Russian Academy of Sciences, Kosygina st. 4, Moscow, 119991, Russia

E-mail: aquamirabilis@yandex.ru

Abstract. Gas-filled composite materials based on polyethylene were obtained. It was assumed that introduction of porosity in polyethylene will improve the biodegradability of synthetic materials. The morphological and structural changes were estimated, physical and mechanical properties, stability in water and soil of these materials were determined. It is stated that filling the polymer matrix with pores increases the ability to degrade in nature.

1. Introduction

Polyethylene comes forward in capacities of production and consumption in comparison with all other kinds of polymers in Russia. It is resistant to chemical and biological factors, meets high operational requirements. But after use it stays the same in landfills for hundreds of years, emitting harmful substances. The problem to improve the ability of polyethylene and other industrial synthetic polymeric materials (polypropylene, polyvinylchloride, polystyrene, polyethylene terephthalate) to degrade in natural conditions takes an important place among scientific researches [1, 2].

One of the ways to solve this problem is to make compounds of synthetic polymers with biodegradable components that are able to initiate fragmentation of a material or even decay of a polymer chain. Scientists develop this trend by making different kinds of composite materials, based on polymer matrix. Some researchers propose to add natural polymers [3-5]. Other scientists offer biocomposites containing finely divided particles of natural origin [6-8]. In some researches vegetable waste products are added to the polymer that helps to get not only biodegradable but inexpensive materials [7,9]. A synthetic matrix provides fine operational properties; an addition of natural filler provides a possibility to degrade after use. Structure, physical and mechanical properties, stability in water environment and soil of these compositions should be investigated [5].

The resistance of polymers to biological factors is due to the high degree of crystallinity and big sizes of molecules [10]. Addition of nature components makes the structure heterogeneous, decreasing resistance to different destructive factors. The process of filled polymer destruction starts with the natural component first, then the polymeric matrix declines. The polymeric framework, involved in

IOP Conf. Series: Materials Science and Engineering **286** (2017) 012009 doi:10.1088/1757-899X/286/1/012009

biological processes, decomposes in terms of oxidation and hydrolysis. These reactions cause fragmentation of the polymer because of macromolecules destruction and chain length reduction.

The aim of this work was to make gas-filled compositional materials, to investigate their structure and properties and reveal the influence of the porous structure on the ability to degrade in nature; then to choose compositions with optimal properties for adding natural component to it.

2. Experimental part

2.1. Objects and methods

The research considers compositions on the basis of low density polyethylene. A blowing agent - hydrocerol was added to make the structure porous. Wood flour with average particle size of 0.08, 0.14, 0.2 mm and starch with particle size 0.1 mm were added as natural disperse fillers.

Compositions of the polyethylene with hydrocerol were proposed (table 1). The hydrocerol contents were: 0, 1, 3, 5, 10 and 15wt%. Compositions were made by mechanical blending on the laboratory micro-rollers, warmed to 120°C. Then the compositions were foamed by using laboratory manual hydraulic press at 160°C (decomposition temperature for hydrocerol) without loading. Such a method allowed getting flat gas-filled plates.

Table 1. Compositions of test samples.

Number of the composition	Quantity of polyethylene (%)	Quantity of hydrocerol (%)
1	99	1
2	97	3
3	95	5
4	90	10
5	85	15

The density of samples was determined by the hydrostatic weighing method. Parameters of the structure were investigated by the optical microscope (Axio Imager Z2m Carl Zeiss, Germany). Thermophysical parameters (polyethylene melting temperature, gas formation temperature and the degree of crystallinity) were measured by differential scanning calorimetry on micro-calorimeter (DSM-10m, Russia). Listed above studies were carried out with the equipment of the Center of Shared Usage «New Material and Technologies of Emanuel Institute of Biochemical Physics». Water absorption was measured by samples weight change in distilled water. The ability to degrade in the environment was measured by samples weight change in soil.

2.2. Results and discussion

The thermogram of heating samples of polyethylene, polyethylene with hydrocerol is presented in figure 1. It shows two peaks that give an idea of temperatures of melting and gas formation. The first peak corresponds to the process of polyethylene melting (106°C). The second peak corresponds to the process of gas formation (160°C). It did not appear in case of pure polyethylene. With the help of the computer program «Peak Explorer» squares of peaks were determined and the degree of cristallinity was measured. The investigation of samples with the different content of hydrocerol showed that thermal effect of hydrocerol decomposition is directly proportional to its content in the polymeric mass. The degree of cristallinity stays almost the same (28%) in spite of hydrocerol containing change.

IOP Conf. Series: Materials Science and Engineering **286** (2017) 012009 doi:10.1088/1757-899X/286/1/012009

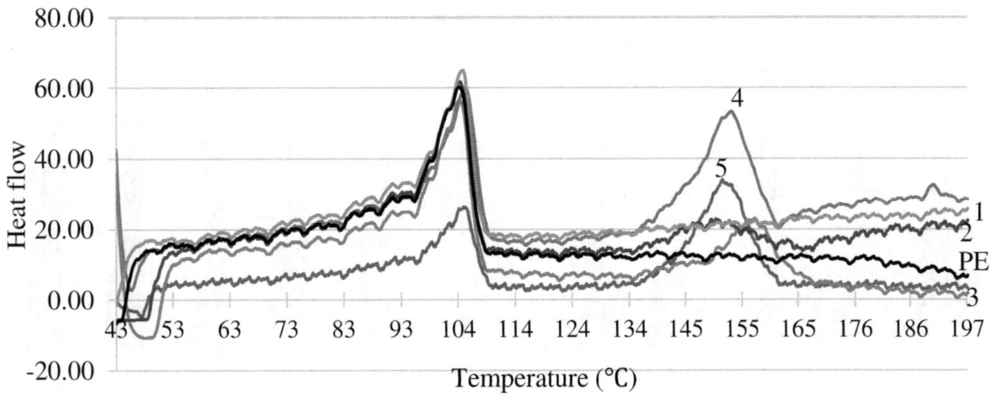

Figure 1. Thermograms of melting and gas formation: thermogram of heating samples of polyethylene (PE), polyethylene with 1% (1), 3% (2), 5% (3), 10% (4) and 15% (5) of hydrocerol.

At the next stage, the structure of samples was investigated. Microphotographs of the gas-filled samples in case of magnification 100x are shown in figure 2. The investigation of cross section of samples presents that pores appear in the polyethylene matrix. In samples with the low content of blowing agent pores and cells have clear regular spherical shape with smooth walls, isolated from each other and from the external environment. Within an increase of hydrocerol content a number of cells is growing, their shapes are getting irregular, walls between them are thinning and destroying. The cells interpenetrate into one another, connect with each other and with the external environment.

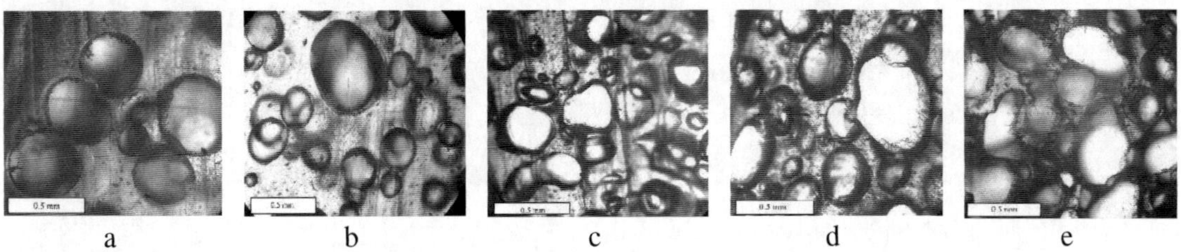

Figure 2. Microphotographs of the gas-filled samples of compositions with different hydrocerol content, a – 1%, b – 3%, c – 5%, d – 10%, e – 15%.

Diameter of pores and thickness of walls between it were measured, making the assumption, that all pores have a clear and spherical form. The average diameter of pores varies from 0.21 to 0.40 mm, few pores had bigger diameters (0.61-0.80 mm) and few pores had larger diameters (0 – 0.20 mm). The average thickness of walls of pores varies from 0.01 to 0.40 mm; walls with thickness of up to 0.10 mm prevail. The results of this measurement are given in figure 3, a - pores, b - walls. A number of pores per 1 mm^2 was calculated. It varies between from 9.6 to 12.5.

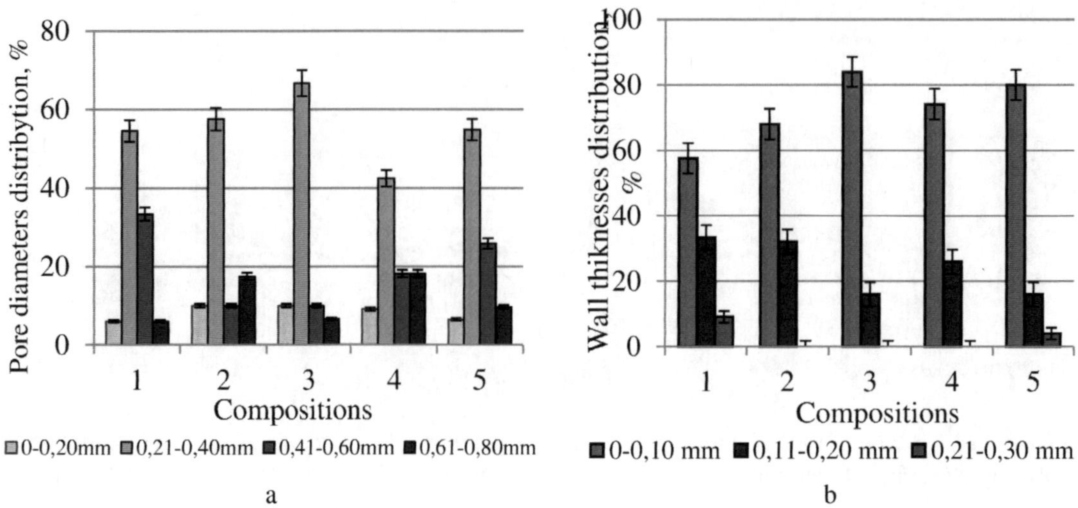

Figure 3. Size distribution of elements of the structure, a – cells and pores, b – walls.

The structure of the gas-filled materials corresponds to schemes given in figure 4. A scheme, pictured in figure 3a, is more likely to match gas-filled polyethylene with the hydrocerol content of 1 – 10%. This is a porous structure with closed, isolated cells. A scheme, pictured in figure 3b, is more likely to match samples with the hydrocerol content of 15%. These samples contain many connecting and interpenetrating cells.

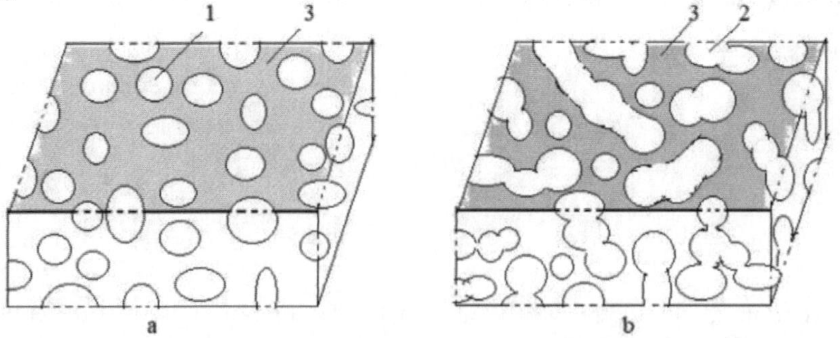

Figure 4. Types of the structure: a – with isolated cells, b – with interconnected cells; 1 – cell, 2 – pore, 3 – polymer.

Such factors as the number of cells per 1mm^2, a type of the structure, a shape of cells are crucial for operational and consumer properties. On the one hand, gas-filled polymeric materials are valued for its low density, fine isolating properties [11]. On the other hand, these materials may be characterized by low strength properties. So, it is important, to determine density, tensile strength and elongation at break of the samples.

The density of test samples, depending on hydrocerol content, was determined and compared to the pure polyethylene (table 2). As seen from the table, the density of gas-filled materials is lower than the one of polyethylene. There is no direct correlation between the density and the composition of the samples. The sample with hydrocerol content of 10% has the lowest density (0.33 g/cm^3), but the sample with hydrocerol content of 15% shows an increase to 0.55 g/cm^3. It may be explained by the fact that a solid residue accumulates during the reaction. It is known that only 40% of hydrocerol decomposes with emission of gas (CO_2) during gas-blowing. The other 60% of hydrocerol produces a solid residue.

IOP Conf. Series: Materials Science and Engineering **286** (2017) 012009 doi:10.1088/1757-899X/286/1/012009

Table 2. Density of test samples under investigation.

Hydrocerol content (%)	Density (g/cm^3)
0	0.92
1	0.58
3	0.73
5	0.57
10	0.33
15	0.55

Tensile strength and elongation at break of polyethylene (PE) and the test samples were investigated to learn mechanical and operational properties (table 3). As seen from the table, filling the structure with gas does not make much influence on the strength of polyethylene samples. Nevertheless, elongation at break decreased by about 10%. It may be due to forming thin walls of pores.

Table 3. Tensile strength and elongation at break of test samples.

Compositions	Tensile strength (MPa) ±0.5MPa	Elongation at break (%) ±2%
PE	10.0	74
1	9.7	65
2	9.5	64
3	9.4	67
4	9.1	63
5	9.1	65

Gas bubbles make the structure heterogeneous, cause formation of defects, which, in its turn, initiate formation of nascent pores, microscopic cracks. Such a structure is not equilibrium. It will be able to degrade faster into microscopic pieces under environmental conditions, namely microorganisms, oxygen, moisture, water, temperature changes and so on.

When gas-filled materials get to environmental conditions, moisture starts to penetrate in it through pores. And the moisture is the enabling environment for development of microorganisms. On this step the idea was to estimate how different hydrocerol contents influence water absorption. Main factors, influencing the speed of water absorption, are presented below:

-the chemical structure of a polymer and gas, namely permeability of this material for water;

-the size of cells or pores, which means that the large number of small cells exert stronger resistance to penetration of moisture than the little number of big cells;

-presence of open cells on the surface or compacted surface crust.

Results of investigating of water absorption of samples of polyethylene and gas-filled materials during eleven days are given in figure 5. As seen on the picture samples of polyethylene and material with hydrocerol content 1%, 3% absorb less water (not more than 4%).With an increase of hydrocerol content the structure becomes a system with a number of connecting and interpenetrating cells that leak and accumulate moisture.

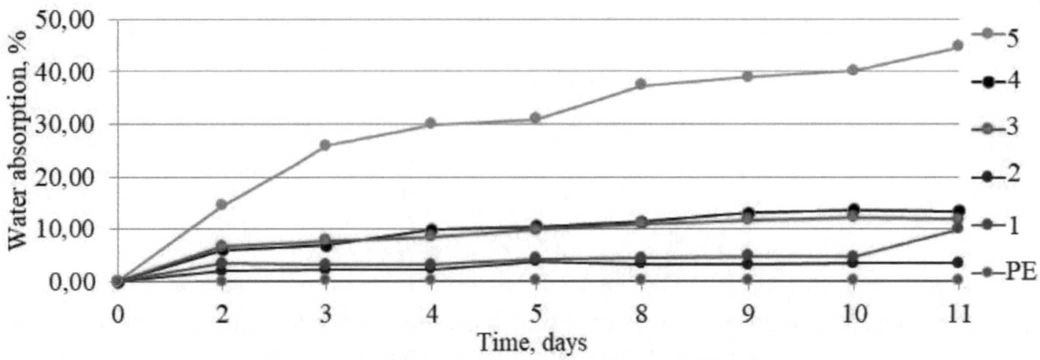

Figure 5. Water absorption of polyethylene and gas-filled samples during 11 days.

The samples were kept in soil to estimate the ability of the material to degrade in a landfill. The change in mass as compared with the initial mass was measured. Course of degradation was controlled by periodic weighing of the test samples. The samples were expected to demonstrate weight loss because of fragmentation. Results of investigating changes in mass of test samples during thirty six months are given on figure 6.

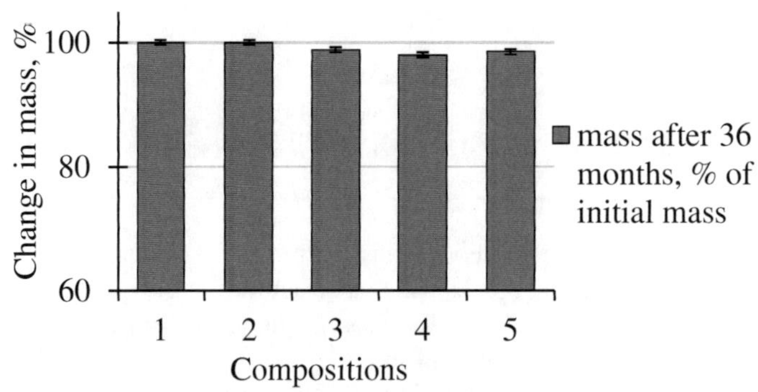

Figure 6. Changes in mass of gas-filled samples after 36 months in soil.

Addition of the finely divided particles of natural fillers is made to samples with initial hydrocerol content 5 and 15%. This will lead to accelerating processes of biological destruction of the materials. Microphotographs of several gas-filled composite samples with wood flour (WF) and starch in case of magnification 100x samples are given in figure 7.

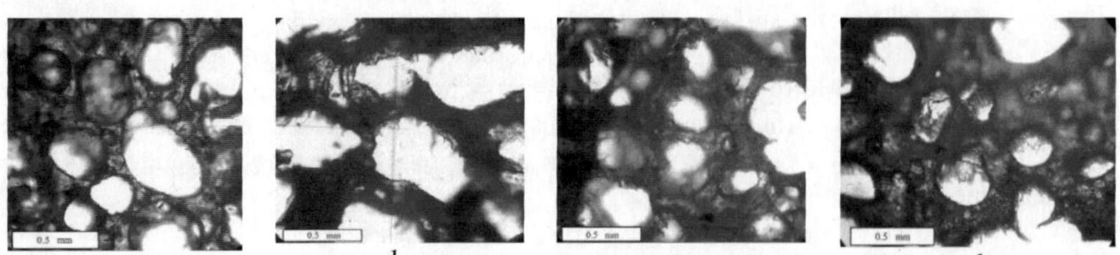

a b c d

Figure 7. Microphotographs of the gas-filled compositions with addition of 15% of natural fillers, a – with 5% of hydrocerol, WF of 0.14 mm; b – with 5% of hydrocerol, WF of 0.20 mm; c – with 15% of hydrocerol, WF of 0.08 mm 5%; d – with 15% of hydrocerol, starch.

The form of pores and cells was distorted when the particles were introduced. Addition of components, that are hygroscopic, swelling in water, leads to significant increase of water absorption.

Penetration of moisture, soil and microorganisms into the material is becoming easier. The particles can be washed out of the matrix by water or assimilated by microorganism.

Addition of fillers will lead to decrease of tensile strength because the structure would be more heterogeneous. It is suggested to produce goods for short-term use on the base of these compositions (packaging materials, packaging elements, elements of medical or cosmetology tools).

3. Conclusion

Addition of hydrocerol to polyethylene leads to creation of porous structure. It is stated, that as the content of hydrocerol in the composition increases, the number of cells and pores in the samples becomes bigger, the thickness of pore walls decreases. As a result, mechanical and operational properties decrease.

The presence of pores in the polymer matrix promotes the penetration of moisture and microorganisms into it. Such materials will decompose in landfills under the influence of physical, mechanical and biological factors.

Addition of a natural filler makes the structure of gas-filled material more heterogeneous and defective. Materials, containing wood flour or starch, will also destroy under the influence of biological factors. Thus, the presence of pores and natural filler particles in the matrix of synthetic polymer is vital for its biodegradability.

References

[1] Arkatkar A, Arutchelvi J, Bhauri S, Uppara P V and Doble M 2009 *J. The Open Environ. Enj. J.* **2** 68-80
[2] Harding K G, Gounden T and Pretorius S 2017 *J. Procedia Manuf.* **7** 106-110 doi:10.1016/j.promfg.2016.12.027
[3] Rogovina S Z, Alexanyan C V and Prut E V 2011 *J. of App. Pol. Sc.* **121** 1850-59 doi:10.1002/app.33477
[4] Olkhov A A, Tyubaeva P M, Staroverova O V, Mastalygina E E, Popov A A, Ischenko A A and Iordanskii A L 2016 *AIP Conf. Proc.* **1736** (US: AIP Publishing) doi: 10.1063/1.4949673
[5] Mastalygina E E, Varyan I A, Kolesnikova N N and Popov A A 2016 *AIP Conf. Proc.* **1736** (US: AIP Publishing) doi: 10.1063/1.4949672
[6] Mastalygina E E, Kolesnikova N N, Popov A A and Olkhov A A 2015 *AIP Conf. Proc.* **1683** (US: AIP Publishing) doi: 10.1063/1.4932833
[7] Zykova A K, Pantyukhov P V, Kolesnikova N N and Monakhova T V 2017 *J. Polym. Environ.* doi: 10.1007/s10924-017-1039-9
[8] Zykova A, Pantyukhov P and Popov A 2017 *J. Polym. Enj. And Scien.* doi: 10.1002/pen.24626
[9] Pantyukhov P, Kolesnikova N and Popov A 2016 *J. Polym. Compos.* doi: 10.1002/pc.23315
[10] Ashwin Kumar A, Karthick K and Arumugam K P 2011 *J. of Ghem. Enj. and App.* **2** 164-167
[11] Toninski M 2015 *J. Additiv. for Polyolefins* 147-152 doi:10.1016/B978-0-323-35884-2.00013-2

IOP Conf. Series: Materials Science and Engineering **286** (2017) 012010 doi:10.1088/1757-899X/286/1/012010

Examination of the junctures of aluminium and dielectric structures using subminiature eddy-current transducers

S F Dmitriev[1], V N Malikov[1], A M Sagalakov[1], A V Ishkov[2], A O Katasonov[1]

[1]Altai State University, 61, Lenina avenue, Barnaul, 656049, Russia
[2]Altai State Agrarian University, 98, Krasnoarmeyskiy avenue, Barnaul, 656049, Russia

E-mail: mirotnas@gmail.com

Abstract. The article presents the structure of a subminiature eddy-current transducer of a transformer type, which is based on the pyramid-shaped core providing the localization of the magnetic field on the area of 2500 μm². It also describes the major transducer parameters, which provide such localization of the magnetic field. The article describes the structure of the measuring system and the measurement procedure, which allows determining the thickness of different objects depending on the signal. The measuring system includes two subminiaturized eddy-current transducers controlled by software. The output signal of the transducers is processed in the hardware filtration systems. The article contains data demonstrating dependency of the amplitude part of the signal on the different thickness of objects, and states the objects experimentally determined size limits that precondition the practicability of these measurements. Such objects as conductive coatings, laminated structures (metal-polyethylene), solid conductive objects (aluminium) and non-conductive dielectric coatings were examined. The article demonstrates the change of the signal amplitude of the eddy-current transducer depending on the type of the monitored material. The pattern of this change allows determining a type of the studying object and its thickness on a small area of the monitored item.

1. Introduction

Conductive coatings find wide application in some brunches of industry, such as pipelines, spacecraft and aircraft, through their advantages in corrosion protection, electrical contact, wearing resistance and thermal isolation [1]. The thermal barrier coatings are applied in the surface construction of the airplane engine; it effectively protects the substrate metal of the blades from high temperature [2]. In any case, inhomogeneous distribution in thickness is supposed to be the problem that influence coating performance [3]. In addition, there is a problem in the incongruity of the parameters between the substrate metal and the coating, and in the corrosion under the coatings. The thickness of the coating influences on the coating term of the work, surface stress and consumption of the material. Therefore, a smart method of measuring the coating thickness is required for both process control and checking up of the key conductive structures.

Nowadays thermal, microwave, ultrasonic and eddy current (or EC) inspection techniques are applied in the measurements of coating thickness [4,5]. EC has many advantages over the other ones: it has high resistance to harsh conditions, measurements can be carried out in a noncontact way, and its cost is low [6]. To estimate the thickness of a coating with determined conductivity Takashi et al. [7]

conversed about the possibility of estimating the thickness of the alloy with the Ni-based coating sprayed on stainless steels of austenitic type 304 applying test with a measured EC frequency (SFEC). The coating thickness was estimated within an accuracy of 22 μm with inverse modeling methods. Barbosa [8] proposed a formula that obtains the coating thickness of galvanized steel from the impedance of a solenoid containing a sample of wires. Tai et al., Yang[5,9] and Tai [10] applied pulsed EC and SFEC techniques, trying to determine the conductivity and thickness of coating on a metallic substrate in conditions of either the coating or the substrate is magnetic.

Mandache and Lefebvre [11] explored the behaviour of the lift-off point of intersection (LOI) in different test conditions. LOI is used for measuring the thickness of conductive layers over the ferromagnetic substrates.

Thickness gauges based on the eddy-current method are applied to study the thickness of the electrically conductive sheets, plates, films and coatings; cylindrical and spherical gas cartridges; pipe walls. It also determines the air-gap distance between the plates made from the identical material. The purpose of the work was to assess the possibility of the amplitude eddy-current method in order to find the thickness of the conductive or dielectric coating, placed on the conductive base, as well as to estimate the error of such measurement. The research of the conduction demonstrated the possibility of the amplitude eddy-current technique application to determine the local thickness of the conductive objects, represented by several alternating conductive and non-conductive layers and solid conductive objects. An eddy-current transducer (ECT) is intended for on-site studies of the various coatings thickness, and to bring out the effect of different coatings on the output low frequencies.

2. Specifications, material choices and design

A subminiaturized eddy-current transducer [12, 13] is designed for the local examination of the different coatings thickness in experiments and to determine how different coatings influence the output signal value.

The actuating winding of the subminiature transducer consists of 10 winds and its diameter is 0.12-0.13 mm. The measuring winding includes 130 winds and its diameter is 0.05-0.08 mm. The compensation winding is included into the scheme to minimize the influence of the actuating winding on the received signal. The compensation winding is connected with the measuring winding so that the researcher could deduct the voltage of the actuating winding. It consists of 20 winds. A copper wire, that is 0.005 mm thick, is used to reel the winds. The windings are wound round the pyramid shaped core. The proposed shape of the core is favourable for the area of the magnetic field. The core is made of ferrite with initial magnetic conductivity equal to H/m .

The characteristics of the constructed transducers allow one to efficiently localize the magnetic field within 2500 μm^2 and larger and ensure a large (up to 5 mm) depth of penetration into the investigated object in operation at relatively low frequencies.

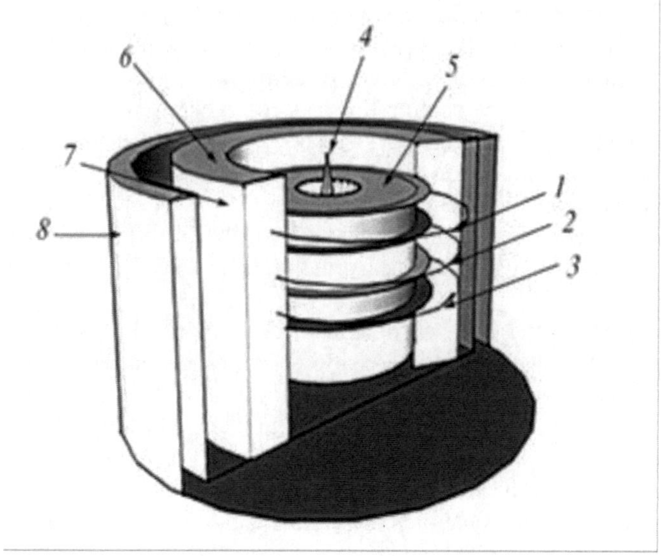

Figure 1. A block diagram of the eddy-current transducer.

The eddy-current transducer (figure 1) is a transformer with measuring (1), exciting (2), and compensation (3) windings and a magnetic circuit 4, which is located inside the cylindrical platform 5 with tracks that are cut on the external side for windings. The platform is impregnated with a compound 6 at a temperature of 200 °C to prevent the disintegration of the windings in the process of application of the ferrite screen 7, which is used to locate the electromagnetic field in the monitored object. Outside the transducer, there is a corundum washer 8, which protects the core 4 from contacting the tested object.

The core consists of a tetrahedral pyramid (height 1 mm), with a square base (sides of 0.2 mm). The measuring winding is situated on the top of the pyramid, which improves the localization of the magnetic field. Such transducer permit effective localization of the magnetic field, so that defects as small as 250 μm may be detected. In addition, the magnetic field penetrates into the sample to a considerable depth when working at relatively low frequencies.

The pyramid shape of the core was received with the help of two main methods. The first one is as follows: thin triangles (0.1 mm) were cut of the alloy 81NMA [14]. After that the plates were put together and their upper parts were ground off. Such kind of an approach has its disadvantages: the upper parts turned out to be uneven which resulted in the creation of a weak the magnetic field.

The second way to make a pyramid shaped core was grinding off the thin triangles and their matching into a pyramid.

Wire coils were placed on the pyramid shaped cores. The coils were impregnated with compound. The temperature of impregnation was 200 °C, the diameter of the wire was 1.5×10^{-6} meters. The measuring winding was placed at the end of the top of the pyramid, the diameter of the winding was 0.05 mm., the number of winds varied from 100 up to 200. In the middle of the pyramid there was an actuating winding that consisted of one wind. The compensating winding was situated on the mobile frame and included 100 wires. The mobile frame enables it to move freely along the oxide core from the bottom up to the actuating winding.

On the whole, 10 ECTs were made and their calibration was carried out in order to deduct the voltage induced in the measuring winding from the actuating winding. ECTs of different types were made in order to get the magnetic field of different voltage. Therefore, the cores of different sizes were created. The interrelation between the diagonal of the bottom and the edge of the pyramid varied from 1:1 to 1:10 (figure 2).

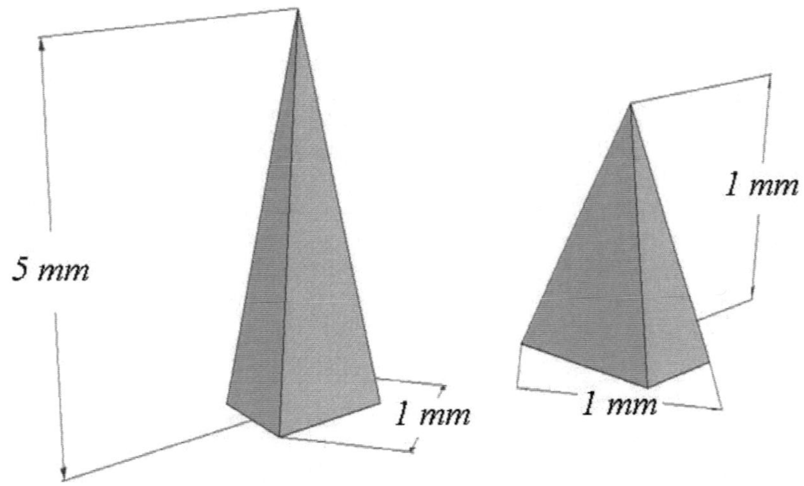

Figure 2. The cores of different sizes.

The ECTs constructed on the basis of the cores, having similar interrelation between the diagonal of the base (400 μm) and the length of the rib (4 mm), have been calibrated on semi-conductors with the certain conductivity. Thus, the ECTs have identical geometric parameters of the cores and the same number of the winds of the actuating (1 wind), measuring (200 winds) and compensation (200±40) windings.

The developed measuring system is used in the following way. The software of the PC supervises the operation of the generator, which produces a chain of rectangular voltage pulses with the repetition rate "f" - it is necessary for the functioning of eddy-current transducers. The voltage pulses are transmitted from the generator output to two series integrators. Then they are going further to the input of the power amplifier. At the output of the amplifier, the pulses are applied to the exciting inductance coils of the ECT. The signal passes through two rows of high-quality low-pass filters, then - two-series selective amplifiers and reaches the amplitude detector. Then the signal goes via an analog-to-digital converter to a PC. Through the contemporaneous control of the generated signal frequency on the exciter coil, the filtration system frequency, and selective amplification, the signal contains either the information on the distribution of electrical conductivity within the object and the data about the probable defects of the object. The program allows changing the operational frequency of the measuring system, so that the received signal can be reliably registered.

3. Experimental results

In order to test the measuring system, an aluminum coating of various thickness located on a 3 mm thick copper base was scanned. Figure 3 demonstrates the dependence of the signal magnitude on the thickness of the aluminum coating on a non-ferromagnetic substrate. If the coating thickness is increased to 1200 μm, a decrease in the signal magnitude from 28.2 to 22 mV is observed. In the range from 750 to 1500 μm the signal magnitude is much smaller than the signal magnitude from the solid unit. It can be explained by the insufficient thickness of the coating. Nevertheless, when the thickness of the coating is between 1500 and 2500 μm, the magnitude of the signal is stable. The next experiment illustrates the change in signal from the thickness of a laminated coating with alternating foil and polyethylene layers placed on a copper substrate.

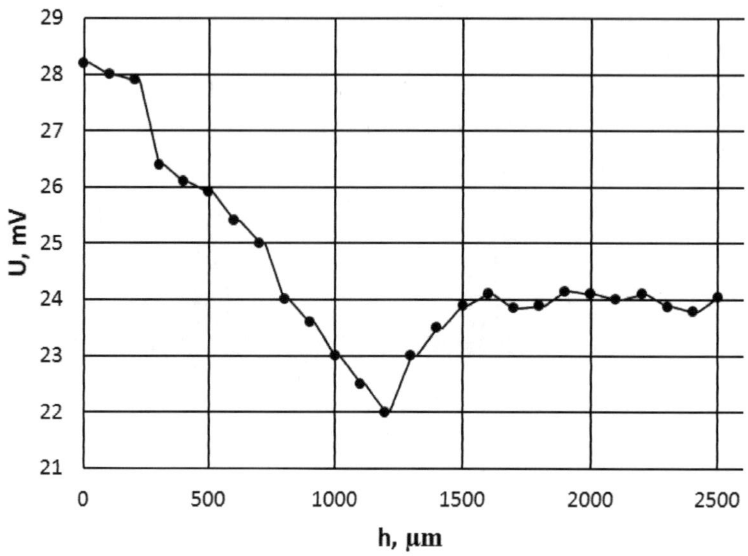

Figure 3. Dependency of the signal value on the aluminium coating on the copper base.

The next experiment illustrates the change of the signal depending on the thickness of the laminated coating with alternating foil and polyethylene layers placed on a copper substrate. Figure 4 presents the dependency of the intensity of the signal on the thickness of laminated coating. If the thickness is within the range of 0 to 100 µm, the signal value from the base falls from 29 to 26 mV; whereas if the thickness is within the range of 100 to 240 µm, the signal change pattern is flatter. In the range of 240 to 400 µm, the signal changes from 26 to 23 mV; this is possible due to the contribution of the laminated coating signal and decrease of the copper base signal contribution.

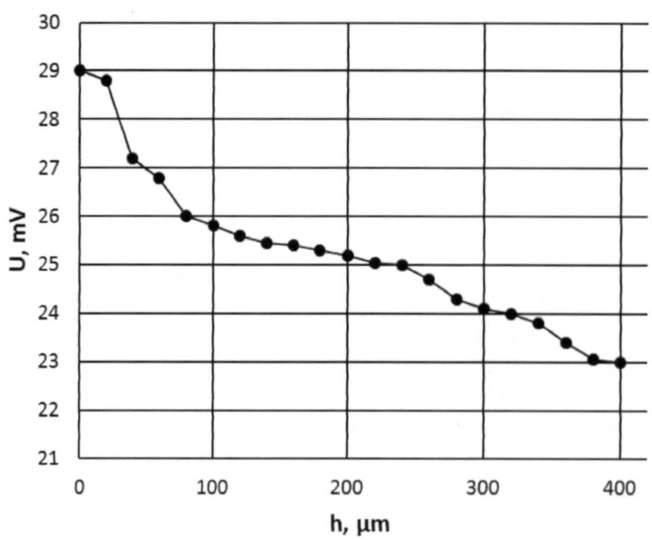

Figure 4. Dependency of the signal intensity, while scanning a laminated structure.

The final experiment considered a copper plate and a dielectric coating as the studied objects. As a dielectric coating, the varnish coating was used. A copper sample consisted of a unitary copper block.

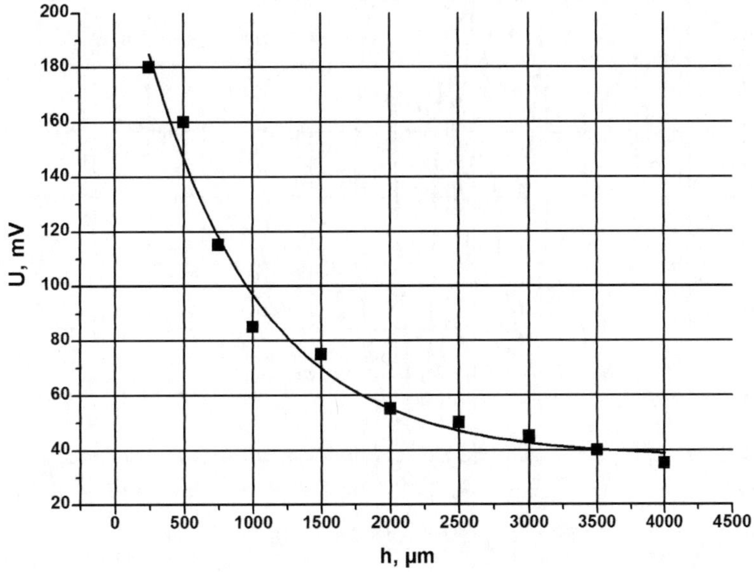

Figure 5. Dependency of the eddy-current transducer response on the thickness of the dielectric coating.

A set of experiments with copper-samples and a varnish coating were conducted. A layer of varnish was put on pre-cut samples of copper. As can be seen from the dependence of the signal amplitude on the dielectric coating thickness on figure 5, the output signal value decreases rapidly in parallel with increasing coating thickness. This dependency can be fit as an exponential function:

$$y = A_1 e^{-\frac{x}{t}} + y_0 \, (1) \qquad (1)$$

To demonstrate the operational capability of the suggested method, a structure consisting of the alternating aluminum foil and paper layers (100 μm thick) was used. A hollow parallelepiped was placed between the layers as a model defect. Its walls were 300 μm thick.

In figure 6, there is a spectral picture, observed when the sensor is moved above the layered medium, which has an inner defect. The signal level from the measuring winding characterizes the values of conductivity on the survey plot. For the fundamental frequency at 1000 Hz the level of the voltage, sent to the measuring winding, was (130±2) mV. Area 1 and area 2 on the diagram where the voltage level falls to 115 mV, correspond to the walls of the parallelepiped. These changes in the signal amplitude account for 11 percent from the signal level corresponding to the nondefective zone of the sample. Thus the signal amplitude in the nondefective zone does not exceed 4 mV, it accounts for 3 percent from the signal level corresponding to the nondefective zone of the sample.

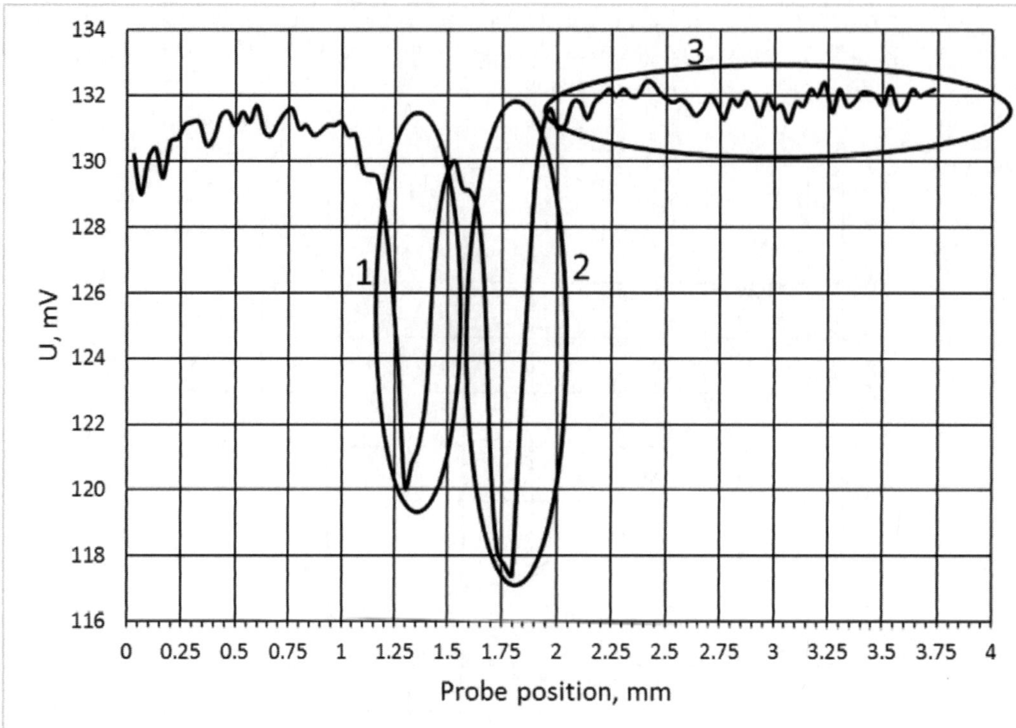

Figure 6. Dependency of the eddy-current transducer response, observed as the sensor is used to examine the layered medium with a defect. 1, 2 – the sides of the parallelepiped, 3 – a nondefective zone of the sample.

4. Conclusions

On the basis of the eddy-current transducer of the transformer type, a measuring system was developed. It allows estimating the possibility of the eddy-current method application to measure the thickness of the conductive and dielectric coatings located on the conductive base. The developed measuring system was applied to examine the objects (conductive and non-conductive coatings) located on the conductive base, as well as to measure the thickness of the solid conductive objects. It was determined that the thickness of the coating influences the signal of the eddy-current transducer. In perspective, this will allow using the amplitude control method of such class of objects for exact local measurements of the thickness of conductive and non-conductive coatings as well as other objects. Owing to the received dependencies of the eddy-current transducer signal on various coatings, it becomes possible to use the developed system in the process of diagnostic testing of composite hardening coatings. Therefore, the results of the experiment demonstrate the wide range of possibilities provided by the eddy-current method in the process of the examination of defects concealed in metal-dielectric-metal structures.

References

[1] Ptchelintsev A, Halleux B 1998 *Rev. Sci. Instrum* **69(3)** 1488
[2] Wen S Q, He A J, Wang H 2009 *Gas Turbine Experim. Res* **22(1)** 59
[3] Zhang H, Tian G Y, Simm A and Alamin M *2013 Proc. 8th Int. Symp. Precision Eng. Meas. Instrum(Chengdu, China)* vol 8759 p 875919-1
[4] Ghasr M T, Horst M J, Lechuga M, Rapoza R 2015 *IEEE Trans. Instrum. Meas* **64(10)** 2802
[5] Tai C C, Rose J H, Lechuga M, Moulder J C 1996 *Rev. Sci. Instrum* **67(11)** 3965
[6] Halleux B, Ptchelintsev A, Limburg Stirum B 1997 *Rev. Sci. Instrum* **68(9)** 3533

[7] Takahashi Y 2012 *Int. J. Appl Electromagn. Mech* **39(1)** 419
[8] Barbosa C F 2014 *IEEE Trans. Magn* **50(7)** 1
[9] Tai C C 2000 *Rev. Sci. Instrum* **71(8)** 3161
[10] Yang H C, Tai C C 2002 *Meas. Sci. Technol* **13(8)** 1259
[11] Lefebvre J, Mandache C 2008 *Int. J. Appl. Electromagn. Mech* **27(1-2)** 1
[12] Malikov V, Davydchenko M, Dmitriev S, Sagalakov A 2016 *Proc. International Conf. Mech Eng, Autom Cont Sys (Tomsk, Russia)* 7414951
[13] Dmitriev S, Ishkov A, Katasonov A, Malikov V, Sagalakov A 2017 *MATEC Web of Conferences* **106** 03007
[14] Dmitriev S, Ishkov A, Katasonov A, Malikov V, Sagalakov A 2014 *Instruments and Experimental Techniques* **57(6)** 751

IOP Conf. Series: Materials Science and Engineering **286** (2017) 012011 doi:10.1088/1757-899X/286/1/012011

Nanofiber-bonded cloth materials based on poly-3-hydroxybutyrate with antibacterial properties for medical purposes

P M Tyubaeva[1,3], A A Olkhov[1,2,3], S G Karpova[2], A L Iordansky[3] and A A Popov[1,3]

[1] Department of Chemistry and Physics, Plekhanov Russian University of Economics, 36 Stremyanny Lane, Moscow, 117997, Russia
[2] Emanuel Institute of Biochemical Physics Russian Academy of Sciences, 4 Kosygina Street, Moscow, 119334, Russia
[3]Semenov Institute of Chemical Physics Russian Academy of Sciences, 4 Kosygina Street, Moscow, 119334, Russia

E-mail: polina-tyubaeva@yandex.ru

Abstract. Different transdermal systems based on solid polymer matrices or gels containing functional substances with antiseptic (antibacterial) properties have application to the therapy of many infectious diseases and cancer. Today the most promising type of matrices with antiseptic characteristics are the nano- and microfiber nonwoven materials. Fibers on the biopolymer (poly(3-hydroxybutyrate)) basis were obtained using the electrospinning method. In the present work, the effects of iron (III) complex with tetraphenylporphyrin and its influence on bactericidal and antibacterial properties of the ultrathin PHB fibers were investigated.

1. Introduction

The previous experimental studies of biopolymer materials revealed the necessity of development of special materials with specific structure. The aim of the research work was to develop the method of preparation of nanostructured fibrous material based on the biopolymer poly(3-hydroxybutyrate) (PHB), which would be suitable for manufacturing of innovative drugs, overpouches, capsules with specific properties, and as well as for bandages, insulants and barrier materials.

Much attention is paid to the development of systems based on biopolymers with such characteristics as biocompatibility, thromboresistance and bioresorption [1]. Nowadays one of the most commonly used biopolymers is PHB, which is frequently applied to such therapeutic systems. One of the most perspective matrices with antiseptic properties are the nano- and microfiber nonwoven materials, obtained by electrospinning. The electrospinning method is based on the orientation of a polymer solution as a thin viscous jet in the field of mechanical and electrostatic forces with the subsequent formation of fibers with a nanoscale diameter [2].

2. Materials and Methods

Natural biodegradable polymer PHB 16F series, which was obtained by microbiological synthesis from company BIOMER® (Germany), was used as the polymeric matrix for the formation of the ultrathin fibers. The PHB's average molecular weight was 2.06x105. Meso-tetraphenylporphyrin iron (III) chloride complex (FeCl(TPP)) was used as a modifying additive for the development of fibrous matrices with antiseptic properties. The forming solutions of PHB and PHB with FeCl(TPP) was prepared in chloroform. The PHB concentration in the solution was 7% wt. The mass content of FeCl(TPP) in the forming solution was 1, 3 and 5% wt. respectively in ratio to PHB mass. The forming solutions of PHB with FeCl(TPP) were prepared at a temperature of 60 °C with the use of automatic magnetic stirrer. Fibers were obtained by the electrospinning method with the use of monocapillary laboratory setup with the diameter of capillary 0.1 mm, the current electric voltage was 12 kV, the distance between the electrodes was 18 cm, and the electrical conductivity of the solution was 10 mmhos/cm.

It is possible to vary the main technological parameters, for instance, voltage, diameter of capillary or distance between the electrodes in dependence to characteristics of forming biopolymer solution and content of additional components. Such possibilities show that the details of the process of electrospinning method provide the potential of producing different kinds of non-woven materials for medical purposes with specific properties (depending on the certain parameters of the process and the electrical conductivity, viscosity and concentration of the forming solution of PHB) [3].

In the current research, the conceptual approach was applied to the development of biopolymer fibrous matrices with a wide range of properties for medical purpose. Moreover, such matrices are based on the biopolymer with the specified structure obtained by the spinning in an electrostatic field from a melt of a natural biopolymer poly-3-hydroxybutyrate (PHB), a bactericidal agent - porphyrin complex (Fe-3-ClTPhP) and they have antimicrobial and microbial activity.

The main goal of this study is to obtain such specific nanostructured materials for medical purposes, which can provide the maximum surface density in minimum area contact with the tissues of the body, in accordance with the requirements of physical and mechanical performance and hygiene standards.

3. Results and Discussion

Various parameters were studied to characterize the structure of the nonwoven material. Physical, mechanical, chemical properties of the material were determined. Assessment and comparison of some technological parameters allowed classifying the materials, which were obtained by electrospinning method, in dependence to their morphology and location of ultrathin fibers. Among basic indicators of fiber organization in the material some properties were identified as the determinant, such as density of fiber structure (□), the index of orientation of the fibers (ϕ), material intensity (function of superficial filling, expressed as the mass per unit volume) (δ). The average values for these parameters are given in Table 1.

Table 1. Indicators of structure of nonwoven materials based on PHB with different contents of bactericidal additive.

Samples with the content of bactericidal additive FeCl(TPP)	Average fiber diameter, d (μm)	Density of fiber structure, γ (%)	Index of orientation of the fibers, ϕ	Material intensity, δ (g/m^2)
7% PHB	2.0-2.5	74	0.26	37.7
7% PHB + 1%		93	0.34	36.6
7% PHB + 3%		96	0.72	36
7% PHB + 5%	2.5-3.0	89	0.67	32

As it is shown in Figure 1 and Figure 2, the addition of FeCl(TPP) into a solution of PHB leads to significant changes in the morphology of the fiber. At first, original fibers of PHB (Figure 1) the alternation of cylindrical and spindle-shaped thickenings are observed along entire length. The presence of thickenings in the fiber structure can be explained by the low values of electrical conductivity and surface tension of the polymer forming solution. The average diameter of the cylindrical sections of the fibers is 1 - 6 um, and spindle elements have a maximum diameter ~ 10 μm by length 20 - 30 μm. The addition of 1% FeCl(TPP) leads to the emerging of the uniform fibers in the material's structure with the average diameter 1.5; 3; 5 μm for each concentration of FeCl(TPP). The increase of FeCl(TPP) concentration of provides the formation of fibers with the average diameter no less than 3 μm (Figure 2).

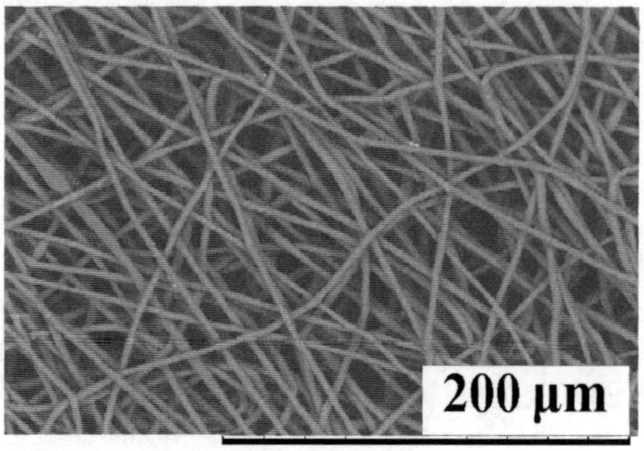

Figure 1. Microphotography of fibrous material PHB.

Figure 2. Microphotography of fibrous material PHB+3% FeCl(TPP).

The study of fibrous materials by using differential scanning calorimetry (DSC) showed that the PHB degree of crystallinity increases dramatically with increasing concentration of the FeCl(TPP) in the mixed composition.

The study, which determines the annealing of materials, was carried out to establish the existence of mesomorphic structures in the material, which is capable of further crystallization after annealing. It was postulated that the bactericidal agent fills these regions in the material.

It was confirmed the presence of supramolecular structure of ultrathin fibers based on PHB with FeCl(TPP), which causes a high antimicrobial properties of the resulting material. By the method of DSC was determined thermal transitions and the values of melting and crystallization enthalpies. The obtained data allow to estimate the changes in the speed of crystallization, the nature of the arrangement of polymer units in ultrafine fiber, which determines the specific features of the thermal transitions, which were recorded for samples of the material.

From the obtained data, it was concluded that porphyrine has a plasticizing effect on the crystallization of polymer PHB during the process of electrospinning. The changes in crystallinity degree of polymer are shown in Figure 3.

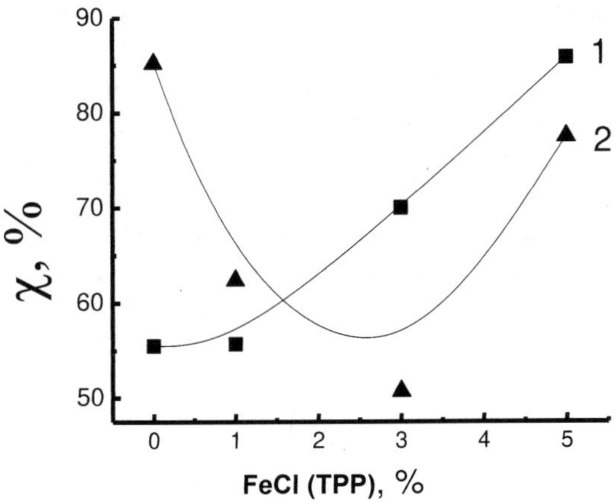

Figure 3. The change in crystallinity degree of PHB with different concentrations FeCl(TPP) in fibers: 1 – initial fiber, 2 – after annealing at the 140^{0}C.

The method of electron paramagnetic resonance was used to study amorphous regions of the fibers, which allows assessing phenomena detected by DSC and the ratio of polymer phases, providing high accuracy in the evaluation of supramolecular structure of polymer units in the fibers of the material, its location in a germicidal complex.

Moreover, adding of FeClTPhP higher content (3-5% wt.) leads to the increase in intermolecular distance, mobility of the chains and resulting in efficiency of the fiber orientation process. As result, the ratio of crystalline and mesomorphic regions in polymer increases.

4. Conclusion

The crystalline and amorphous regions in biodegradable polymers and their compositions show mutual correlation and this is an important issue in modern polymer materials science. An important result of the research was the data from biological tests. The data obtained for the sowing of the test cultures (S. aureus p 209 (Staphylococcus aureus), S. typhimurium (Salmonella typhimurium), E. coli 1257 (Escherichia coli)) point out the prospects of using the FeCl(TPP) for the impregnation of nano- and microfiber nonwoven materials based on the PHB with antiseptic properties for medical properties. During the research, an experimental batch of ultra-fine fibers for medical purposes with different diameter, density and geometry of the spatial position was obtained. The mechanical, chemical-physical, thermal properties, phase distribution in the macromolecular structure were determined. Analysis of these indicators allowed assessment of the possibility to establish the set of characteristics in the fixed range of values, which were taken into account according to requirements to medical materials. The results of the study of ultrafine fibers based on PHB with the bactericidal agent demonstrate the principle of the changes of various properties depending on the component composition of the molding solution.

Acknowledgements

This study was financed by a grant from the Plekhanov Russian University of Economics.

The authors express an acknowledgement to the common use centre "New materials and technologies" at the Emanuel Institute of Biochemical Physics of Russian Academy of Sciences.

References

[1] Chasin M and Langer R 1990 *Biodegradable polymers as drug delivery systems* (New York: Marcel Dekker) p 347

[2] Filatov Yu N 1997 *Electropermanent fibrous materials under the editorship* (M.: Oil and gas) p 298

[3] Olkhov A A, Tyubaeva P M, Staroverova O V, Mastalygina, E E, Popov A A, Ischenko A A and Iordanskii A L 2016 Process optimization electrospinning fibrous material based on polyhydroxybutyrate *AIP Conference Proceedings* **1736** 020098-1 - 020098-4

[4] Olkhov A A, Lobanov A A, Staroverova O V, Tyubaeva P M, Zykova A K, Pantyukhov P V, Popov A A and Iordanskii A L 2017 Fibrous materials on polyhydroxybutyrate and ferric iron (III)-based porphyrins basis: Physical-chemical and antibacterial properties *IOP Conference Series: Materials Science and Engineering* **175 I 1** 012008

[5] Karpova S G, Iordanskii A L, Motyakin M V, Olkhov A A, Staroverova O V, Lomakin S M, Shilkina N G, Rogovina S Z and Berlin A A 2015 Structural-dynamic Characteristics of matrixes based on ultrathin poly(3-hydroxybutyrate) fibers prepared via electrospinning *Polymer Science* A **57** **2** 131-138

IOP Conf. Series: Materials Science and Engineering **286** (2017) 012012 doi:10.1088/1757-899X/286/1/012012

Alumosilicate ceramic proppants based on natural refractory raw materials

T V Vakalova[1], L P Devyashina[1], M A Burihina[1], A S Kisner[1] and N V Pashenko[1]

[1]Tomsk Polytechnic University, 30 Lenin Avenue, Tomsk 634050, Russia,

E-mail: tvv@tpu.ru

Abstract. The sintering-strengthening effect of the additions of the highly ferrous bauxite (with Fe_2O_3 content of 20-25 % in the calcined state) in the compositions with refractory clays was established. It was found that in the temperature range 1350-1500°C the additions of bauxite in amounts of 10-40% have a fluxing effect due to the iron oxide introduced with bauxite in compositions with clay. An increasing the bauxite additive in the amount of 50-70% ensures its strengthening effect by increasing the total content of the mullite of the prismatic habit in the firing products of composites with clay. Preliminary clay and bauxite calcination at 900 °C and an increase in the content of bauxite additive up to 50-70% in compositions with clay allow to produce aluminosilicate proppants with a bulk density of 1.62-1.65 g/cm^3 and compressive strength up to 52 MPa.

1. Introduction

Alumosilicate ceramics based on natural raw materials, combining high refractoriness, chemical resistance and increased strength, have not lost their relevance in the total volume of modern ceramic and refractory materials. The perspective of such ceramics is also the fact that the main raw materials for its production are available natural and technogenic raw materials, such as refractory clays and kaolins, bauxites, rocks containing hydrates and silicates of alumina, refractory and metallurgical waste.

The availability of a raw material base for the production of aluminosilica ceramic materials determines the further traditional use of such materials in ferrous and non-ferrous metallurgy, glass and cement industries, the construction industry and others [1-5]. Recently, alumina-silicate ceramics has been used in non-traditional applications in the oil and gas extraction industry as propping-up materials for extracting difficult-to-recover oil and gas by hydraulic fracturing [6-11]. This is a mechanical method of influencing the productive oil and gas bearing layer, in which the rock breaks along the lines of the line of strength due to the impact on the formation of the pressure created by injecting into the formation of fracturing fluid. After rupture under pressure, the created cracks increase, and their connection with the system of natural cracks arises. Granular material (proppant) in size from 2 to 0.2 mm is transported to the cracks formed by the fracturing fluids and fix the cracks in the opened state after removal of the excess pressure. The conditions of service of proppants determine their basic functional properties, according to which they must withstand high reservoir pressures and withstand the corrosive action of the corrosive medium (acid gases, saline solutions). The complex of required physico-mechanical and chemical properties of proppants largely determines the efficiency of

[1] Corresponding author: tvv@tpu.ru

hydraulic fracturing. Therefore, improving the quality of proppants is one of the most important reserves to improve the technical and economic indicators of oil and gas production, which entails the need to improve the technology for manufacturing these proppants.

After years of research, preference in the production of proppants is given to ceramic technology, because ceramic proppants combine high strength, relatively low bulk density, chemical resistance and high conductivity. Medium and high strength materials, preferably aluminous (in the case of heavy proppants) or aluminosilicate (in the case of lightweight proppants) composition can be used as ceramic proppants.

Further developments are under way to expand the raw material base of ceramic proppants based on wollastonite, forsterite, clinoenstatite and other minerals. However, strict requirements to the raw materials for high-quality ceramic proppants, combining high strength with reduced bulk density have not been established either in domestic or in foreign practice of obtaining ceramic proppants. Possible ways of solving the problem are to increase the proportion of crystalline phases in the ceramic matrix (preferably with a fibrous-needle structure [12-14]), to ensure the most dense packing of grains during granulation and the minimum porosity of the granular material after firing.

Of the total variety of natural aluminosilicate raw materials, the most common are basic kaolins, bauxites and refractory clays, as well as various additives of natural and technogenic genesis, enhancing the sintering process of the material and increasing raw material strength for the production of aluminosilicate proppants.

Previously, the authors established physicochemical parameters for assessing the suitability of refractory clay raw materials for the production of aluminosilicate proppants, which include mineralogical (the content of kaolinite is not less than 65% by weight, the content of free quartz is not more than 15% by weight), the chemical (Al_2O_3 content in the calcined state, not less than 35% by weight, preferably 40 - 45% by weight, and the content of alkaline oxides is not more than 1.2% by weight).

The purpose of this work was the development of compositions and technological parameters for the production of aluminosilicate ceramic proppants based on the refractory clay-containing raw materials - argillite refractory clay and high-iron bauxite.

2. Materials and methods

As raw materials, refractory clay and bauxite were used in the work. The study of the physicochemical features of clay was carried out with the aim of elucidating the prospects of its use as the main component of the ceramic masses, and bauxite as a natural reinforcing additive in compositions with clay. According to the chemical-mineralogical composition, clay is a highly basic raw material (the content of Al_2O_3 in the calcined state is 48.3 mass %), with an average content of coloring oxides (Fe_2O_3 + TiO_2 - 5.5 mass%), kaolinite composition with hydrargillite (gibbsite) admixture in clay part, hematite and quartz – in the non-plastic part.

Bauxite is a kind of hematite-gibbsite-kaolinite bauxite with a content of up to 62% Al_2O_3 with an a low silica content (up to 10 wt%) and a low content of alkali and alkaline-earth oxides (less than 2% by weight) in the chemical composition.

Both types of investigated raw materials are clay-containing starting materials of predominantly kaolinite composition (in the case of clay) or an admixture of kaolinite (clay bauxite). Therefore, one of the ways to increase the strength characteristics of ceramic materials based on them is the addition of clay-containing additives [15-19]. These additives neutralize the negative softening effect of cristobalite, which is formed during the polymorphic transformation of impurity quartz and the crystallization of amorphous silica released from the structure of kaolinite. This will increase the total content of the crystalline phase and reduce the amount of unstable vitreous phase in the structure of the calcined material, which, in the future, will favorably enhance the hardening of the calcined material.

To improve the mechanical strength of aluminosilicate proppants while reducing the temperature of their firing, the possibility of improving the functional properties of the ceramic materials of the

IOP Conf. Series: Materials Science and Engineering **286** (2017) 012012 doi:10.1088/1757-899X/286/1/012012

system (RO, RO_2, R_2O_3) - Al_2O_3-SiO_2 was examined by activating the processes of mullite synthesis and sintering of aluminosilicate ceramics on the basis of raw materials with mineral oxide additives.

As additives-mineralizers, the following oxides were chosen: 2% MnO_2 - in the form of manganese concentrate; 5% Fe_2O_3 - in the form of pyritic cinders; 2% MgO - in the form of caustic magnesite. Mineralizing additives in an amount of 2-5 % over 100% were introduced into the aluminosilicate raw materials (clay and bauxite) calcined at 900 ° C and 980 ° C.

Chemical analysis of the materials sample was carried out as per GOST 2642. The mineralogical composition of raw materials was determined by a set of physicochemical analysis techniques (X-ray and thermal methods). X-ray study was carried out on a diffractometer «Shimadzu XRD-7000S» with Cu Kα-radiation at the tube voltage of 40 kV and the current of 25 mA, the goniometer rotation speed was 4°/min.

The thermal analysis of the samples was performed on combined SDT Q600 TGA/DSC/DTA analyzer under the following imaging conditions: the heating rate was 10 °C/min, the surrounding medium was air, crucibles were from corundum, and weighed amount was 20-40 mg.

The samples were molded from calcined clay rock by semi-dry pressing under the pressure of 10 MPa; the solution of 0.3% carboxymethylcellulose was involved as the binder. The sintering of the compacted samples was carried out in the temperature range of 1350–1450 °C with the step of 50 degrees with holding at the maximum temperature for 2 hours.

Structural and phase changes were evaluated by X-ray analysis. The microstructure of the starting and processed materials were investigated using JEOL 6000 scanning electron microscope.

3. Results and discussion

A comparative analysis of the effect of the oxide additives on the sintering process of refractory clay showed that the optimum reinforcing additive is the addition of 5% Fe_2O_3, which provides an increase in strength from 230 to 300 MPa (1.3 times) of molded clay samples heat-treated at 900°C and calcined at 1350 °C (Figure 1). In the case of high-iron bauxite, the obtained data (Figure 2) indicate that the optimal pre-calcination temperature of bauxite is 980 °C, the optimum sintering temperature of the samples is 1300 °C, since there is a danger of vitrification of the surface of the samples and the formation of cakes and fusions in the furnace at the burning temperature of 1350 °C.

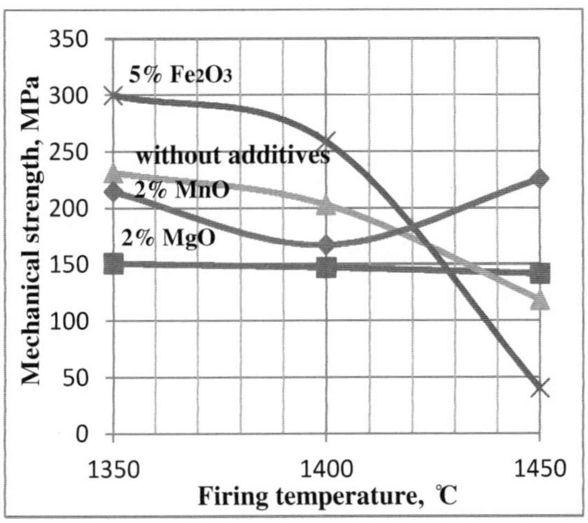

Figure 1. The influence of additives and of preliminary calcination temperature (900 °C) of the argillaceous clay on the its sinterability at 1350-1450 °C

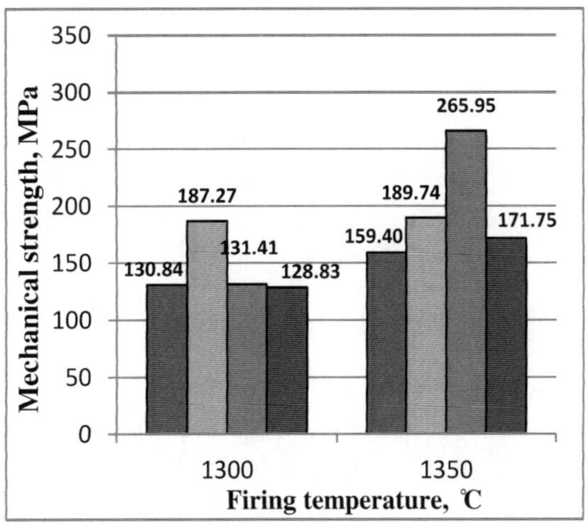

Figure 2. The influence of preliminary calcination temperature (980°C) on the high-iron bauxite sinterability at 1350-1450 °C

IOP Conf. Series: Materials Science and Engineering **286** (2017) 012012 doi:10.1088/1757-899X/286/1/012012

An evaluation of the effect of the oxide additives on the process of sintering of bauxite rock showed that the optimum reinforcing additive is the 2% MnO, which increases the strength from 130 to 187 MPa (by a factor of 1.5) from bauxite samples thermally treated at 980 °C and calcined at 1300 °C.

The strengthening effect of these additives is due to the formation of defective solid solutions for the introduction and / or replacement of iron in corundum in connection with the proximity of the ionic radius of Al^{3+} (0.057 nm) and the ionic radius of Mn^{4+} (0.052nm) and Fe^{3+}(0.067 nm), which causes a significant activation of the Al_2O_3 lattice and, as a result, improvement of its sintering. Moreover, the possibility of the influence of Mn^{2+} and Fe^{2+} cations on the appearance of melts and on their properties, which promote sintering and recrystallization of corundum and mullite, is not excluded.

Thus, according to the degree of activation of the sintering process of the clay and the bauxite, ferruginous additives are of interest for using them in aluminosilicate proppants from this raw material.

The production of aluminosilicate proppants based on clay with additions of bauxite was carried out according to the traditional technological scheme, which includes heat treatment of raw materials at a temperature of 900 °C for the purpose of dehydration of clay and bauxite minerals, wet grinding of the heat-treated products to a particle size of less than 63 μm, drying the slurry, mixing and homogenizing the components, granulating to a bulk density of 0.9 to 1.0 g/cm^3 (dry), drying and firing granules, sieving into commodity fractions of proppants. On the granules with a fraction of 20/40 (0.4-0.8 mm), compressive strength and bulk density were determined (Table 1).

Table 1. Properties of proppants fraction 20/40 from clay-based compositions with the addition of bauxite, fired at 1450 -1500°C (crush pressure – 52 MPa)

number	Clay / bauxite wt %	Temperature			Properties of proppants		Chemical composition proppants	
		heat treatment		Firing granules	Bulk density, g/cm^3	Crush resistance[2], %	Oxide content, wt.%	
		Clay	bauxite				Al_2O_3	Fe_2O_3
Clay without additives								
A	100/0	900	-	1450	1.50	25.6	48.1	2.3
				1500	1.54	23.5		
Bauxite additive								
AB50	50/50	900	900	1450	1.61	14.7	54.6	12.2
				1500	1.62	9.8		
AB70	30/70	900	900	1450	1.63	9.7	57.5	16.2
				1500	1.65	9.5		

4. Conclusion

It has been established that the addition of high-iron bauxite in an amount of 10 to 40% does not ensure the production of proppants based on argillic clay with a strength at a fracture pressure of 52 MPa. The increase in the content of bauxite additives from 50 to 70% by weight in compositions with refractory clay ensures the production of aluminosilica proppants at a firing temperature of 1450-1500 °C with a bulk density of 1.62-1.65 g/cm^3 and strength, withstanding the destructive pressure of 52 MPa.

[2] According to the requirements of the API (American Petroleum Institute) - the fraction of broken granules at a pressure of 52 MPa fraction 20/40 - no more than 10%.

Acknowledgement
The research is carried out at Tomsk Polytechnic University within the framework of Tomsk Polytechnic University Competitiveness Enhancement Program grant, Project Number TPU CEP_IHTP_73\2017.

References
[1] Migal V P, Margishvili A P and Skurikhin V V Mullitic products for blast furnace blast furnaces 2010 *Ref. and Tech. Ceramics* **3** 39–46
[2] Shvetsov R N, Tretyakova Yu Yu and Pilipchatin L D Development of a non-shrinkage mullite-corundum mass for lining the steel casting ladle 2008 Ref. and Tech. Ceramics 10 30–6
[3] Franken M and Van Berkel P Refractories for the aluminum industry. Melting furnaces and mixers 2009 Ref. and Tech. Ceramics 6 35–42
[4] Azarov S M, Azarova T A, Rat'ko N V and et al. Multilayer porous aluminosilicate ceramic 2009 *Steklo i Keramika* **3** 23-25
[5] Maslennikova G N Some trends in the development of aluminosilicate ceramics 2001 *Glass and Ceramics* **2** 10–4
[6] Liang F, Sayed M,-Muntasheri G and et.al A comprehensive review on proppant technologies 2016 *Petroleum* **2** 26-39
[7] McDaniel G, Abbott J, Mueller F and et.al Changing the shape of fracturing: new proppant improves fracture conductivity 2010 *Proceedings*–SPE Annual Technic. Conference and Exhibition **6** 4764-4784
[8] Edelman J, Maghrabia M, Semary A and et.al. Rod-shaped proppant provides superior proppant flowback control in the Egyptian Eastern Desert 2013 *Society of Petroleum Engineers*–SPE Middle East Unconventional Gas Conference and Exhibition 659-665
[9] Liu Z, Zhao J, Li Y and et.al. Lowerature sintering of bauxite-based fracturing proppants containing CaO and MnO2 additives 2016 *Materials Letters* **171** 300-303
[10] Usova Z Yu and Pogrebenkov V M Ceramic proppant based on aluminum borate 2012 *Ref. and Tech. Ceramics* **1–2** 28–32
[11] Biryukova A A, Tikhonova T A, Akchulakova S T and et al. Influence of fluorine-containing additives on synthesis and properties of mullite ceramics based on aluminosilicate raw materials of Kazakhstan 2016 *Complex use of min. raw mat.* **1** 80-87
[12] Vakalova T V, Pogrebenkov V M and Chernousova O A Structure-phase transitions upon firing of new ceramic feedstock, topaz-containing rocks 2002 *Steklo i Keramika* **6** 24-27
[13] Vakalova T V, Pogrebenkov V M, Chernousova O A and Khabas T A Special features of phase formation in fireclay in the presence of topaz 2001 *Steklo i Keramika* **4** 13-17
[14] Pogrebenkova V V, Vakalova T V, Gorbatenko V V and Grekhova M V Features of phaseformation of mullite-corundum materials in mixtures of kaolin with a fluoriding component 2010 *Ref. and Indust. Ceramics* **51 (3)** 197–201
[15] Temuujin J, MacKenzie K J D and Schmucker D M Phase evolution in mechanically treated mixtures of kaolinite and alumina hydrates (gibbsite and boehmite) 2000 *J. Eur. Ceram. Soc.* **20 (4)** 413-421
[16] Oliveira M and Ferreira J M F Structural and mechanical characterisation of MgO-, CaO- and BaO- doped aluminosilicate ceramics 2003 *Materials Science and Engineering* **344 (1-2)** 35-44
[17] Kawai S and Yoshida M Preparation of mullite from kaolin by dry grinding 1990 *J. Ceram. Soc. Japan* **98** 363-369
[18] Allahverdi M, Allaire C and Afshar S Effect of $BaSO_4$, CaF_2, and AlF_3 as well as Na_2O on aluminosilicates having a mullite-like composition 1997 *Canadian Ceramics* **66 (3)** 223-230
[19] Ma X, Yao X, Hua S and Chen Y Effects of MnO2 and Fe2O3 on microstructure and crush resistance of alumina matrix fracturing proppant 2009 *Journal of the Chinese Ceramic Society* **37 (2)**, 280-284

IOP Conf. Series: Materials Science and Engineering **286** (2017) 012013 doi:10.1088/1757-899X/286/1/012013

Spectral locus interpolation with splines in optical instruments

I O Zharinov[1] **and O O Zharinov**[2]

[1] Department of Computer Design of On-Board Computer Equipment, Saint Petersburg National Research University of Information Technologies, Mechanics and Optics, 49, Kronverksky ave., Saint Petersburg, 197101, Russia

[2] Department of Problem-Oriented Computing Complexes, Saint Petersburg State University of Aerospace Instrumentation, 69, Bolshaya Morskaia str., Saint Petersburg, 190000, Russia

E-mail: mpbva@mail.ru

Abstract. The article deals with the problem of representation of the spectral locus curve on the CIE xy chromaticity diagram. The issue is that spectral locus is specified at discrete set of points corresponding to given set of wavelengths of monochromatic colors. Thus, the problem appears which way to calculate chromaticity coordinates for any other wavelength. Three different methods how to interpolate the spectral locus were considered. The method we proposed to build an interpolant curve is based on using of Bezier curves. This approach allows to calculate chromaticity coordinates for any arbitrary wavelength directly on the CIE xy chromaticity diagram. Obtained analytical expressions for the spectral locus curve will increase the evaluation accuracy of the luminous radiation dominant wavelength value which defines the hue and the saturation value of the emitting color and other things.

1. Introduction

In 1931 the International Commission on Illumination (CIE) established [1] the spectral tristimulus values $\bar{x}(\lambda),\bar{y}(\lambda),\bar{z}(\lambda)$ of the constant power spectral emissions for the wavelengths $\lambda[i]$ of the monochromatic colors, from 380 nm to 780 nm. These tristimulus values were tabulated with the step $\Delta\lambda$=5 nm, i=0, 1, ..., $N_\lambda-1$. The total number N_λ of the tabulated spectral tristimulus values $\bar{x}(\lambda),\bar{y}(\lambda),\bar{z}(\lambda)$ is $N_\lambda=81$.

According to that tabulated data for each wavelength value $\lambda[i]$ of the monochromatic color its correspondent chromaticity coordinates ($x[i]$,$y[i]$,z[i]) can be defined directly. Although the coordinate $z[i]$ is usually omitted in colorimetric computations because it can be easily defined with the other two coordinates: $z[i]=1-x[i]-y[i]$. The chromaticity coordinates set $\{x[i],y[i]\}$ for $\lambda\in[380,780]$ with a step $\Delta\lambda$ forms a point specified contour on the XY-plane, these points are nodes for interpolation procedure [2, 3] that enables to obtain the spectral locus continuous curve for the monochromatic colors which defines the color locus area. The straight line between two endpoints on the XY-plane corresponding the wavelengths 380 nm and 780 nm is called the line of purple colors (LPC).

An interpolation of a set of points $\{x[i],y[i]\}$ must provide [4,5] a possibility to obtain the interim values of the (x,y)-chromaticity coordinates of monochromatic colors with an arbitrary wavelength step which is not equal to 5 nm according to the CIE standards. Combining those values a continuous and «smooth» spectral locus graph can be obtained. A task to interpolate the tabulated spectral locus is very important for increasing an accuracy of colorimetric computations.

2. Spectral locus interpolation approaches

The primary requirement for an interpolation method is the first derivative continuity of the interpolating function in the specified nodal points ($x[i]$,$y[i]$) except for the endpoints ($x[0]$,$y[0]$) and ($x[80]$,$y[80]$). One of the effective approaches to solve the interpolation problem is the approach when a spline function is built [6,7]. In the colorimetry there are three different approaches to build a graph which interpolates the spectral locus (figure 1):

1) interpolation of the tabulated CIE *XYZ* tristimulus values — when interpolation is done with three tabulated sequences of nodal points $(\lambda[i],\bar{x}(\lambda[i])),(\lambda[i],\bar{y}(\lambda[i])),(\lambda[i],\bar{z}(\lambda[i]))$ separately and then the (x,y)-chromaticity coordinate values of the monochromatic colors are obtained;

2) interpolation of the (x,y)-chromaticity coordinates — interpolation procedure is performed with two sequences of the nodal points ($\lambda[i]$,$x[i]$), ($\lambda[i]$,$y[i]$) separately;

3) the interpolation of the tabulated spectral locus nodal points ($x[i]$,$y[i]$) directly on *XY*-plane to built a curve with some distinctive form features.

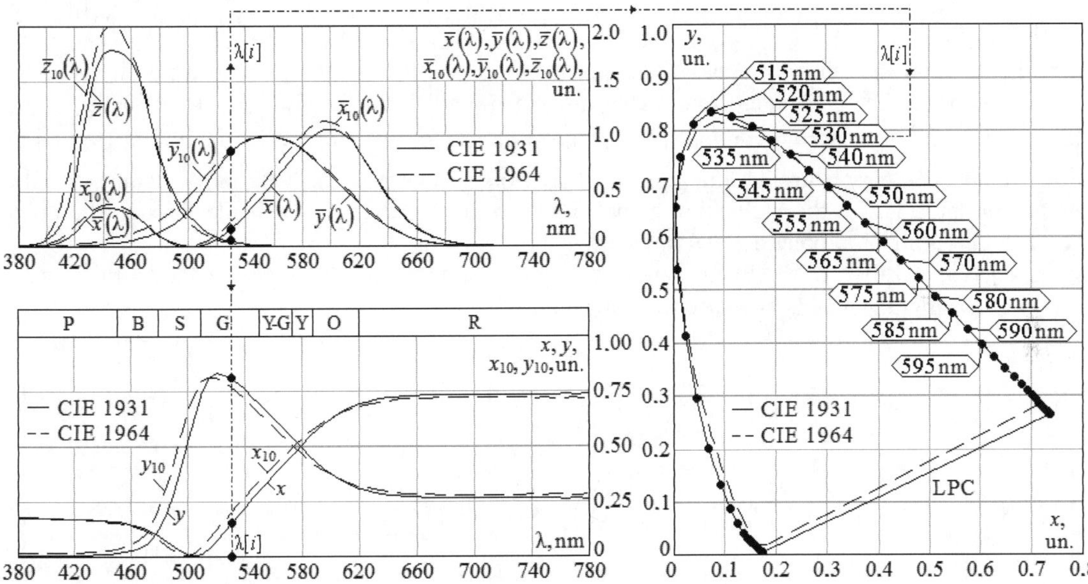

Figure 1. The colorimetric curves accepted by the CIE (The CIE 1931 — solid line; the CIE 1964 — dotted line, the variable index is «10»), P – purple, B – blue, S – sky blue, G – green, Y-G – yellowish green, Y – yellow, O – orange, R – red color.

3. The solution of the interpolation problem for the spectral locus

To solve the interpolation problem using the first and the second approaches mentioned above the standard method of spline interpolation can be used, cubic spline interpolation technique in patricular [8]. Main disadvantage of the cubic spline technique is that resulting interpolant curve usually has undesirable «oscillations».

There is an alternative and original approach to interpolate the spectral locus curve represented as a trajectory which is drawn on the *XY*-plane by chromaticity coordinates point when it moves according to wavelength change $\lambda[i]$ from 380 nm to 780 nm. To interpolate spectral locus directly on *XY*-plane, Bezier interpolation [9,10] technique seems to be useful tool. Locally specified Bezier curves may be calculated and then jointed with each other in the interpolation nodes to create a joint Bezier spline for spectral locus curve. For that purpose quadratic and cubic Bezier curves are used.

An interpolating function which is used to build locus curve between wavelengths from $\lambda[i]$ to

IOP Conf. Series: Materials Science and Engineering **286** (2017) 012013 doi:10.1088/1757-899X/286/1/012013

$\lambda[i+1]$ on the basis of the quadratic Bezier curve must be represented as:

$$\begin{cases} x(\lambda)=t_2^2(\lambda)x[i]+2t_1(\lambda)t_2(\lambda)x_Q[i]+t_1^2(\lambda)x[i+1] \\ y(\lambda)=t_2^2(\lambda)y[i]+2t_1(\lambda)t_2(\lambda)y_Q[i]+t_1^2(\lambda)y[i+1], \end{cases} \quad (1)$$

where $x_Q[i]$, $y_Q[i]$, $x_{P_1}[i]$, $y_{P_1}[i]$, $x_{P_2}[i]$, $y_{P_2}[i]$ – control point coordinates $Q[i]$, $P_1[i]$ and $P_2[i]$ on the XY-plane which define the interpolating function properties in each interval;

$$t_1(\lambda)=(\lambda-\lambda[i])/(\lambda[i+1]-\lambda[i]);\ t_2(\lambda)=1-t_1(\lambda)=(\lambda[i+1]-\lambda)/(\lambda[i+1]-\lambda[i]).$$

An interpolating function which is used to build locus curve between wavelengths from $\lambda[i]$ to $\lambda[i+1]$ on the basis of the cubic Bezier curve must be represented as:

$$\begin{cases} x(\lambda)=t_2^3(\lambda)x[i]+3t_1(\lambda)t_2^2(\lambda)x_{P_1}[i]+3t_1^2(\lambda)t_2(\lambda)x_{P_2}[i]+t_1^3(\lambda)x[i+1] \\ y(\lambda)=t_2^3(\lambda)y[i]+3t_1(\lambda)t_2^2(\lambda)y_{P_1}[i]+3t_1^2(\lambda)t_2(\lambda)y_{P_2}[i]+t_1^3(\lambda)y[i+1]. \end{cases} \quad (2)$$

The computation of the control point coordinates $Q[i]$, $P_1[i]$ and $P_2[i]$ includes the precomputation of the angle of the tangent in each nodal point $(x[i],y[i])$, defining desired interpolating curve using the formula:

$$\alpha[i]=angle(x[i+1]-x[i-1],y[i+1]-y[i-1]),\ i=1,2,\dots,N_\lambda-2, \quad (3)$$

where the function $angle(x,y)$ computes the angle(in radians) of the directing vector which starts in the XY-plane coordinate origin and terminates in the point defined with the coordinates (x,y), $\alpha[i]\in[0,2\pi)$. In the i-th interpolation interval the starting tangent with the inclination angle $\alpha[i]$ to the X axis must be made through the point $(x[i],y[i])$ of the interpolation interval beginning and the terminating tangent with the inclination angle $\alpha[i+1]$ to the X axis must be made through the point $(x[i+1],y[i+1])$ of the interpolation interval ending.

The selection method of the Bezier curve type and its construction method are shown in figure 2 using as an example two parts of a spectral locus graph.

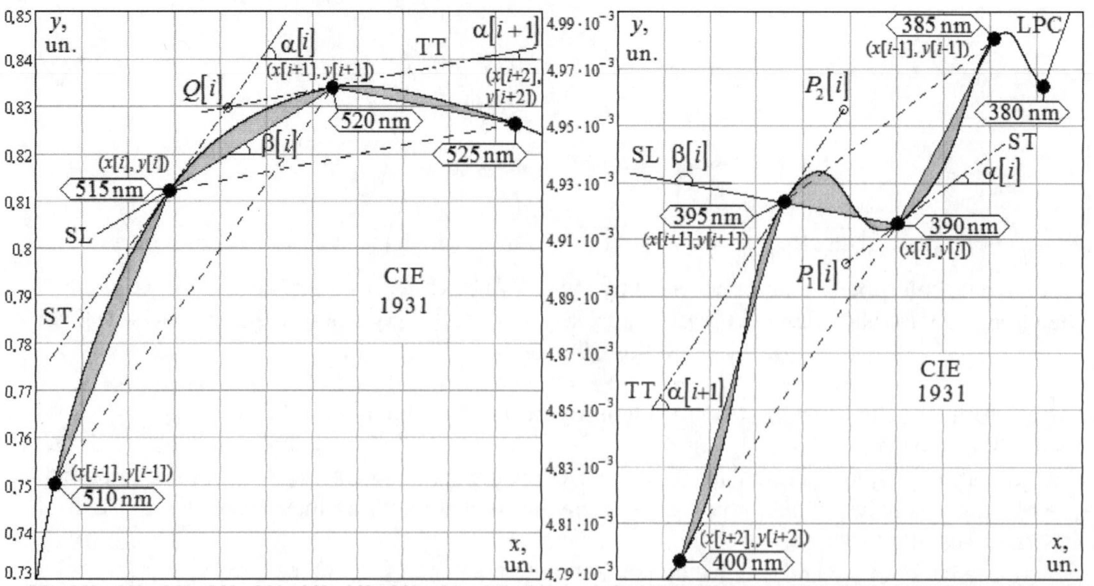

a) b)

Figure 2. Selection method of the Bezier curve type to interpolate the spectral locus on the i-th interval: a) Quadratic Bezier curve, b) Cubic Bezier curve.

A wavelength $\lambda[i]$ of 515 nm corresponds to the nodal point $(x[i],y[i])$ through which a starting tangent is drawn as shown in figure 2,a. The starting tangent is a line which is parallel to the line (shown in figure 2,a as a dotted line) which connects the point ($x[i$-1],$y[i$-1]) for λ[i-1]=510 nm and the point ($x[i$+1],$y[i$+1]) for λ[i+1]=520 nm. The terminating tangent which intersects the spectral locus in the point ($x[i$+1],$y[i$+1]) for λ[i+1]=520 nm is a line which is parallel to the line (shown in figure 2,a as a dotted line) which connects the point ($x[i]$,$y[i]$) for λ[i]=515 nm and the point ($x[i$+2],$y[i$+2]) for λ[i+2]=525 nm. In figure 2 the starting (ST) and terminating tangents (TT) are shown by dot and dash lines.

The wavelength λ[i] of 390 nm corresponds to the nodal point ($x[i]$,$y[i]$) in figure 2,b through which the starting tangent is drawn. The starting tangent is a line which is parallel to the line (shown in figure 2,b as a dotted line) which connects the point ($x[i$-1],$y[i$-1]) for λ[i-1]=385 nm and the point ($x[i$+1],$y[i$+1]) for λ[i+1]=395 nm. The terminating tangent which intersects the spectral locus in the point ($x[i$+1],$y[i$+1]) for λ[i+1]=395 nm is a line which is parallel to the line (shown in figure 2,b as a dotted line) passing through two points ($x[i]$,$y[i]$) for λ[i]=390 nm and ($x[i$+2],$y[i$+2]) for λ[i+2]=400 nm.

To select the Bezier curve type the supporting line inclination angle must be calculated between the nodal points {$x[i]$,$y[i]$} and {$x[i$+1],$y[i$+1]}:

$$\beta[i] = angle\left(x[i+1]-x[i],\, y[i+1]-y[i]\right), i=0,1,\ldots,N_\lambda-2. \quad (4)$$

For the spectral locus ending nodal points in the expressions (3) and (4) the edge conditions must be assumed: α[0]=β[0], α[N_λ-1]=β[N_λ-2]. In that way in each of the i-th interpolation interval which correspond to the wavelength λ[i] three parameters must be assigned: the angle α[i] of the starting tangent inclination to the desired curve in the interval beginning; the angle β[i] of the supporting line (SL) inclination between the nodal points {$x[i]$,$y[i]$} and {$x[i$+1],$y[i$+1]}; the angle α[i+1] of the terminating tangent inclination to the desired curve in the interval ending.

The reciprocal ratios of the parameters α[i], β[i], α[i+1] determine the choice of the Bezier curve type to interpolate a part of the curve in a particular interval: a quadratic Bezier curve with a single control point $Q[i]$; a cubic Bezier curve with two control points $P_1[i]$ and $P_2[i]$.

The control point $P_1[i]$ lies on the starting tangent at a distance of $L_1[i]$ from the nodal point with coordinates ($x[i]$,$y[i]$). The control point $P_2[i]$ lies on the terminating tangent at a distance of $L_1[i]$ from the point ($x[i$+1],$y[i$+1]). The control points in figure 2 designated with a mark «o», the nodal points — with a mark «•». The control points coordinates $P_1[i]$ and $P_2[i]$ for cubic curve (2) must be preliminarily calculated with the formulae:

$$x_{P_1}[i] = x[i] + L_1[i]cos\left(\alpha[i]\right),\ y_{P_1}[i] = y[i] + L_1[i]sin\left(\alpha[i]\right), \quad (5)$$

$$x_{P_2}[i] = x[i+1] - L_2[i]cos\left(\alpha[i+1]\right),\ y_{P_2}[i] = y[i+1] - L_2[i]sin\left(\alpha[i+1]\right). \quad (6)$$

When the control points $P_1[i]$ and $P_2[i]$ are found on the different sides from the supporting line which is drawn through the nodal points ($x[i]$,$y[i]$) and ($x[i$+1],$y[i$+1]) the preliminary control points must be finally set in the positions (5), (6) and in this interval a cubic Bezier curve must be drawn. In other case, when preliminarily calculated control points $P_1[i]$ and $P_2[i]$ are found on the same side (both in the upper part, in the lower part, in the left part or in the right part) from the supporting line, then calculating of a cubic Bezier curve become pointless. In this case one control point $Q[i]$ must be assigned instead of two preliminarily calculated. This new point is the point where the starting and terminating tangents are crossed. As the result, a quadratic Bezier curve must be build.

The control point coordinates $Q[i]$ which are used to interpolate spectral locus by a quadratic curves (2) must be calculated with the following formulae:

$$x_Q[i] = \begin{cases} \dfrac{b[i+1]-b[i]}{tg\left(\alpha[i]\right)-tg\left(\alpha[i+1]\right)}, & tg\left(\alpha[i]\right) \neq tg\left(\alpha[i+1]\right) \\ \left(x[i]+x[i+1]\right)/2, & tg\left(\alpha[i]\right) = tg\left(\alpha[i+1]\right) \end{cases}, \quad (7)$$

$$y_Q[i] = tg\left(\alpha[i]\right)x_Q[i] + b[i],\ b[i] = y[i] - tg\left(\alpha[i]\right)x[i]. \quad (8)$$

The control points position (5), (6) and (7), (8) on the starting and terminating tangents enables to

IOP Conf. Series: Materials Science and Engineering **286** (2017) 012013 doi:10.1088/1757-899X/286/1/012013

combine the Bezier curves in the nodal points to ensure the continuity of the interpolating function first derivative. The decision function $n[i]$ which enables a preferred selection of the quadratic (2) or the cubic (3) interpolated curve in the i-th interpolation interval may be represented as:

$$n[i]=\left(tg\left(\beta[i]\right)\left(x_{P_1}[i]-x[i]\right)+y[i]-y_{P_1}[i]\right)\cdot\left(tg\left(\beta[i]\right)\left(x_{P_2}[i]-x[i]\right)+y[i]-y_{P_2}[i]\right). \quad (9)$$

If $n[i]>0$ a quadratic Bezier curve must be selected and if $n[i]\leq 0$ — a cubic Bezier curve must be selected. As a result an interpolating function system which defines the spectral locus on the XY-plane must be represented as:

$$\begin{cases} x(\lambda)=\begin{cases} t_2^2(\lambda)x[i]+2t_1(\lambda)t_2(\lambda)x_Q[i]+t_1^2(\lambda)x[i+1], n[i]>0 \\ t_2^3(\lambda)x[i]+3t_1(\lambda)t_2^2(\lambda)x_{P_1}[i]+3t_1^2(\lambda)t_2(\lambda)x_{P_2}[i]+t_1^3(\lambda)x[i+1], n[i]\leq 0 \end{cases} \\ y(\lambda)=\begin{cases} t_2^2(\lambda)y[i]+2t_1(\lambda)t_2(\lambda)y_Q[i]+t_1^2(\lambda)y[i+1], n[i]>0 \\ t_2^3(\lambda)y[i]+3t_1(\lambda)t_2^2(\lambda)y_{P_1}[i]+3t_1^2(\lambda)t_2(\lambda)y_{P_2}[i]+t_1^3(\lambda)y[i+1], n[i]\leq 0 \end{cases} \end{cases}. \quad (10)$$

Definition of the distances $L_1[i]$ and $L_2[i]$ may impact significantly the interpolating function incurvature in the intervals where a cubic Bezier curve is formed. The spectral locus interpolation problem in the given method can be solved with the following expressions: $L_1[i]=L[i-1]L[i]/(L[i-1]+L[i+1])$, $L_2[i]=L[i+1]L[i]/(L[i-1]+L[i+1])$, $i=0,1,\ldots,N_\lambda-2$.

The value $L[i]$ is a Euclidian distance between the points $(x[i],y[i])$ and $(x[i+1],y[i+1])$ on the XY-plane. For the starting and terminating spectral locus interpolation intervals the following distance is used: $L_1[0]=L_2[0]=L[0]/2$, $L_1[N_\lambda-2]=L_2[N_\lambda-2]=L[N_\lambda-2]/2$.

In figure 2 the nodal points are connected with straight solid lines and that connection method gives a possibility to evaluate visually the proposed interpolation method efficiency in comparison with the method of linear interpolation on XY-plane. The spectral locus interpolating curves obtained with three different methods are shown in figure 3 at a larger scale in order to reveal some features of each method of interpolation.

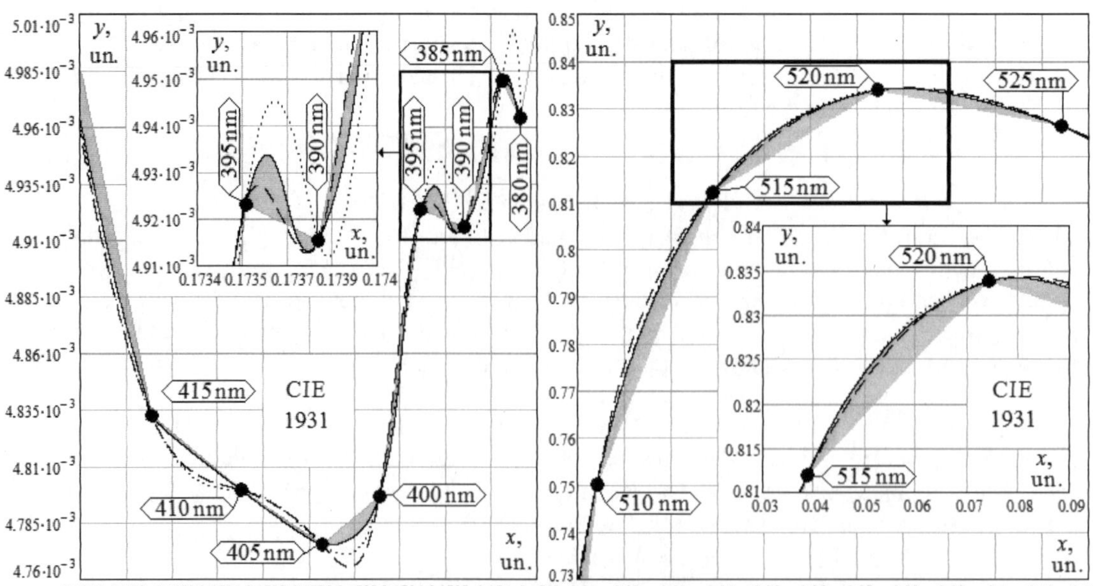

a) b)

Figure 3. The spectral locus interpolation results: a) in the band 380 nm - 420 nm, b) in the band 510 nm - 525 nm (the dotted curve is obtained with the method of CIE XYZ tristimulus values interpolation by cubic splines, the pointed curve — the chromaticity coordinates data interpolation by cubic splines, the solid curve — the proposed interpolation method on XY-plane using Bezier splines).

IOP Conf. Series: Materials Science and Engineering **286** (2017) 012013 doi:10.1088/1757-899X/286/1/012013

The most important point of interest is an analysis of spectral intervals where the position of the spectral locus nodal points tabulated in the CIE standard corresponds to the curve parts of the function maximum curvature: at the curve beginning in the spectral band from 380 nm to 420 nm; in the curve part which corresponds to the wavelength from 515 nm to 525 nm. The analysis of figure 3 shows that the application of the spline interpolation method in the tristimulus values data (method 1) and the chromaticity coordinates data (method 2) causes «space oscillations» in the interpolating function at the beginning of the visible spectral band wavelength 380 nm – 415 nm. For the spectral interval 515–525 nm where curvature variations of the spectral locus are significantly less, all three methods seem to be nearly equivalent.

4. Conclusion

As a result for the spectral locus interpolation, the authors propose the system (10) which defines analytical expressions for interpolating function allowing to calculate (x,y)-chromaticity coordinates for monochromatic colors with any given wavelength λ in the visible band 380 nm – 780 nm.

The solution to the spectral locus interpolation problem is obviously useful for increasing the accuracy of the quantitative colorimetric computations. In particular, the analytical expressions for the spectral locus representation will increase the evaluation accuracy of the luminous radiation dominant wavelength value which defines the hue and the saturation value of the emitting color and other things.

References

[1] Ibraheem N A, Hasan M M, Khan R Z and Mishra P K 2012 *ARPN Journal of science and technology* **3** 265-275

[2] Masaoka K 2010 *Optics Letter* **12** 2031-33

[3] Wyman Ch, Sloan P, Shirley P 2013 *Journal of Computer Graphics Techniques* **2** 1-11

[4] Abbas M, Jamal E, Ali J M 2011 *Applied Mathematical Sciences* **37** 1817-32

[5] Fitter H N, Pandey A B, Patel D D and Mistry J M 2014 *Procedia Engineering* **97** 1155-66

[6] Hasegawa A Y, Tormena C, Parpinelli R S 2014 *International Journal of Computer Science and Applications* **2** 1-18

[7] Pal S, Ganguly P, Biswas P K 2007 *Pattern Recognition* **40** 2730-41

[8] Madi M 2004 *International Journal of Applied Mathematics and Computer Science* **1** 33-41

[9] Juhasz I, Hoffmann M 2008 *Journal of Computational and Applied Mathematics* **2** 413-424

[10] Ahn Y J, Hoffmann Ch, Rosen P 2014 *Journal of Computational and Applied Mathematics* **255** 887-897

Evaluation of parameters of color profile models of LCD and LED screens

I O Zharinov[1] **and O O Zharinov**[2]

[1] Department of Computer Design of On-Board Computer Equipment, Saint Petersburg National Research University of Information Technologies, Mechanics and Optics, 49, Kronverksky ave., Saint Petersburg, 197101, Russia
[2] Department of Problem-Oriented Computing Complexes, Saint Petersburg State University of Aerospace Instrumentation, 69, Bolshaya Morskaia str., Saint Petersburg, 190000, Russia

E-mail: mpbva@mail.ru

Abstract. The purpose of the research relates to the problem of parametric identification of the color profile model of LCD (liquid crystal display) and LED (light emitting diode) screens. The color profile model of a screen is based on the Grassmann's Law of additive color mixture. Mathematically the problem is to evaluate unknown parameters (numerical coefficients) of the matrix transformation between different color spaces. Several methods of evaluation of these screen profile coefficients were developed. These methods are based either on processing of some colorimetric measurements or on processing of technical documentation data.

1. Introduction

The luminescence color of each screen pixel is programmed with the code *RGB* (*R* – Red, *G* – Green, *B* – Blue) each component of the code defines desirable luminescence intensity of red, green and blue respectively [1]. Mathematical description of color profile of a screen is defined with the screen profile coefficients (SPC) X_r, X_g, X_b, Y_r, Y_g, Y_b, Z_r, Z_g, Z_b. The screen profile matrix binds the values of the *RGB* code and the color coordinate values *XYZ* of the image shown on the screen [2]:

$$\begin{bmatrix} X \\ Y \\ Z \end{bmatrix} = \begin{bmatrix} X_r & X_g & X_b \\ Y_r & Y_g & Y_b \\ Z_r & Z_g & Z_b \end{bmatrix} \begin{bmatrix} R \\ G \\ B \end{bmatrix}, \begin{bmatrix} R \\ G \\ B \end{bmatrix} = \begin{bmatrix} X_r & X_g & X_b \\ Y_r & Y_g & Y_b \\ Z_r & Z_g & Z_b \end{bmatrix}^{-1} \begin{bmatrix} X \\ Y \\ Z \end{bmatrix}. \quad (1)$$

The color gamut of the screen based on three-component color formation principle is approximated by the triangle on either *xy*- or *u'v'*-chromaticity diagram. The CIE (International Commission on Illumination) introduced the following formulae to translate from the color coordinates system *XYZ* to the (*x*,*y*)- or (*u'*,*v'*)-chromaticity coordinates system [2]:

$$x = \frac{X}{X+Y+Z},\ y = \frac{Y}{X+Y+Z},\ X = \frac{xY}{y},\ Z = (1-x-y)\frac{Y}{y},$$

$$u' = \frac{4X}{X+15Y+3Z},\ v' = \frac{9Y}{X+15Y+3Z},\ X = \frac{9Yu'}{4v'},\ Z = \frac{3Y(4-u')}{4v'} - 5Y,$$

IOP Conf. Series: Materials Science and Engineering **286** (2017) 012014 doi:10.1088/1757-899X/286/1/012014

$$u' = \frac{2x}{6y - x + 1{,}5},\ v' = \frac{4{,}5y}{6y - x + 1{,}5},\ x = \frac{4{,}5u'}{3u' - 8v' + 6},\ y = \frac{2v'}{3u' - 8v' + 6}. \quad (2)$$

The screen manufacturers do not specify the numerical values of the screen profile coefficients X_r, X_g, X_b, Y_r, Y_g, Y_b, Z_r, Z_g, Z_b in the technical documentation. Thus, a technique of evaluation of these values is necessary.

2. Color spaces transformation

Basically all mathematical expressions which are used to compute the evaluations of the SPC contain the basic conversions (1), (2), which bind the color coordinate values of the different color spaces. The figure 1 shows the color space transformation general rules which are essential to evaluate the SPC.

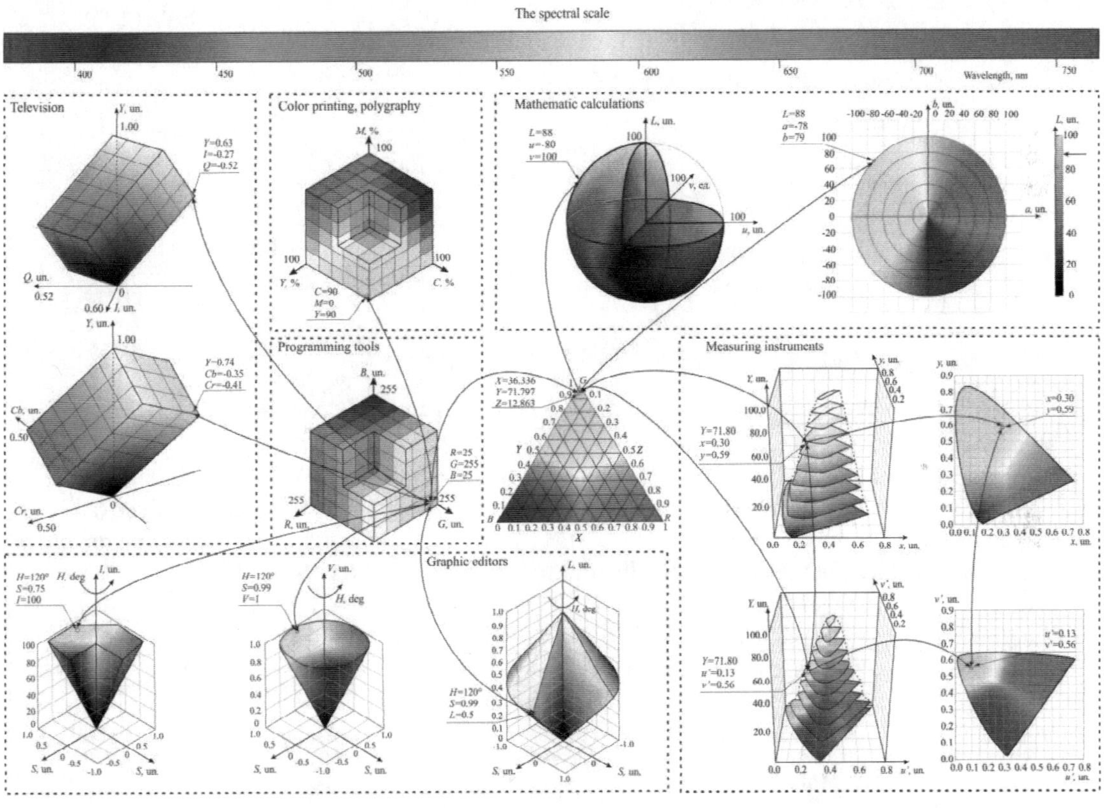

Figure 1. The color space transformation rules.

The following methods are proposed to compute the evaluations of the SPC:
- on the basis of colorimetric computations using data of manufacturer's technical documentation;
- on the basis of colorimetric experiments using the measuring instruments which allow to obtain the chromaticity coordinate values in different color spaces [3-5].

3. The evaluation of the SPC using the *xy* plane technical documentation data

Usually manufacturer specifies in the technical documentation some values of chromaticity coordinates on the *xy* chromaticity diagram, namely coordinates (x_R, y_R), (x_G, y_G), (x_B, y_B) of vertices of the color triangle, and also chromaticity coordinates of the white point (x_W, y_W). There is the way to obtain SPC values via these data by solving the following equations:

$$x_R\left(\hat{X}_r + \hat{Y}_r + \hat{Z}_r\right) = \hat{X}_r = \frac{x_R}{y_R}\hat{Y}_r,\ y_R\left(\hat{X}_r + \hat{Y}_r + \hat{Z}_r\right) = \hat{Y}_r = \frac{y_R}{x_R}\hat{X}_r,\ x_G\left(\hat{X}_g + \hat{Y}_g + \hat{Z}_g\right) = \hat{X}_g = \frac{x_G}{y_G}\hat{Y}_g,$$

$$y_G\left(\hat{X}_g+\hat{Y}_g+\hat{Z}_g\right)=\hat{Y}_g=\frac{y_G}{x_G}\hat{X}_g,\ x_B\left(\hat{X}_b+\hat{Y}_b+\hat{Z}_b\right)=\hat{X}_b=\frac{x_B}{y_B}\hat{Y}_b,\ y_B\left(\hat{X}_b+\hat{Y}_b+\hat{Z}_b\right)=\hat{Y}_b=\frac{y_B}{x_B}\hat{X}_b,$$

$$x_W\left(\hat{X}_r+\hat{X}_g+\hat{X}_b+\hat{Y}_r+\hat{Y}_g+\hat{Y}_b+\hat{Z}_r+\hat{Z}_g+\hat{Z}_b\right)=\hat{X}_r+\hat{X}_g+\hat{X}_b,$$

$$y_W\left(\hat{X}_r+\hat{X}_g+\hat{X}_b+\hat{Y}_r+\hat{Y}_g+\hat{Y}_b+\hat{Z}_r+\hat{Z}_g+\hat{Z}_b\right)=\hat{Y}_r+\hat{Y}_g+\hat{Y}_b. \qquad (3)$$

The system (3) contains eight equations and nine variables: $\hat{X}_r$, $\hat{Y}_r$, $\hat{X}_g$, $\hat{Y}_g$, $\hat{X}_b$, $\hat{Y}_b$, $\hat{Z}_r$, $\hat{Z}_g$, $\hat{Z}_b$. To solve the simultaneous (3) we need to use the ninth equation:

$$\hat{Y}_r+\hat{Y}_g+\hat{Y}_b=1\ , \qquad (4)$$

which describes the screen white color balance.

The joint solution formula of the equations (3) and (4) may be presented as:

$$\hat{X}_r=\frac{x_R}{y_R}\hat{Y}_r,\ \hat{X}_g=\frac{x_G}{y_G}\hat{Y}_g,\ \hat{X}_b=\frac{x_B}{y_B}\hat{Y}_b,\ \hat{Y}_r=1-\hat{Y}_g-\hat{Y}_b,\ \hat{Y}_g=\left(\frac{x_W}{y_W}-\frac{x_R}{y_R}+\left(\frac{x_R}{y_R}-\frac{x_B}{y_B}\right)\hat{Y}_b\right)\Bigg/\left(\frac{x_g}{y_g}-\frac{x_R}{y_R}\right),$$

$$\hat{Y}_b=\frac{\dfrac{1-x_W}{y_W}+\dfrac{x_R-1}{y_R}+\dfrac{x_Wy_R-y_Wx_R}{x_Gy_R-y_Gx_R}\cdot\dfrac{y_G}{y_W}\cdot\left(\dfrac{1-x_R}{y_R}+\dfrac{x_G-1}{y_G}\right)}{\dfrac{x_Ry_B-y_Rx_B}{x_Gy_R-y_Gx_R}\cdot\dfrac{y_G}{y_B}\cdot\left(\dfrac{1-x_G}{y_G}+\dfrac{x_R-1}{y_R}\right)+\dfrac{1-x_R}{y_B}+\dfrac{x_R-1}{y_R}},$$

$$\hat{Z}_r=\hat{Y}_r\left(\frac{1-x_R-y_R}{y_R}\right),\ \hat{Z}_g=\hat{Y}_g\left(\frac{1-x_G-y_G}{y_G}\right),\ \hat{Z}_b=\hat{Y}_b\left(\frac{1-x_B-y_B}{y_B}\right). \qquad (5)$$

4. The evaluation of the SPC using the plane *u'v'* technical documentation data

The chromaticity coordinate values (u'_R,v'_R), (u'_G,v'_G), (u'_B,v'_B) of the triangle color gamut vertices in the red *R*, green *G* and blue *B* colors respectively and also the chromaticity coordinate values (u'_W,v'_W) of the screen white point which the manufacturer specifies in the technical documentation are the initial data to evaluate the SPC. In this case the evaluations of the SPC may be presented as:

$$\hat{X}_r=\frac{9u'_R}{4v'_R}\hat{Y}_r,\ \hat{X}_g=\frac{9u'_G}{4v'_G}\hat{Y}_G,\ \hat{X}_b=\frac{9u'_B}{4v'_B}\hat{Y}_B,\ \hat{Z}_g=\frac{3\hat{Y}_g\left(4-u'_G\right)}{4v'_G}-5\hat{Y}_g,\ \hat{Z}_b=\frac{3\hat{Y}_b\left(4-u'_B\right)}{4v'_B}-5\hat{Y}_b,$$

$$\hat{Y}_g=\left(det\begin{pmatrix}u'_W & u'_R\\ v'_W & v'_R\end{pmatrix}-\hat{Y}_b\,det\begin{pmatrix}u'_B & u'_R\\ v'_B & v'_R\end{pmatrix}\right)\Bigg/det\begin{pmatrix}u'_G & u'_R\\ v'_G & v'_R\end{pmatrix},\ \hat{Y}_r=1-\hat{Y}_g-\hat{Y}_b,\ \hat{Z}_r=\frac{3\hat{Y}_r\left(4-u'_R\right)}{4v'_R}-5\hat{Y}_r,$$

$$\hat{Y}_b=\frac{v'_B}{v'_W}\cdot det\begin{pmatrix}det\begin{pmatrix}u'_G & u'_R\\ v'_G & v'_R\end{pmatrix} & det\begin{pmatrix}4-u'_G & v'_G\\ 4-u'_R & v'_R\end{pmatrix}\\ det\begin{pmatrix}u'_W & u'_R\\ v'_W & v'_R\end{pmatrix} & det\begin{pmatrix}4-u'_W & v'_W\\ 4-u'_R & v'_R\end{pmatrix}\end{pmatrix}\Bigg/det\begin{pmatrix}det\begin{pmatrix}u'_G & u'_R\\ v'_G & v'_R\end{pmatrix} & det\begin{pmatrix}4-u'_G & v'_G\\ 4-u'_R & v'_R\end{pmatrix}\\ det\begin{pmatrix}u'_B & u'_R\\ v'_B & v'_R\end{pmatrix} & det\begin{pmatrix}4-u'_B & v'_B\\ 4-u'_R & v'_R\end{pmatrix}\end{pmatrix} \qquad (6)$$

5. The evaluation of the SPC using the measurement results obtained in the *xy* parameters

When the manufacturer does not specify the chromaticity coordinates (x_R,y_R), (x_G,y_G), (x_B,y_B) of the triangle color gamut vertices and the chromaticity coordinates (x_W,y_W) of the white point in the screen technical documentation the evaluations of the SPC can be obtained by the experimental measurements of the screen colorimetric features in the parameters of the color plane *xy*. In this case the evaluations of the SPC may be presented as:

$$\begin{bmatrix} \hat{X}_r \\ \hat{X}_g \\ \hat{X}_b \\ \hat{Y}_g \\ \hat{Y}_b \\ \hat{Z}_r \\ \hat{Z}_g \\ \hat{Z}_b \end{bmatrix} = \begin{bmatrix} d_{11} & d_{12} & d_{13} & d_{14} & d_{15} & d_{16} & d_{17} & d_{18} \\ d_{21} & d_{22} & d_{23} & d_{24} & d_{25} & d_{26} & d_{27} & d_{28} \\ d_{31} & d_{32} & d_{33} & d_{34} & d_{35} & d_{36} & d_{37} & d_{38} \\ d_{41} & d_{42} & d_{43} & d_{44} & d_{45} & d_{46} & d_{47} & d_{48} \\ d_{51} & d_{52} & d_{53} & d_{54} & d_{55} & d_{56} & d_{57} & d_{58} \\ d_{61} & d_{62} & d_{63} & d_{64} & d_{65} & d_{66} & d_{67} & d_{68} \\ d_{71} & d_{72} & d_{73} & d_{74} & d_{75} & d_{76} & d_{77} & d_{78} \\ d_{81} & d_{82} & d_{83} & d_{84} & d_{85} & d_{86} & d_{87} & d_{88} \end{bmatrix}^{-1} \begin{bmatrix} x_1 R_1 \\ R_1(1-y_1) \\ x_2 R_2 \\ R_2(1-y_2) \\ x_3 R_3 \\ R_3(1-y_3) \\ x_4 R_4 \\ R_4(1-y_4) \end{bmatrix}, \tag{7}$$

where: $\hat{Y}_r = 1 - \hat{Y}_g - \hat{Y}_b$, $d_{11} = R_1(1-x_1)$, $d_{12} = G_1(1-x_1)$, $d_{13} = B_1(1-x_1)$, $d_{14} = x_1(R_1 - G_1)$, $d_{15} = x_1(R_1 - B_1)$, $d_{16} = -x_1 R_1$, $d_{17} = -x_1 G_1$, $d_{18} = -x_1 B_1$, $d_{21} = R_1 y_1$, $d_{22} = G_1 y_1$, $d_{23} = B_1 y_1$, $d_{24} = (R_1 - G_1)(1-y_1)$, $d_{25} = (R_1 - B_1)(1-y_1)$, $d_{26} = R_1 y_1$, $d_{27} = G_1 y_1$, $d_{28} = B_1 y_1$, $d_{31} = R_2(1-x_2)$, $d_{32} = G_2(1-x_2)$, $d_{33} = B_2(1-x_2)$, $d_{34} = x_2(R_2 - G_2)$, $d_{35} = x_2(R_2 - B_2)$, $d_{36} = -x_2 R_2$, $d_{37} = -x_2 G_2$, $d_{38} = -x_2 B_2$, $d_{41} = R_2 y_2$, $d_{42} = G_2 y_2$, $d_{43} = B_2 y_2$, $d_{44} = (R_2 - G_2)(1-y_2)$, $d_{45} = (R_2 - B_2)(1-y_2)$, $d_{46} = R_2 y_2$, $d_{47} = G_2 y_2$, $d_{48} = B_2 y_2$, $d_{51} = R_3(1-x_3)$, $d_{52} = G_3(1-x_3)$, $d_{53} = B_3(1-x_3)$, $d_{54} = x_3(R_3 - G_3)$, $d_{55} = x_3(R_3 - B_3)$, $d_{56} = -x_3 R_3$, $d_{57} = -x_3 G_3$, $d_{58} = -x_3 B_3$, $d_{61} = R_3 y_3$, $d_{62} = G_3 y_3$, $d_{63} = B_3 y_3$, $d_{64} = (R_3 - G_3)(1-y_3)$, $d_{65} = (R_3 - B_3)(1-y_3)$, $d_{66} = R_3 y_3$, $d_{67} = G_3 y_3$, $d_{68} = B_3 y_3$, $d_{71} = R_4(1-x_4)$, $d_{72} = G_4(1-x_4)$, $d_{73} = B_4(1-x_4)$, $d_{74} = x_4(R_4 - G_4)$, $d_{75} = x_4(R_4 - B_4)$, $d_{76} = -x_4 R_4$, $d_{77} = -x_4 G_4$, $d_{78} = -x_4 B_4$, $d_{81} = R_4 y_4$, $d_{82} = G_4 y_4$, $d_{83} = B_4 y_4$, $d_{84} = (R_4 - G_4)(1-y_4)$, $d_{85} = (R_4 - B_4)(1-y_4)$, $d_{86} = R_4 y_4$, $d_{87} = G_4 y_4$, $d_{88} = B_4 y_4$.

6. The evaluation of the SPC using the measurement results obtained in the *u'v'* parameters

In this case the evaluations of the SPC must be presented as:

$$\begin{bmatrix} \hat{X}_r \\ \hat{X}_g \\ \hat{X}_b \\ \hat{Y}_g \\ \hat{Y}_b \\ \hat{Z}_r \\ \hat{Z}_g \\ \hat{Z}_b \end{bmatrix} = \begin{bmatrix} d'_{11} & d'_{12} & d'_{13} & d'_{14} & d'_{15} & d'_{16} & d'_{17} & d'_{18} \\ d'_{21} & d'_{22} & d'_{23} & d'_{24} & d'_{25} & d'_{26} & d'_{27} & d'_{28} \\ d'_{31} & d'_{32} & d'_{33} & d'_{34} & d'_{35} & d'_{36} & d'_{37} & d'_{38} \\ d'_{41} & d'_{42} & d'_{43} & d'_{44} & d'_{45} & d'_{46} & d'_{47} & d'_{48} \\ d'_{51} & d'_{52} & d'_{53} & d'_{54} & d'_{55} & d'_{56} & d'_{57} & d'_{58} \\ d'_{61} & d'_{62} & d'_{63} & d'_{64} & d'_{65} & d'_{66} & d'_{67} & d'_{68} \\ d'_{71} & d'_{72} & d'_{73} & d'_{74} & d'_{75} & d'_{76} & d'_{77} & d'_{78} \\ d'_{81} & d'_{82} & d'_{83} & d'_{84} & d'_{85} & d'_{86} & d'_{87} & d'_{88} \end{bmatrix}^{-1} \begin{bmatrix} 15u'_1 R_1 \\ R_1(9-15v'_1) \\ 15u'_2 R_2 \\ R_2(9-15v'_2) \\ 15u'_3 R_3 \\ R_3(9-15v'_3) \\ 15u'_4 R_4 \\ R_4(9-15v'_4) \end{bmatrix}, \tag{8}$$

where: $\hat{Y}_r = 1 - \hat{Y}_g - \hat{Y}_b$, $d'_{11} = R_1(4-u'_1)$, $d'_{12} = G_1(4-u'_1)$, $d'_{13} = B_1(4-u'_1)$, $d'_{14} = 15u'_1(R_1 - G_1)$, $d'_{15} = 15u'_1(R_1 - B_1)$, $d'_{16} = -3u'_1 R_1$, $d'_{17} = -3u'_1 G_1$, $d'_{18} = -3u'_1 B_1$, $d'_{21} = R_1 v'_1$, $d'_{22} = G_1 v'_1$, $d'_{23} = B_1 v'_1$, $d'_{24} = (R_1 - G_1)(9-15v'_1)$, $d'_{25} = (R_1 - B_1)(9-15v'_1)$, $d'_{26} = 3R_1 v'_1$, $d'_{27} = 3G_1 v'_1$, $d'_{28} = 3B_1 v'_1$, $d'_{31} = R_2(4-u'_2)$, $d'_{32} = G_2(4-u'_2)$, $d'_{33} = B_2(4-u'_2)$, $d'_{34} = 15u'_2(R_2 - G_2)$, $d'_{35} = 15u'_2(R_2 - B_2)$, $d'_{36} = -3u'_2 R_2$, $d'_{37} = -3u'_2 G_2$, $d'_{38} = -3u'_2 B_2$, $d'_{41} = R_2 v'_2$, $d'_{42} = G_2 v'_2$, $d'_{43} = B_2 v'_2$,

$d'_{44}=(R_2-G_2)(9-15v'_2)$, $d'_{45}=(R_2-B_2)(9-15v'_2)$, $d'_{46}=3R_2v'_2$, $d'_{47}=3G_2v'_2$, $d'_{48}=3B_2v'_2$, $d'_{51}=R_3(4-u'_3)$, $d'_{52}=G_3(4-u'_3)$, $d'_{53}=B_3(4-u'_3)$, $d'_{54}=15u'_3(R_3-G_3)$, $d'_{55}=15u'_3(R_3-B_3)$, $d'_{56}=-3u'_3R_3$, $d'_{57}=-3u'_3G_3$, $d'_{58}=-3u'_3B_3$, $d'_{61}=R_3v'_3$, $d'_{62}=G_3v'_3$, $d'_{63}=B_3v'_3$, $d'_{64}=(R_3-G_3)(9-15v'_3)$, $d'_{65}=(R_3-B_3)(9-15v'_3)$, $d'_{66}=3R_3v'_3$, $d'_{67}=3G_3v'_3$, $d'_{68}=3B_3v'_3$, $d'_{71}=R_4(4-u'_4)$, $d'_{72}=G_4(4-u'_4)$, $d'_{73}=B_4(4-u'_4)$, $d'_{74}=15u'_4(R_4-G_4)$, $d'_{75}=15u'_4(R_4-B_4)$, $d'_{76}=-3u'_4R_4$, $d'_{77}=-3u'_4G_4$, $d'_{78}=-3u'_4B_4$, $d'_{81}=R_4v'_4$, $d'_{82}=G_4v'_4$, $d'_{83}=B_4v'_4$, $d'_{84}=(R_4-G_4)(9-15v'_4)$, $d'_{85}=(R_4-B_4)(9-15v'_4)$, $d'_{86}=3R_4v'_4$, $d'_{87}=3G_4v'_4$.

7. The evaluation of the SPC based on the measurement results obtained in the *XYZ* parameters

Some colorimetric measurement instruments allow obtaining coordinates in the color space *XYZ*. Using such instruments, the evaluation of the SPC may be presented as:

$$\begin{bmatrix}\hat{X}_r\\ \hat{X}_g\\ \hat{X}_b\\ \hat{Y}_r\\ \hat{Y}_g\\ \hat{Y}_b\\ \hat{Z}_r\\ \hat{Z}_g\\ \hat{Z}_b\end{bmatrix}=\begin{bmatrix}R_1 & G_1 & B_1 & 0 & 0 & 0 & 0 & 0 & 0\\ 0 & 0 & 0 & R_1 & G_1 & B_1 & 0 & 0 & 0\\ 0 & 0 & 0 & 0 & 0 & 0 & R_1 & G_1 & B_1\\ R_2 & G_2 & B_2 & 0 & 0 & 0 & 0 & 0 & 0\\ 0 & 0 & 0 & R_2 & G_2 & B_2 & 0 & 0 & 0\\ 0 & 0 & 0 & 0 & 0 & 0 & R_2 & G_2 & B_2\\ R_3 & G_3 & B_3 & 0 & 0 & 0 & 0 & 0 & 0\\ 0 & 0 & 0 & R_3 & G_3 & B_3 & 0 & 0 & 0\\ 0 & 0 & 0 & 0 & 0 & 0 & R_3 & G_3 & B_3\end{bmatrix}^{-1}\begin{bmatrix}X_1\\ Y_1\\ Z_1\\ X_2\\ Y_2\\ Z_2\\ X_3\\ Y_3\\ Z_3\end{bmatrix}. \quad (9)$$

8. The evaluation of the SPC using the measurement results obtained in the color space *Yxy* parameters

Some colorimetric measurement instruments allow obtaining coordinates in the color space *Yxy*. Using such instruments one can evaluate the SPC as follows:

$$\begin{bmatrix}\hat{X}_r\\ \hat{X}_g\\ \hat{X}_b\\ \hat{Y}_r\\ \hat{Y}_g\\ \hat{Y}_b\\ \hat{Z}_r\\ \hat{Z}_g\\ \hat{Z}_b\end{bmatrix}=\begin{bmatrix}R_1 & G_1 & B_1 & 0 & 0 & 0 & 0 & 0 & 0\\ 0 & 0 & 0 & R_1 & G_1 & B_1 & 0 & 0 & 0\\ 0 & 0 & 0 & 0 & 0 & 0 & R_1 & G_1 & B_1\\ R_2 & G_2 & B_2 & 0 & 0 & 0 & 0 & 0 & 0\\ 0 & 0 & 0 & R_2 & G_2 & B_2 & 0 & 0 & 0\\ 0 & 0 & 0 & 0 & 0 & 0 & R_2 & G_2 & B_2\\ R_3 & G_3 & B_3 & 0 & 0 & 0 & 0 & 0 & 0\\ 0 & 0 & 0 & R_3 & G_3 & B_3 & 0 & 0 & 0\\ 0 & 0 & 0 & 0 & 0 & 0 & R_3 & G_3 & B_3\end{bmatrix}^{-1}\begin{bmatrix}Y_1x_1/y_1\\ Y_1\\ Y_1(1-x_1-y_1)/y_1\\ Y_2x_2/y_2\\ Y_2\\ Y_2(1-x_2-y_2)/y_2\\ Y_3x_3/y_3\\ Y_3\\ Y_3(1-x_3-y_3)/y_3\end{bmatrix}. \quad (10)$$

9. The evaluation of the SPC using the measurement results obtained in the color space *Yu'v'* parameters

Some colorimetric measurement instruments allow us to obtain coordinates in the color space *Yu'v'*. In this case, to evaluate the SPC three colors encoded as $R_1G_1B_1$, $R_2G_2B_2$, $R_3G_3B_3$ must be projected randomly on the screen and then the respective coordinate values $(Y_1,u_1',v_1'),(Y_2,u_2',v_2'),(Y_3,u_3',v_3')$ measurements must be obtained. The evaluation of the SPC must be presented as:

$$\begin{bmatrix} \hat{X}_r \\ \hat{X}_g \\ \hat{X}_b \\ \hat{Y}_r \\ \hat{Y}_g \\ \hat{Y}_b \\ \hat{Z}_r \\ \hat{Z}_g \\ \hat{Z}_b \end{bmatrix} = \begin{bmatrix} R_1 & G_1 & B_1 & 0 & 0 & 0 & 0 & 0 & 0 \\ 0 & 0 & 0 & R_1 & G_1 & B_1 & 0 & 0 & 0 \\ 0 & 0 & 0 & 0 & 0 & 0 & R_1 & G_1 & B_1 \\ R_2 & G_2 & B_2 & 0 & 0 & 0 & 0 & 0 & 0 \\ 0 & 0 & 0 & R_2 & G_2 & B_2 & 0 & 0 & 0 \\ 0 & 0 & 0 & 0 & 0 & 0 & R_2 & G_2 & B_2 \\ R_3 & G_3 & B_3 & 0 & 0 & 0 & 0 & 0 & 0 \\ 0 & 0 & 0 & R_3 & G_3 & B_3 & 0 & 0 & 0 \\ 0 & 0 & 0 & 0 & 0 & 0 & R_3 & G_3 & B_3 \end{bmatrix}^{-1} \begin{bmatrix} 9Y_1u_1'/4v_1' \\ Y_1 \\ \left(3Y_1\left(4-u_1'\right)/4v_1'\right)-5Y_1 \\ 9Y_2u_2'/4v_2' \\ Y_2 \\ \left(3Y_2\left(4-u_2'\right)/4v_2'\right)-5Y_2 \\ 9Y_3u_3'/4v_3' \\ Y_3 \\ \left(3Y_3\left(4-u_3'\right)/4v_3'\right)-5Y_3 \end{bmatrix}. \quad (11)$$

10. Conclusion

There are several different methods of evaluation of screen profile coefficients of either LCD or LED display. The choice of computation method of the SPC depends on availability and features of colorimetric instruments [6-10].

The mathematical method computational complexity which is applied for all mentioned above methods to evaluate the screen profile coefficients is practically equal. Obtained results may be used by engineers who are designing modern indication equipment based on LCD- or LED-panels.

References

[1] Lee B-S, Bala R, Sharma G 2007 *Journal of Electronic Imaging* **4** 1-13
[2] Ibraheem N A, Hasan M M, Khan R Z and Mishra P K 2012 *ARPN Journal of science and technology* **3** 265-275
[3] Menesatti P, Angelini C, Pallottino F, Antonucci F, Aguzzi Y and Costa C 2012 *Sensors* **12** 7063-79
[4] Chhajed S, Xi Y, Li Y-L, Gessmann Th and Schubert E F 2005 *Journal of applied physics* **5** 1-8
[5] Kostishin M O, Shukalov A V, Zharinov I O, Zharinov O O and Ershov A N 2016 *Indian Journal of Science and Technology* **29** 99451
[6] Huang W, Li J-M, Yang L-M, Jin Zh-L, Zhong Zh-G and Liu Y 2011 *Optics & Laser technology* **43** 214-217
[7] Zargaryants G S, Mikhailov O M 2008 *Light & Engineering* **3** 69-77
[8] Chiu T-L, Lee J-H 2010 *Optics communications* **283** 373-378
[9] Hernandez-Andres J, Lee R L, Romero Jr and Romero J 1999 *Applied optics* **27** 5703-09
[10] Wang Ch-Fu, Tang Ch-W, Huang B-J 2010 *IEEE Trans. on power electronic* **2** 417-428

IOP Conf. Series: Materials Science and Engineering **286** (2017) 012015 doi:10.1088/1757-899X/286/1/012015

The research of statistical properties of colorimetric features of screens with a three-component color formation principle

I O Zharinov[1] **and O O Zharinov**[2]

[1]Department of Computer Design of On-Board Computer Equipment, Saint Petersburg National Research University of Information Technologies, Mechanics and Optics, 49, Kronverksky ave., Saint Petersburg, 197101, Russia
[2]Department of Problem-Oriented Computing Complexes, Saint Petersburg State University of Aerospace Instrumentation, 69, Bolshaya Morskaia str., Saint Petersburg, 190000, Russia

E-mail: mpbva@mail.ru

Abstract. The problem of the research is concerned with quantitative analysis of influence of technological variation of the screen color profile parameters on chromaticity coordinates of the displayed image. Some mathematical expressions which approximate the two-dimensional distribution of chromaticity coordinates of an image, which is displayed on the screen with a three-component color formation principle were proposed. Proposed mathematical expressions show the way to development of correction techniques to improve reproducibility of the colorimetric features of displays.

1. Introduction

Developers of software for a color multi-function display units (MFDU) in avionics indication equipment have to solve the problem of the image color palette selection, which is displayed on the LCD (liquid crystal display), or LED (light emitting diode) screen panel [1-3]. The color palette selection involves the necessity to comply the requirements of the aviation industry technical documentation and the necessity to show to the observer the image with the optimal visual characteristics of perception. Quality of avionics indication equipment is evaluated using a lighting test with the measuring instruments. Final result of these tests is the color palette — an array of records encoded with the *RGB* code (*R* – Red, *G* – Green, *B* – Blue) for each involved color [2, 4-6].

The practice of the color MFDU development reveals that once formed color palette, which is approved with the results, and the lighting test protocol of a particular color MFDU type is used in the software of the following developments of this type of equipment. This software engineering approach based on the application of the maximally uniform and standardized successful solutions has some significant disadvantages [7-10]. Mainly problem is caused by unavoidable technological variation of the screen color profile parameters. As the result, random shifts of chromaticity coordinates appear for the same once approved color palette. Because of this reason, the research of statistical properties of colorimetric features of screens with a three-component color formation principle is an actual problem.

2. Color space transformation rules used in avionics

Results of tests of the color MFDU manufactured on the basis of the LCD panels revealed that the same *RGB* code which determines the image displayed color leads to different *XYZ*-tristimulus values and (*x*,*y*)-chromaticity coordinates in the different samples of color multi-function displays.

For each pair of the color MFDU samples that use the software with the same color palette according to the transformation rules [2] for the color spaces *XY*, *XYZ* and the *RGB* color space the following expressions are correct:

$$\begin{bmatrix} X_{r_2} & X_{g_2} & X_{b_2} \\ Y_{r_2} & Y_{g_2} & Y_{b_2} \\ Z_{r_2} & Z_{g_2} & Z_{b_2} \end{bmatrix}^{-1} \begin{bmatrix} X_2 \\ Y_2 \\ Z_2 \end{bmatrix} = \begin{bmatrix} R \\ G \\ B \end{bmatrix} = \begin{bmatrix} X_{r_1} & X_{g_1} & X_{b_1} \\ Y_{r_1} & Y_{g_1} & Y_{b_1} \\ Z_{r_1} & Z_{g_1} & Z_{b_1} \end{bmatrix}^{-1} \begin{bmatrix} X_1 \\ Y_1 \\ Z_1 \end{bmatrix} \Rightarrow \begin{bmatrix} X_2 \\ Y_2 \\ Z_2 \end{bmatrix} = \begin{bmatrix} X_{r_2} & X_{g_2} & X_{b_2} \\ Y_{r_2} & Y_{g_2} & Y_{b_2} \\ Z_{r_2} & Z_{g_2} & Z_{b_2} \end{bmatrix} \begin{bmatrix} X_{r_1} & X_{g_1} & X_{b_1} \\ Y_{r_1} & Y_{g_1} & Y_{b_1} \\ Z_{r_1} & Z_{g_1} & Z_{b_1} \end{bmatrix}^{-1} \begin{bmatrix} X_1 \\ Y_1 \\ Z_1 \end{bmatrix} =$$

$$= \begin{bmatrix} \breve{X}_r & \breve{X}_g & \breve{X}_b \\ \breve{Y}_r & \breve{Y}_g & \breve{Y}_b \\ \breve{Z}_r & \breve{Z}_g & \breve{Z}_b \end{bmatrix} \begin{bmatrix} X_1 \\ Y_1 \\ Z_1 \end{bmatrix} = \begin{bmatrix} \dfrac{x_2 Y_{xy_2}}{y_2} \\ Y_{xy_2} \\ (1 - x_2 - y_2)\dfrac{Y_{xy_2}}{y_2} \end{bmatrix} = \begin{bmatrix} \breve{X}_r & \breve{X}_g & \breve{X}_b \\ \breve{Y}_r & \breve{Y}_g & \breve{Y}_b \\ \breve{Z}_r & \breve{Z}_g & \breve{Z}_b \end{bmatrix} \begin{bmatrix} \dfrac{x_1 Y_{xy_1}}{y_1} \\ Y_{xy_1} \\ (1 - x_1 - y_1)\dfrac{Y_{xy_1}}{y_1} \end{bmatrix} \Rightarrow$$

$$\Rightarrow \begin{cases} \dfrac{x_2 Y_{xy_2}}{y_2} = \breve{X}_r \dfrac{x_1 Y_{xy_1}}{y_1} + \breve{X}_g Y_{xy_1} + \breve{X}_b (1 - x_1 - y_1)\dfrac{Y_{xy_1}}{y_1} \\ Y_{xy_2} = \breve{Y}_r \dfrac{x_1 Y_{xy_1}}{y_1} + \breve{Y}_g Y_{xy_1} + \breve{Y}_b (1 - x_1 - y_1)\dfrac{Y_{xy_1}}{y_1} \\ (1 - x_2 - y_2)\dfrac{Y_{xy_2}}{y_2} = \breve{Z}_r \dfrac{x_1 Y_{xy_1}}{y_1} + \breve{Z}_g Y_{xy_1} + \breve{Z}_b (1 - x_1 - y_1)\dfrac{Y_{xy_1}}{y_1} \end{cases}, \qquad (1)$$

Where X_1, Y_1, Z_1 are the displayed image color coordinates of the first color MFDU sample; X_2, Y_2, Z_2 - the displayed image color coordinates of the second color MFDU sample; x_2, y_2, Y_{xy_2} - the chromaticity coordinates and the image relative brightness component on the first color MFDU screen; x_2, y_2, Y_{xy_2} - the chromaticity coordinates and the image relative brightness component on the second color MFDU screen; $X_{r_i}, X_{g_i}, X_{b_i}, Y_{r_i}, Y_{g_i}, Y_{b_i}, Z_{r_i}, Z_{g_i}, Z_{b_i}$ - the LCD screen profile matrix coefficients used in the color MFDU; *RGB* – the image sample color code.

Analysis of the LCD panels technical documentation reveals that the color difference of the (*x*,*y*)-coordinates of an image displayed in the different color MFDU samples with the same screen type and the same program *RGB* code may reach up to 0.1 unit of chromaticity value, which is visible for an observer and measurable with instruments.

3. Evaluation of the chromaticity coordinates shift

The system (1) binds the chromaticity coordinates and the image relative brightness components with the same color palette *RGB*. The difference in the (*x*,*y*)-chromaticity coordinate values can be explained by the different values of the LCD panel profile coefficients and can be calculated by solving the system (1) as follows:

$$x_2 = \frac{\breve{X}_r \dfrac{x_1}{y_1} + \breve{X}_g + \breve{X}_b (1 - x_1 - y_1)\dfrac{1}{y_1}}{(\breve{X}_r + \breve{Y}_r + \breve{Z}_r)\dfrac{x_1}{y_1} + (\breve{X}_g + \breve{Y}_g + \breve{Z}_g) + (\breve{X}_b + \breve{Y}_b + \breve{Z}_b)(1 - x_1 - y_1)\dfrac{1}{y_1}}, \qquad (2)$$

$$y_2 = \frac{\breve{Y}_r \frac{x_1}{y_1} + \breve{Y}_g + \breve{Y}_b (1 - x_1 - y_1)\frac{1}{y_1}}{(\breve{X}_r + \breve{Y}_r + \breve{Z}_r)\frac{x_1}{y_1} + (\breve{X}_g + \breve{Y}_g + \breve{Z}_g) + (\breve{X}_b + \breve{Y}_b + \breve{Z}_b)(1 - x_1 - y_1)\frac{1}{y_1}}. \quad (3)$$

The expressions (2) and (3) define the chromaticity coordinates conversion rule on the *XY*-plane for an image programmed with the same *RGB* code but displayed on two color MFDU with different LCD panel matrix profiles.

4. LCD panel color reproduction stochastic model

Stochastic model of the LCD panel color reproduction includes their colorimetric characteristics technological dispersion in mass production, and it may be represented as:

$$\begin{bmatrix} X \\ Y \\ Z \end{bmatrix} = \begin{bmatrix} X_r + \Delta \cdot \xi_{X_r} & X_g + \Delta \cdot \xi_{X_g} & X_b + \Delta \cdot \xi_{X_b} \\ Y_r + \Delta \cdot \xi_{Y_r} & Y_g + \Delta \cdot \xi_{Y_g} & Y_b + \Delta \cdot \xi_{Y_b} \\ Z_r + \Delta \cdot \xi_{Z_r} & Z_g + \Delta \cdot \xi_{Z_g} & Z_b + \Delta \cdot \xi_{Z_b} \end{bmatrix} \begin{bmatrix} R \\ G \\ B \end{bmatrix}, \quad (4)$$

where $\xi_{X_r} \in [-X_r/2;+X_r/2], \xi_{X_g} \in [-X_g/2;+X_g/2], \xi_{X_b} \in [-X_b/2;+X_b/2], \xi_{Y_r} \in [-Y_r/2;+Y_r/2]$,

$\xi_{Y_g} \in [-Y_g/2;+Y_g/2], \xi_{Y_b} \in [-Y_b/2;+Y_b/2], \xi_{Z_r} \in [-Z_r/2;+Z_r/2], \xi_{Z_g} \in [-Z_g/2;+Z_g/2]$,

$\xi_{Z_b} \in [-Z_b/2;+Z_b/2]$, – uniformly distributed random values, Δ – the technological dispersion parameter explained with the screens production quality of the same manufacturer.

The geometric representation of the color reproduction stochastic model is shown in figure 1.

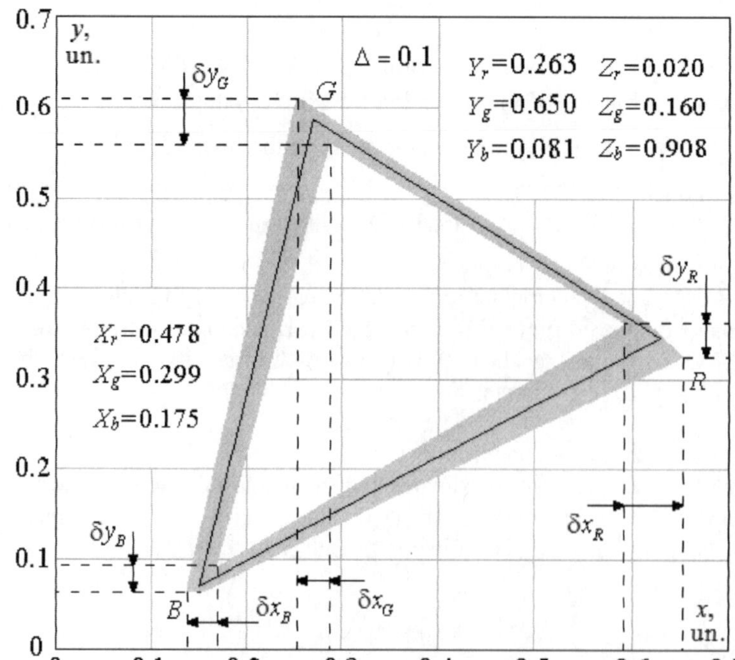

Figure 1. The color gamut variability of the mass production of screens.

5. Simulation results

There are two-dimensional histogram sections (shaped as distorted ellipses) shown in figure 2, which describe the chromaticity coordinates distribution $p(x,y)$ for the different technological dispersion values: Δ=0.2; Δ=0.1; Δ=0.05; Δ=0.01. The point (marked in figure 2 with a mark «+») which has the (*x,y*)-coordinates calculated with the formula (4) when Δ=0 corresponds to the ellipse central point which is the same for all the values of Δ.

The threshold level *L* for the horizontal histogram section of the *p*(*x,y*) chromaticity coordinates distribution is calculated using the following equation:

$$L=\arg\min_{L}\left(\sum_{k_1=0}^{N-1}\sum_{k_2=0}^{N-1}I\left(H_{k_1,k_2}>L\right)H_{k_1,k_2}-P_D\right),\ I(q)=\begin{cases}0,q<0\\1,q>0\end{cases}, \tag{5}$$

where H_{k_1,k_2} - intersection relative frequency of the (*x,y*)-chromaticity coordinates in the histogram subinterval *p*(*x,y*); *I*(*q*) - indicator function; *N* – the number of the *p*(*x,y*) histogram subintervals for each coordinate; P_D– the confidence probability level.

The equation (5) is solved for the *L* with the use of a numerical computing methods in software MathCad 15.0. The subinterval dimensions are 0.0003 x 0.0003 units of chromaticity values on the *XY*-plane. In figure 2 the horizontal section of the histogram *p*(*x,y*) is shown for the confidence probability level P_D=0.999 in the different parts of the color gamut triangle for five different *RGB* code values which are non-zero. In particular, if Δ=0.2 the ellipse area includes almost $5\cdot10^5$ of the (*x,y*)-chromaticity coordinates values.

The section of the histogram *p*(*x,y*) corresponds to the section resulted from the theoretical two-dimensional probability density function for the normally distributed mutually dependent random values (*x,y*):

$$f(x,y)=\frac{1}{2\pi\sqrt{D_xD_y}\sqrt{1-\rho^2}}\exp\left(-\frac{1}{2(1-\rho^2)}\left(\frac{(x-M_x)^2}{D_x}-\frac{2\rho(x-M_x)(y-M_y)}{\sqrt{D_xD_y}}+\frac{(y-M_y)^2}{D_y}\right)\right),$$

$$\rho=\frac{\frac{1}{N}\sum_{i=0}^{N-1}(x_i-M_x)(y_i-M_y)}{\sqrt{D_xD_y}}, \tag{6}$$

where ρ - correlation coefficient of the random values (*x,y*); M_x, M_y - expected value of the random values (*x,y*); D_x, D_y - dispersion of the random values (*x,y*). In the formula (6) the following parameters M_x, M_y, D_x, D_y, ρ can be substituted with the correspondent sample estimates after some colorimetric measurements of LCD panel samples are taken.

The equations of the equal probability density ellipse family formed with horizontal section of the function (6) on the level λ must be represented as:

$$Q(x,y)=\frac{(x-M_x)^2}{D_x}-\frac{2\rho(x-M_x)(y-M_y)}{\sqrt{D_xD_y}}+\frac{(y-M_y)^2}{D_y}=\lambda^2 .$$

Different but constant probabilities for all ellipse points of the probability density function (6) correspond to the various values of the parameter λ, called threshold. The probability that random point (*x,y*) with density function (6) can be found inside the equal probabilities ellipse must be represented as:

$$P_D(\lambda)=\frac{1}{2\pi\sqrt{D_xD_y}\sqrt{1-\rho^2}}\iint_{g(\lambda)}\exp\left(-\frac{1}{2(1-\rho^2)}Q(x,y)\right)dxdy=1-\exp\left(-\frac{\lambda^2}{2(1-\rho^2)}\right),$$

where $\lambda^2=-2(1-\rho^2)\ln(1-P_D)$, function g($\lambda$) specifies points (*x,y*) located inside the ellipse section area.

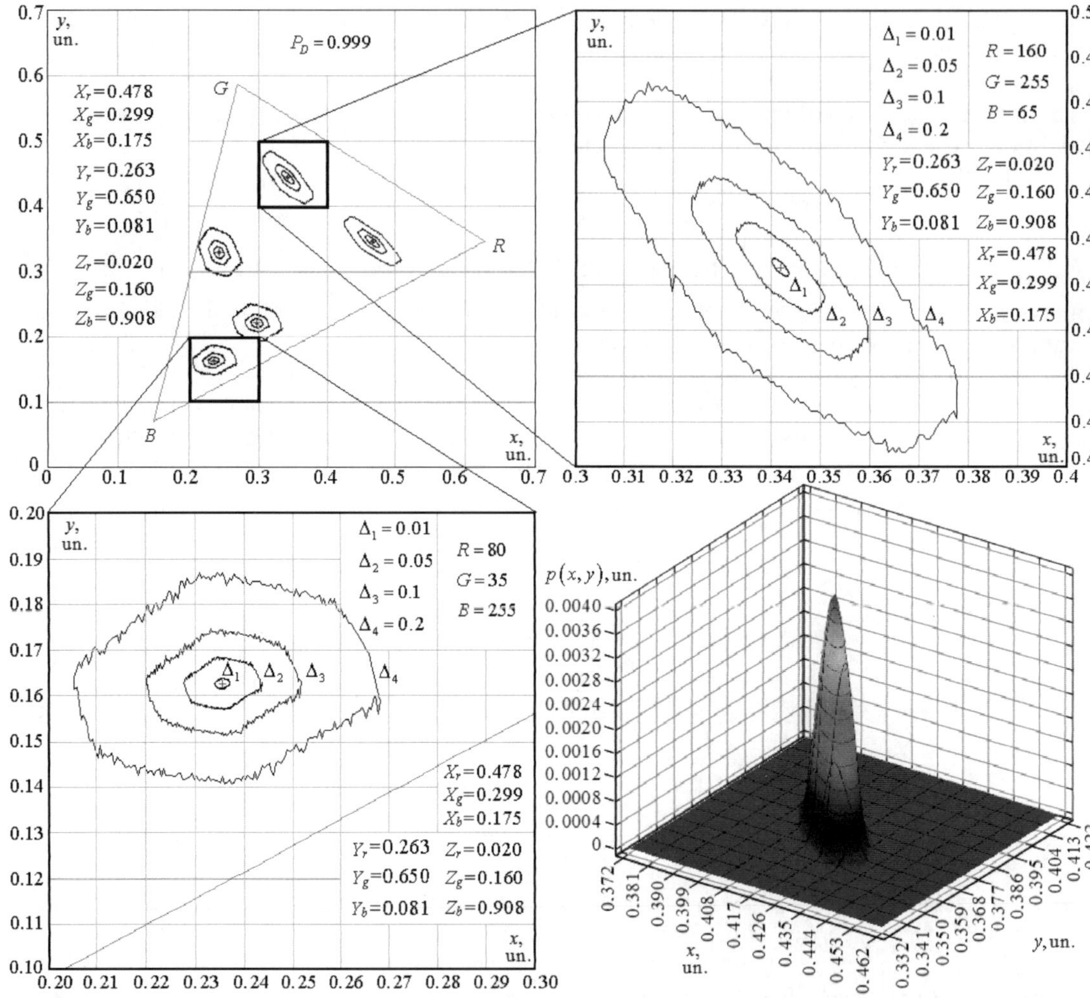

Figure 2. The horizontal section of the histogram *p*(*x*,*y*), projected on the *XY*-plane, and the two-dimensional distribution histogram *p*(*x*,*y*) for the chromaticity coordinates on the *XY*-plane.

The ellipse sections of the theoretical probability density function *f*(*x*,*y*) on the *XY*-plane are shown in figure 3. Horizontal sections of the histogram *p*(*x*,*y*) are also presented in figure 3.

6. Conclusion

The analysis of the figure 3 shows that the (*x*,*y*)-chromaticity coordinates random shift caused by the technological variation of screen color profile parameters is an essential matter to be kept in mind. For example, if Δ=0.01 or higher: the *x*-chromaticity coordinates dispersion is within 0.005 units, the *y*-chromaticity coordinates dispersion is within 0.005 units. The dispersion of 0.005 chromaticity units when Δ=0.01 can be perceived by an observer and can be measured with the modern direct measurement instruments (colorimeters, spectroradiometers). This level of dispersion is acceptable for the quality of the household equipment (monitors, TV-sets, cell phones, i-pad etc.) when the observer looks at only one sample of the display unit. But the value Δ=0.01 is insufficient for manufacturing the aviation display screens when the observer receives simultaneously one the same image from several (up to 6 units) supposed to be identical displays. Today the screen manufacturers provide the screen samples reproducibility with the dispersion level of $0.1 \le \Delta \le 0.2$. Proposed mathematical expressions show the way to development of correction techniques to improve reproducibility of the colorimetric features of displays.

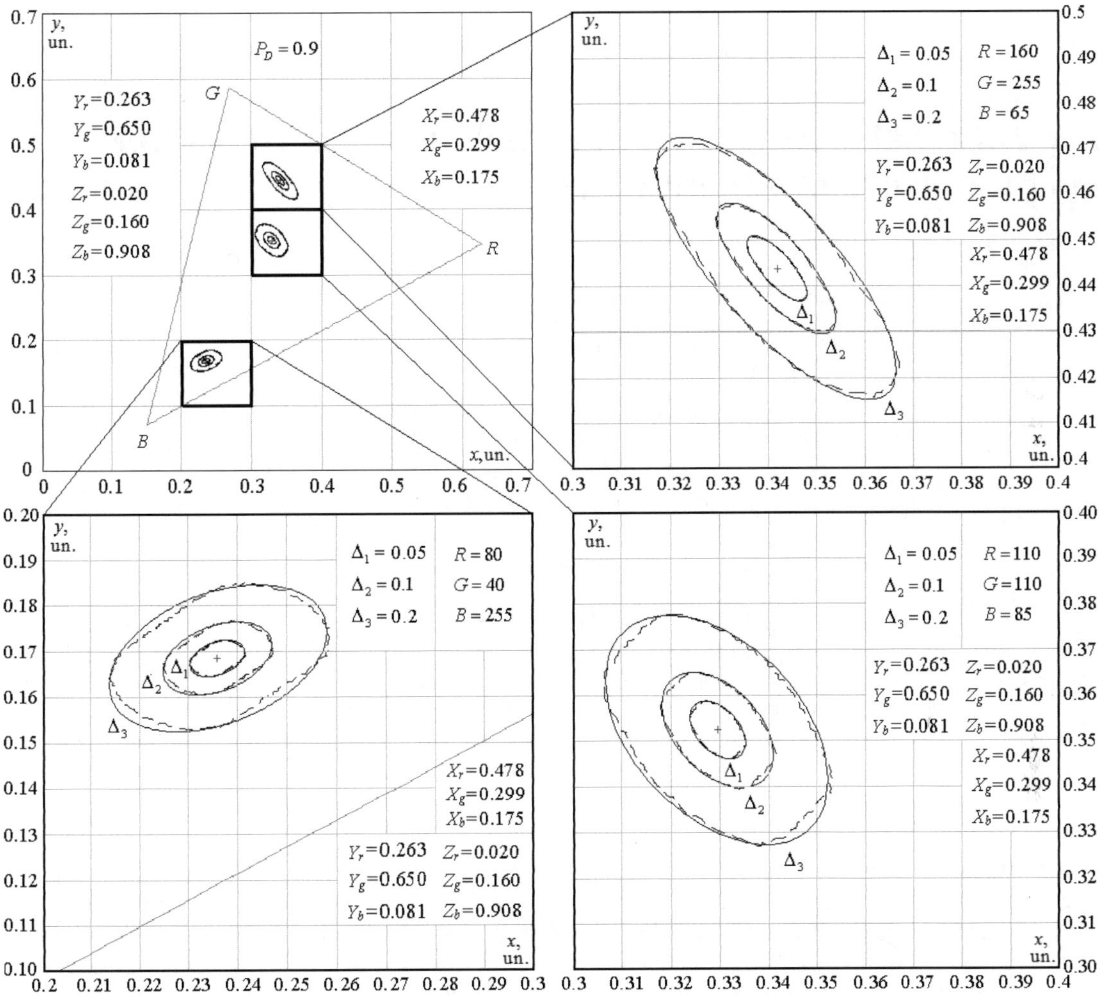

Figure 3. The ellipses of histogram section $p(x,y)$ (dotted line) and the theoretical $f(x,y)$ function of probability density (solid line).

References

[1] Thomas J-B, Hardeberg J Y, Foucherot I and Gouton P 2008 *Color research and application* **6** 449-460

[2] Livada B 2012 *Scientific Technical Review* **3** 70-79

[3] Bradley D M, Gupta R C 2002 *Annals of the Institute of Statistical Mathematics* **3** 689-700

[4] Buonocore A, Pirozzi E, Caputo L 2009 *Statistics & Probability Letters* **19** 2092-97

[5] Murakami H 2014 *Statistica Neerlandica* **4** 267-275

[6] Potuschak H, Muller W G 2009 *Statistical papers* **1** 177-183

[7] Sadooghi-Alvandi S M, Nematollahi A R, Habibi R 2009 *Statistical papers* **1** 171-175

[8] Chiu T-L, Lee J-H 2010 *Optics communications* **283** 373-378

[9] Hernandez-Andres J, Lee R L, Romero Jr and Romero J 1999 *Applied optics* **27** 5703-09

[10] Justel A, Pena D, Zamar R 1997 *Statistics & Probability Letter* **3** 251-259

IOP Conf. Series: Materials Science and Engineering **286** (2017) 012016 doi:10.1088/1757-899X/286/1/012016

The nature of elastoplastic transition in low-carbon steel samples with and without welds

L V Danilova[1,2] and V I Danilov[2,3]

[1] Department of Physics and Technology, Tomsk State University, 36, Lenin Ave, Tomsk, 634050, Russia
[2] Laboratory of Strength Physics, Institute of Strength Physics and Materials Science SB RAS, 2/4, Academic Ave, Tomsk, 634021, Russia,
[3] Department of Welding Production, Yurga Institute of Technology, TPU Affiliate, Leningradskaya St., 26k3, Yurga, 652055, Russia

E-mail: lidaakvo@rambler.ru, dvi@ispms.tsc.ru

Abstract. This paper describes the influence of a weld on elastoplastic transition in low-carbon steels. Depending on the level of internal stresses in the zone of the weld, the formation of Chernov-Lüders bands can occur either exactly in the weld or near it. In the second case, the influence of the weld is detected only when the front of the Chernov-Lüders band reaches it. In both cases, the features of kinetics for the fronts of the Chernov-Lüders bands are caused by the structural and phase inhomogeneity of a heat-affected zone and deposited metal.

1. Introduction

Strength and fracture are the most important characteristics of steels. To select a material for a product it is necessary to consider the material strength, since it determines whether it will resist loads for the required time before the material is fractured, and steels show the highest strength among materials. Another important quality is the availability of material. In this connection, the most common steels used to produce engineering constructions are low-carbon high-grade or low- grade steels. Engineering constructions consist of many elements. These elements can be joined by welds, the strength of which differs from the strength of material, most often toward a lower level. Thus, it is necessary to know the behavior of the base material under loading, as well as the behavior of material in the place of welded joints. The parameters defined as permissible for this structure should be used within elastoplastic transition range at the junction of elements.

Elastoplastic transition in structural low-carbon steels consists of the stages such as microplasticity, during which there are elastic strains and processes leading to irreversible deformation; a yield drop and a yield plateau corresponding to the growth of the Chernov-Lüders band (CLB) through the cross section and its further distribution along the sample [1]. These processes lead to the accumulation of strain in the sample up to 3.5%. This transition ends after the material passes from an elastic stress state to a plastic strain state.

Studying the behavior of material, that is the movement of CLB fronts in the basic material and in the material of welded joints, allows us to determine the limits of permissible stresses and strains for the engineering constructions consisting of these materials.

IOP Conf. Series: Materials Science and Engineering **286** (2017) 012016 doi:10.1088/1757-899X/286/1/012016

2. Methods and materials

The most widely used material in engineering constructions is low-alloy high-grade steel (09G2S) that was chosen to be the object of the study. Specimens of the dog-bone shape with a working part size of 2 × 6 × 40 mm were prepared from the 250 × 250 × 5 mm butt-welded plates, which were subjected to milling and mechanical grinding. In this case, the weld was located in the middle of the working section and perpendicular to the tension axis of the sample. The width of the deposited metal was about 5 mm, and the heat-affected zone surrounding this one was 1.5 mm. To observe CLB fronts, it is necessary to ensure that the working surface of the sample is the diffusively-reflecting one and the samples are subjected to deep etching in a 12% alcohol solution of nitric acid.

A testing machine Walter+Bai AG LFM-125 was used for mechanical uniaxial tensile testing of the samples. During the test, a deformation diagram was recorded in digital form. After analyzing the tests, it can be concluded that the deformation rate of the sample is $8.3 \cdot 10^{-5}$ s^{-1} at the speed of the movable (top) testing machine grip of 0.2 mm/min. The method of digital statistical speckle photography was used during the tensile process to obtain an "in situ" record for the centers of deformation localization [2]. The use of this method leads to the fact that these centers of deformation localization appear as dark contrast bands superimposed on the speckle images of the deformed samples and, in fact, represent the boundaries or fronts of CLB. Comparison of the deformation diagram with the distribution of the deformation localization centers showed that the stages of the tensile curve were in a good agreement with the behavior of the CLB fronts.

3. Results and discussion

The propagation of CLBs in the homogeneous low-carbon steel samples can be called stochastic. However, numerous uniaxial tensile tests show that two CLBs formed often near both testing machine grips [2, 3]. Moreover, each band has two moving fronts diverging in different directions, one of which soon reaches the edge of the working section and then stops. The remaining fronts move towards each other at a constant velocity, at the same time, bands expansion rate remains the constants. This is because the velocity of the CLB front propagation directly depends on the expansion rate of the plastically deformed part in the sample [4]:

$$\sum_{i=1}^{N} \left| V_f^{(i)} \right| = \left(V_f \right) = const, \tag{1}$$

In this formula, $\boldsymbol{N}$ is the number of simultaneously moving CLB fronts, $\boldsymbol{V_f^{(i)}}$ is the propagation velocity of the i-th CLB front, and $\boldsymbol{(V_f)}$ is the expansion rate of the plastically deformed part in the sample, equal to approximately 0.16 mm/s for this case. Thus, if one of the moving CLB fronts stops, the velocity of the remaining one is increases by the velocity of the stopped front. These relations are valid both for any time moment and for any number of moving fronts.

The samples with a weld have a different picture of the CLB formation. The presence of a weld in samples has a significant effect on the formation or distribution of CLBs. Depending on the process, internal stresses in heat-affected zones can be higher or lower compared with the basic metal. Since the stresses in heat-affected zones are higher than in the base metal, the formation of CLB occurs in the decarbonized zone of a weld [5]. In this case, the CLB formation in a weld differs from the CLB formation in the basic metal. If, in the latter case, a CLB grows from the one edge of the working section of the sample to the other one with clear boundaries, then for the first case the picture is different. In the beginning, a large number of small centers of deformation form in the zone of deposited metal, and then a CLB forms during the next 3-8 seconds. This CLB occupies the entire area of deposited metal, but does not have clearly observed fronts. The moving CLB fronts form only in the heat-affected zones of the weld during the first minute after the start of loading. Further propagation of the CLB fronts takes place according to the well-known scenario observed in the samples without a weld [2-4]. The velocity of the fronts is approximately equal to 100-150 μm/s and does not differ much from the velocities in homogeneous samples.

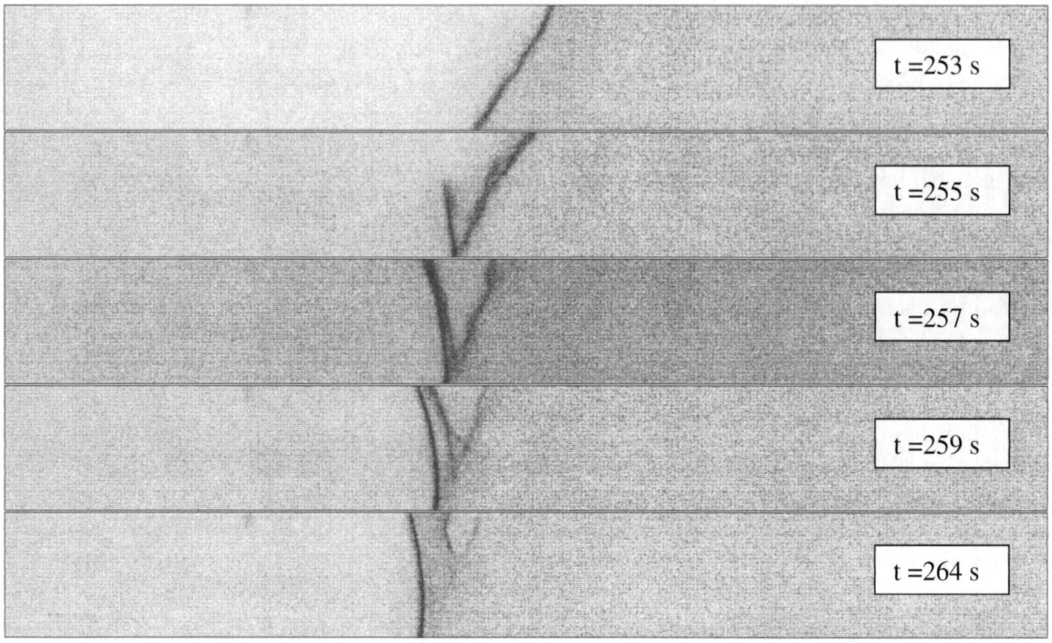

Figure 1. Chernov-Lüders bands behavior in welded joint zones.

In the other case, if the internal stresses in heat-affected zones are lower than in the basic metal [5], another scenario of elastoplastic transition takes place and the presence of a weld has an effect on the CLB distribution. A CLB forms outside the zone of deposited metal, as in the case for the homogeneous sample, near one of the testing machine grips. When one of the fronts reaches the edge of the working section and stops, the velocity of another one increases according to formula (1), and the front continues propagation along the sample. Having reached the boundary of the weld, the CLB front stops and becomes the source of a new CLB nucleation directly in the deposited metal. In this case, the CLB forms in the deposited metal two times faster than in the homogeneous sample under normal conditions [2]. The time required for this process is approximately 3 seconds, and the speed is higher than 2 mm/s. The further deformation of the sample occurs directly in the weld: while the CLB is distributed in the zone of deposited metal, the CLB front remains stationary outside this zone. At the same time, the CLB front moves in the deposited metal at a velocity of about 0.5 mm/s and exceeds the velocity of CLB front propagation in the base material (Figure 1). At the time when the deposited metal is completely deformed, and the CLB front located in it disappears, the stationary CLB front in the heat-affected zone starts moving and continues to be propagated in the base metal with velocity is about 100 μm/s that corresponds to the velocity of front propagation in a homogeneous metal.

4. Conclusions

The existence of a weld in a sample determines the CLB formation. If stress in a heat-affected zone is higher than in the basic metal, a CLB will form directly in the weld. If stress in a heat-affected zone is higher than in the basic metal, a CLB will form outside the area of deposited metal. However, since the weld's area has a large structural and phase inhomogeneity, regardless of where a CLB is formed, it strongly influences on the behavior of deformation. In this case, the difference in the structure of materials leads to the fact that the kinetics of CLB fronts in the deposited and base metal are substantially different, and the velocity of fronts there can differ by an order of magnitude. In addition, a CLB front can stop at the boundaries of zones with a different structural and phase state and become the source for the formation of one or even several CLBs. At the same time, the critical formation processes of moving CLB fronts are localized in heat-affected zones.

Acknowledgments
The work was carried out within the framework of the Program of Fundamental Scientific Research Russian State academies of sciences for 2013-2020.

References

[1] Pelleg J 2013 *Mechanical properties of materials. Springer* (Heidelberg, New York, London)
[2] Gorbatenko V V, Danilov V I and Zuev L B 2017 Plastic flow instability: Chernov–Lüders bands and the Portevin–Le Chatelier effect *Technical Phys.* **36** **3** 395-400
[3] Danilov V I, Gorbatenko V V and Zuev L B 2016 On the kinetics of mobile Chernov–Lüders band fronts *AIP Conf. Proc.* **1783** 020034
[4] Danilov V, Gorbatenko V, Zuev L 2016 Autowaves of localized plastic deformation on the yield plateau and on the work hardening stage *Materials science. Non-equilibrium phase transformation* **3** 21-24
[5] Smirnov A N, Kozlov E V, Ozhiganov E A, Ababkov N V, and Knyazkov V L 2016 The effect of deformation degree of weld joints in carbon steels on the structural-phase state and internal stress fields *Svarka i Diagnostika* **3** 25-8 (in Russian)

A proposal to improve a 3D printing technology of composite materials products

I V Zlobina[1], N V Bekrenev and S P Pavlov

Department "Technical mechanics and machine parts", Yuri Gagarin Saratov State Technical University, 77 Politechnicheskaya street, Saratov, 410054, Russian Federation

E-mail: irinka_7_@mail.ru

Abstract. The objects formed by 3D printing, in particular from nonmetallic materials, have an essential disadvantage not eliminated at the present time - a significant anisotropy of the structure and, as a consequence, of physical and mechanical characteristics. The research of 3DP technology in combination with the influence of microwave electromagnetic field of various power on the formed three-dimensional product has been carried out. It was established that a microwave electromagnetic field with an average specific power of 2450 MHz causes an increase in the homogeneity of the of powder materials` structure, expressed in a decrease of the pore size by 24% and a decrease in their dispersion by almost 30%. As a consequence of the increase in the homogeneity of the structure, the flexural strength of Zp130- powder plates impregnated with cyanoacrylate has increased to 1.77 times. Thus, the use of the microwave electromagnetic field as a final stage in the formation of products made from composite materials is promising and requires additional studies to justify the serial production technology.

1. Introduction

1.1. Status of the question

Development and introduction of perspective transport systems, in particular aerial vehicles, require creation of new high-strength and light materials and improvement of technologies for forming structural elements from them. Analysis of scientific and technical literature, conference materials and exhibitions testifies to the intensive development of production of composite materials based on carbon fibers and glass fabrics and their wide application in the aviation, automotive, shipbuilding, rocketry and space technology. However, composite materials are characterized by a pronounced anisotropy of physicomechanical characteristics, determined by the type and orientation of the reinforcing components.

One of the new ways to reduce this shortcoming is to manage the structure of the material by introducing reinforcing components according to a certain law. This operation can be carried out not only by introducing new components, but also by modifying them by electrophysical influences, which is much easier to implement because the process is easily controlled by regulating technological regimes without interfering with the synthesis of the material. However, the use of traditional technologies for practical implementation of this solution is difficult, which causes the need for

[1] irinka_7_@mail.ru

additional research in the field of creating the scientific basis for technological support for the production of composite products.

1.2. Relevance of research

One of the most topical technological trends currently includes additive technologies, in which layer-by-layer build-up of material in accordance with a given computer solid model practically replaces the traditional processes of cutting, pressing and stamping [1-4]. The main advantage of these technologies is the direct reproduction based on a solid model of a product of any complexity without the use of a complex instrument and expensive equipment. At present, significant progress has been achieved in the use of additive technologies not only in rapid prototyping, but also in the manufacture of structural elements of existing technical systems. At the same time, the technologies for manufacturing products made of polymer materials 3DP, FDM and SLA are quite cheap (the cost of equipment is about $ 1000), but they are imperfect, since products have low strength and durability. Particularly important is the instability of the characteristics from the part to the part and from the installation to the installation. However, scientific research in this area is aimed at developing source materials, methods of components compounding, software that provides maximum accuracy and productivity. One of the alternative ways to improve the technologies of additive production of products from nonmetallic and composite materials may be a directed modification of the structure of existing materials by electrophysical influences. Based on the mentioned above, the solution of the issues of increasing the strength of products made of nonmetallic materials, created based on additive technologies, is an important urgent task.

1.3. Statement of the problem

In the works of Russian and foreign scientists [5, 6], positive effects of the influence of the microwave electromagnetic field on the technological processes of synthesis and modification of nonmetallic materials are reflected, including the positive effect of nonthermal microwave influence on the properties of polymeric materials. We have established when applied to layered carbon plastics with a quasi-isotropic structure, the ultimate bending stress increases by an average of 25%, the interlayer shear stress by 14-15% (confirmed by tests in production conditions), the time to destruction at the ultimate load increases by 1,6 - 2 times [7-10]. The purpose of these studies was the experimental determination of the possibility of improving additive technologies, for example, by 3DP introduction of the structure modification operation in the microwave electromagnetic field.

2. Research methods and equipment

In the experiments we used a 3D printer of the ZPrinter 450 model and a microwave unit with a magnetron OM75S (31) with a frequency of 2450 MHz. We used three microwave power modes: low PI, medium PII and high PIII. The treatment was carried out for 10 and 60 seconds. 3 samples were treated simultaneously. In accordance with 3DP technology there were formed the plate samples of 80 mm long, 8 mm wide and 1.5 mm thick from dielectric powder Zp130. The final impregnation based on Z-BondTM 90 cyanoacrylate was applied to increase strength. The following samples were obtained: I - sample without additional impregnation; II - sample without additional impregnation with microwave processing after formation; III - sample with additional impregnation; IV - sample with microwave treatment after formation and subsequent additional impregnation; V - sample with microwave treatment after additional impregnation; VI - sample with microwave treatment before and after additional impregnation. Accordingly, schemes for introducing microwave modification into the 3D printing process were investigated. The surface of the samples prior to the tests and the surface of the fracture zone were examined using Digital Microscope 2.0 MP 1000X and a computer image analyzer of microstructures AGPM-6M with certified Metallograph software with magnification of x17 and x400, respectively. In addition, the firmness of the obtained samples was studied by the strength of their fracture determined by the digital dynamometer FMI-S10C1 from ALLURIS GmbH & Co. KG with measurement limits of 0-100 N. The authors used a special unit equipped with load strain gauges connected via an analog-digital converter with a computer in which the results were processed in a special Labview software environment. During the testing, we installed the samples like a two-support beam on the tensometric device of the installation and loaded in the middle section with a bending moment with the help of a lever rotated by an electromechanical drive at a constant speed.

3. Results and discussion

As a result of the research, the following main data was obtained.

3.1. The influence of microwave modification on the structure of materials

The study of the structure of the samples at their fracture (Figure 1) showed that the use of microwave treatment after additional impregnation of the formed samples (Schemes V and VI) causes the formation of microcracks in the binder, which is obviously the reason for lowering the ultimate bending stresses. The formation of microcracks can be associated with various thermal and electrophysical characteristics of the powdered material and binder resulting in different interactions with the microwave electromagnetic field and possible changes in the dimensions and deformations of the individual components of the formed composite material. The microstructure of samples with microwave treatment immediately after formation is shown in Figure 1.

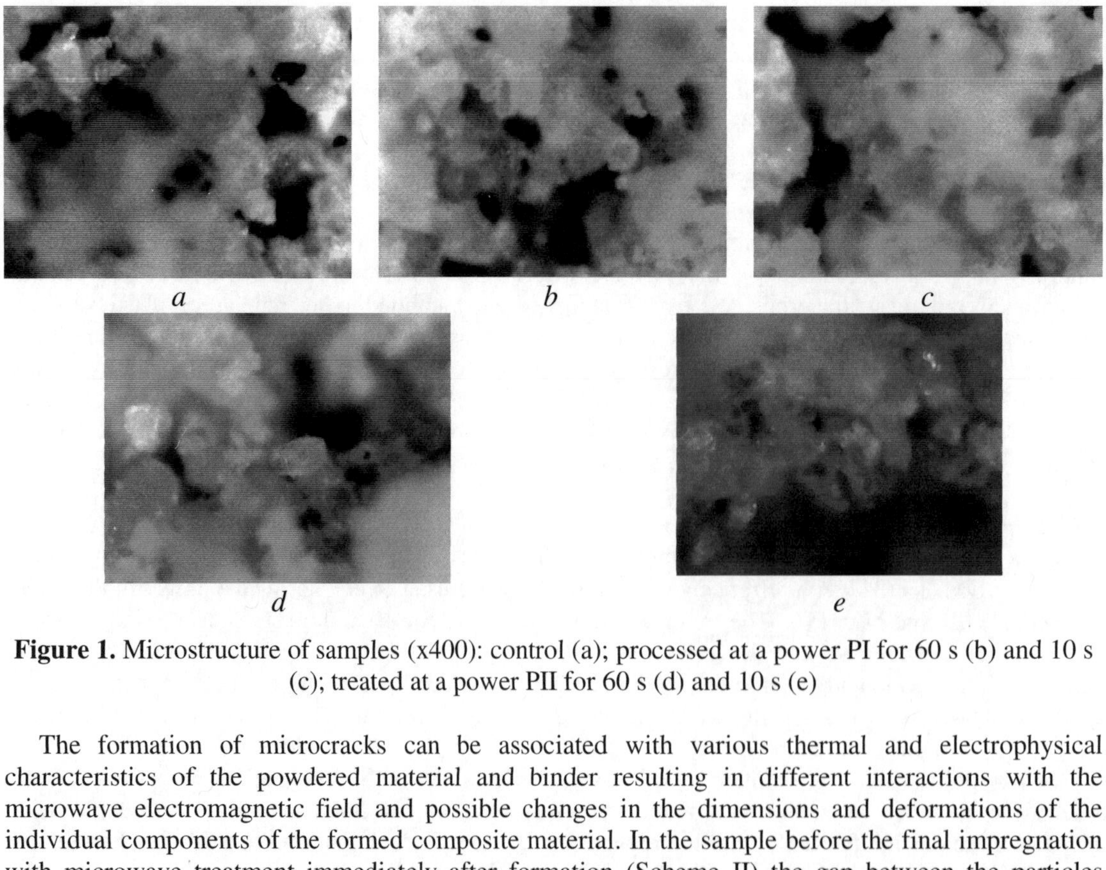

a *b* *c* *d* *e*

Figure 1. Microstructure of samples (x400): control (a); processed at a power PI for 60 s (b) and 10 s (c); treated at a power PII for 60 s (d) and 10 s (e)

The formation of microcracks can be associated with various thermal and electrophysical characteristics of the powdered material and binder resulting in different interactions with the microwave electromagnetic field and possible changes in the dimensions and deformations of the individual components of the formed composite material. In the sample before the final impregnation with microwave treatment immediately after formation (Scheme II) the gap between the particles (macropores) is reduced by 9.2 - 24%, and in the sample with final impregnation without microwave (Scheme III), practically did not change. When treated with impregnation after microwave treatment (Scheme IV), the homogeneity of the structure further increases - the dispersion of pore sizes is reduced by 25-32%. It can be seen that the large microwave power and the longer exposure time in the investigated range (Figure 1c) contribute to the formation of more dense agglomerates of particles with smaller spaces between them. Increasing the homogeneity of the structure and reducing the size of the pores after microwave treatment obviously contributes to an increase in the number of bonds between the particles and an increase in the limiting bending stresses.

Thus, based on the study of the microstructure of the samples studied, the following rational technological route can be proposed with the introduction of the microwave modification operation into the 3D printing process: the formation of a product from a powder on a 3D printer, processing in a microwave electromagnetic field, the final stabilizing impregnation. To identify technological regimes, it is necessary to study the strength characteristics of the formed samples. Impregnation of such a modified structure obviously occurs more evenly, the binder penetrates throughout the volume at approximately the same particle-binder concentration, which further increases the probability of uniform adherence of the particles and the formation of a uniformly strong (isotropic) composition.

3.2. The influence of microwave modification on the physical and mechanical properties of samples
Based on the processing of experimental data for samples with microwave treatment after formation without impregnation, the dependence shown in Figure 2 was obtained.

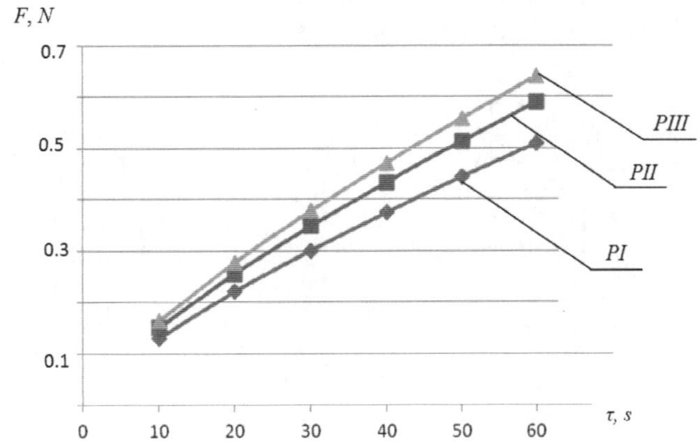

Figure 2. Influence of microwave power $P_{density}$ and time of action τ on the fracture force of samples F.

Based on the known cross-sectional area of the specimens and the determined breaking force, the maximum bending stress acting at the moment of sample destruction was calculated. The greatest effect on increasing bending stresses is achieved at the maximum power density from the investigated range and the maximum time. Where in time has a much greater effect than microwave power. The fact of placing the sample in the microwave electromagnetic field leads to the maximum effect. Here it is important to rearrange the binder and filler at the molecular level and the appearance of new bonds between the components. Figure 3 shows the histogram of the maximum bending stresses σ_{Fmax} arising upon external loading destruction in samples manufactured according to schemes I-VI, in accordance with the procedure described above.

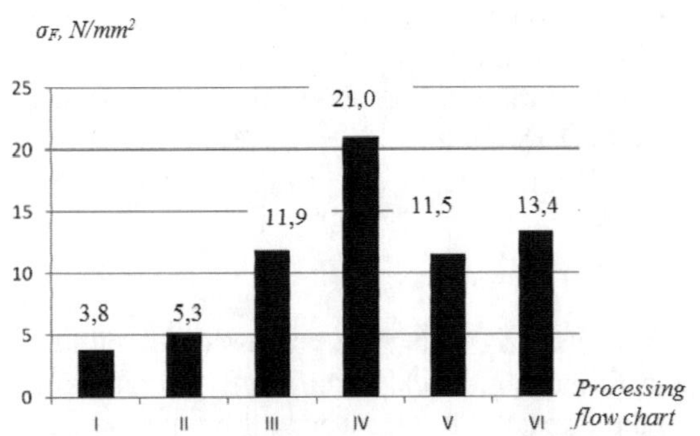

IOP Conf. Series: Materials Science and Engineering **286** (2017) 012017 doi:10.1088/1757-899X/286/1/012017

Figure 3. Influence of the place of microwave treatment in the technological cycle on the bending stresses σF in the samples in comparison with the treatment without microwave exposure.

The histogram shows that the place of microwave treatment in the technological cycle significantly affects the strength of the formed sample, which is characterized by limiting bending stresses. After microwave treatment immediately after forming in a 3D printer, the flexural strength increases by an average of 1.37 times (II). Processing of generated samples in a microwave frequency field of 2,450 MHz with a high specific power (PIII) followed by impregnation (Scheme IV) provides an increase in flexural strength by 1.77 times compared with existing 3DP (Scheme III) technologies. The use of microwave treatment after impregnation has practically no effect on the flexural strength (Scheme V) or decreases it somewhat, which is in full agreement with the above-mentioned results of the change in the microstructure. Two-stage microwave treatment (before and after additional impregnation) gives an increase in the flexural strength of about 13% (Scheme VI), which can be considered almost insignificant in practice.

Figure 4 shows the graphs obtained on a computer installation in the process of loading samples and characterizing its kinetics before the time of their destruction.

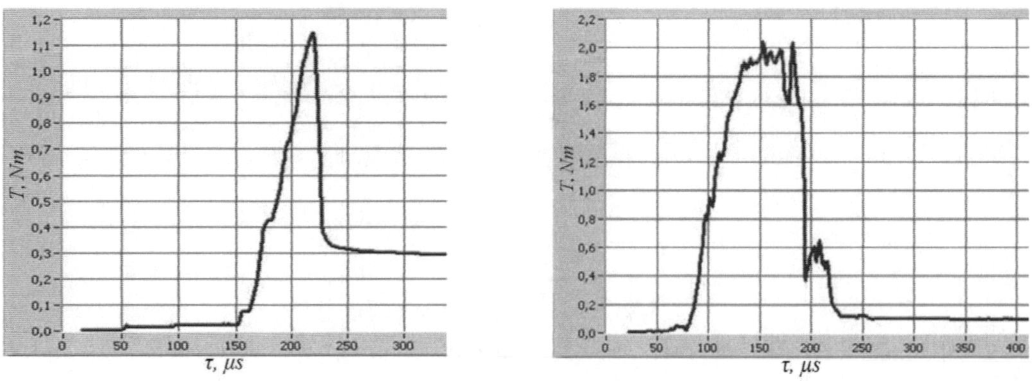

a *b*

Figure 4. Change in the value of the load from the beginning of loading to the moment of destruction of the samples taken after the final impregnation: a - a control sample, b - processed in a microwave electromagnetic field.

During the analysis of the graphs, an increase in the value of the limiting value of the bending moment for the treated sample was established in comparison with the control value from 1.15 Nm to 2 Nm, i.e. by 74 %, which is explained by the changes in the structure of the processed samples described above. It can be seen that in the loading graph of the processed sample, there is a section in the time interval from 150 to 180 μs, at which the recorded load does not increase, i.e. apparently the sample is elastically deformed, absorbing energy due to the increased number of bonds between the agglomerates in the microstructure and not being destroyed. In the control sample, after reaching the maximum load, its recession corresponding to the destruction takes place simultaneously. In this case, one we speak about the absence of deformations and pure brittle fracture. This result is also important for the load-bearing structures of various technical systems, since it increases the time for making decisions for the operator.

4. Conclusion

Based on the foregoing, it can be concluded that the processing of samples generated by 3D printing methods in the microwave electromagnetic field significantly increases their strength parameters. In this case, the effectiveness of the impact is different for different loading schemes (shear or bending) and depends on the supplied microwave power and time. The mechanism of material hardening apparently consists in increasing the number of cross-links of the structural components at the macro and micro levels. To clarify this assumption, it is necessary to conduct research involving electronic and atomic force microscopy.

Acknowledgments
The studies were carried out with the support of RFBR grant No. 17-03-00720 "Methodology of optimizing micro construction of composite materials for objects of complex shape with enhanced dynamic strength layered by electrotechnological methods".

References

[1] Kablov E N 2015 Innovative developments of "VIAM" for the implementation of "Strategic directions for the development of materials and technologies for their processing until 2030" *Aviation materials and technologies* (*Electronic Materials* No. 1) (Moscow) pp 3–33
[2] Kablov E N 2012 Materials and chemical technologies for aviation equipment *Herald of the Russian Academy of Sciences* (vol 82 No. 6) (Moscow: "Publishing house "Science") pp 520–530
[3] Bourell D L, Beaman J J, Leu M C, and Rosen D W 2009 *A Brief History of Additive Manufacturing and the 2009 Roadmap for Additive Manufacturing: Looking Back and Looking* Ahead U. S. Turkey Workshop on Rapid Technologies *(Istanbul, Turkey)*
[4] Ehrenberg R 2013 *The 3-D Printing Revolution: Dreams Made Real One Layer at a Time, Science News* (Washington : Science Service Publication) 183 (5) pp 20–25
[5] Arkhangelskiy Y S 2011 *Reference book on microwave electrothermy: a reference book* (Saratov Scientific Book) p 560
[6] Estel L, Lebaudy Ph, Ledoux A, Bonnet C and Delmotte M 2004 Microwave assisted blow molding of polyethylene-terephthalate (PET) bottles *Proceedings of the Fourth World Congress on Microwave and Radio Frequency Applications* No. 11 (Austin, Texas) p 33
[7] Zlobina I V and Bekrenev N V 2016 *The Influence of Microwave Electromagnetic Field on Mechanical Properties of Composite Materials* Science Intensive Technologies (Moscow: Radio engineering vol 17 No. 2 *)* pp 25–30
[8] Zlobina I V and Bekrenev N V 2016 The influence of electromagnatic field microwave on physical and mechanical characteristics of CFRP (carbon fiber reinforced polymer) structural Solid State Phenomena Materials Engineering and Technologies for Production and Processing II *(Trans Tech Publications Ltd* vol 870*)* pp 101–106
[9] Zlobina I V, Bekrenev N V and Muldasheva G K 2016 *Increasing of the endurance of polymeric construction materials with the multilevel hierarchical structure in the microwave electromagnetic field* AIP Publishing (Melville, New York: AIP Publishing LLC) pp 020236-1–020236-4
[10] Zlobina I V and Bekrenev N V 2017 Investigation of the microstructure of structural layered carbon plastics, modified by electrophysical effects *Vestnik of P A Solovyov Rybinsk State Aviation Technical University No. 1 (40)* (Rybinsk: Publishing house of P A Solovyov Rybinsk State Aviation Technical University) pp 236–242

IOP Conf. Series: Materials Science and Engineering **286** (2017) 012018 doi:10.1088/1757-899X/286/1/012018

Multilevel modeling of damage accumulation processes in metals

K A Kurmoiartseva, P V Trusov and N V Kotelnikova

Department of Mathematical Modeling of Systems and Processes, Perm National Research Polytechnic University, 29 Komsomolsky prospekt, Perm, 614990, Russian Federation

E-mail: kurmoiartseva.k@mail.ru

Abstract. To predict the behavior of components and constructions it is necessary to develop the methods and mathematical models which take into account the self-organization of microstructural processes and the strain localization. The damage accumulation processes and the evolution of material properties during deformation are important to take into account. The heterogeneity of the process of damage accumulation is due to the appropriate physical mechanisms at the scale levels, which are lower than the macro-level. The purpose of this work is to develop a mathematical model for analyzing the behavior of polycrystalline materials that allows describing the damage accumulation processes. Fracture is the multistage and multiscale process of the build-up of micro- and mesodefects over the wide range of loading rates. The formation of microcracks by mechanisms is caused by the interactions of the dislocations of different slip systems, barriers, boundaries and the inclusions of the secondary phase. This paper provides the description of some of the most well-known models of crack nucleation and also suggests the structure of a mathematical model based on crystal plasticity and dislocation models of crack nucleation.

1. Introduction

The increase of the lifetime of components and constructions is one of the most important problems of mechanics and material science, which has been standing for many decades. In this regard, the development of reliable methods and models for predicting the material behavior up to the point of destruction is required. The one of the most perspective directions to solve this problem is the construction of multilevel models based on crystal elastic-visco-plasticity.

A huge number of scientific publications are devoted to the study of damage accumulation processes. The purpose of this work is to develop a mathematical model to analyze the behavior of polycrystalline materials with the description of the damage accumulation processes and fracture in the wide range of thermomechanical effects. By fracture we mean is the multistage and multiscale process of the build-up of micro- and mesodefects. The structure of the model for describing the destruction of polycrystalline materials without a certain ("imposed" a priori) macrolevel damage mode is considered. Herewith, the diffusion and dislocation processes of the nucleation of micropores and microcracks are necessary to include. Then their number and size increases (damage accumulation). The final stage is the coalescence of microdamages and growth of the macroscopic crack.

2. Some dislocation mechanisms of crack nucleation

IOP Conf. Series: Materials Science and Engineering **286** (2017) 012018 doi:10.1088/1757-899X/286/1/012018

The dislocation theory of fracture assumes that plastic deformation plays a decisive role and includes the force criterions of crack nucleation and propagation. This area includes the well-known dislocation models of crack nucleation by Zener, Stroh, Cottrell [1]. The following theoretical directions for the construction of such theory can be specified: 1) models based on the concept of the dislocation pile-up in separate slip planes in front of impediments such as the boundaries of twins, inclusions, etc.; 2) models with the occurrence of cracks related with the interaction of the dislocations located in intersecting slip systems; and 3) models which are not based on locks ("barrier-free models").

The purpose of the current stage of our research is the physical analysis of damage accumulation mechanisms in metals and their mathematical description. We intend to use an explicit description for the changes in the internal structure of material defects in the manufacturing and operating processes. The approach based on the introduction of internal variables is used to construct constitutive models. The models of such class allow, firstly, to analyze deformation processes at various scales and, secondly, to take into account microstructure evolution and damage accumulation.

Let us focus on the description of several well-known crack nucleation models. Microcracks nucleation is the collective effect of dislocation interactions, which is related to their reconstruction accompanied by the conversion of the stored energy into the surface energy. The model of a crack nucleation near a flat dislocation pile-up, the mathematical description of which is given by Stroh [2] will be considered in more detail. The idea of this model is based on the physically reasonable assumption about a occurrence of high local stresses due to a hindered shear, the particular case of which is the pile-up of the n same-sign dislocations as shown in figure 1. Herewith the stresses between the two head dislocations of the pile-up are $\sigma \approx n\tau$, where τ is the stress in the shear plane. The criterion for the formation of a crack is the fulfillment of the condition (1), where γ is the surface energy of the material, G is the shear modulus, ν is the Poisson's ratio, L is the length of the dislocation pile-up which is determined by the relation (2).

$$\tau^2 = \frac{2\pi\gamma G}{8(1-\nu)L} \tag{1}$$

$$L = \frac{Gbn}{\pi(1-\nu)\tau} \tag{2}$$

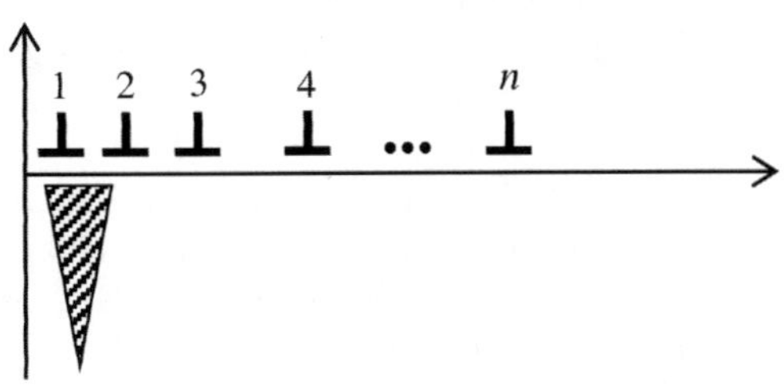

Figure 1. The fracture nucleation scheme in the Stroh's model.

Nevertheless, the original Stroh's model of crack nucleation has several disadvantages. Firstly, according to this model, $10^2 \div 10^3$ dislocations are required, but no such pile-ups are experimentally observed. Secondly, the strong obstacle during the inhibition of such pile-up is necessary. Thirdly, during the evaluation of the energy advantage for a crack nucleation, only the relaxation energy of the already existing stresses is taken into account unlike the work of external forces and the energy of dislocation nuclei. Despite this, the model demonstrates the possibility to describe the nucleation of a crack due to the coalescence of dislocations held in front of a barrier. The mentioned above disadvantages can be overcome by modifying the model [3].

Another crack nucleation model is the mechanism proposed by Gilman and Rozhansky [4], which describes the opening of a crack in the slip plane of dislocations, as it, is typical for HCP metals.

IOP Conf. Series: Materials Science and Engineering **286** (2017) 012018 doi:10.1088/1757-899X/286/1/012018

According to this mechanism, the crack nucleation in the slip plane occurs due to the curvature of atomic planes in the head of a blocked pile-up. The entire area above the pile-up plane lies in the region of compressive stresses while the rest area lies in the region of the tensile stresses; this stress state is similar to the state of a bent beam. Shear along curved planes should cause normal stresses leading to the separation of slip planes. The formation of a microcrack indicates emerging of new free surfaces and, as a consequence, the relaxation of internal stresses near the head dislocation of pile-up. For HCP metals, the value of the specific surface energy is the smallest in the basis plane and it probably plays a decisive role for the nucleation of cracks in this plane. This mechanism was experimentally observed.

Another well-known scheme of crack nucleation is the Cottrell's model of intersecting dislocation pile-ups which is typical for BCC metals [4]. In this scheme, there are no external barriers against slip and the formation of a barrier is caused by the interaction of the head dislocations of the pile-ups with the formation of a sessile dislocation. The realization of the Cottrell mechanism was not experimentally observed.

The scheme of the annihilation model of crack nucleation can be referred to the barrier-free model with the pile-ups of different signs as shown in figure 2 [3]. It is assumed that if the head dislocations of a pile-ups pass one above the other at the distance of $(5\div7)b$ then they annihilate with the formation of vacancies. New vacancies appear by the further inflow and the annihilation of dislocations. The annihilation of a dislocations sequence ($n=5\div10$) forms the sufficient number of vacancies to form a microcrack. In Section 3.3, the criterion of crack nucleation based on the Stroh's model is provided.

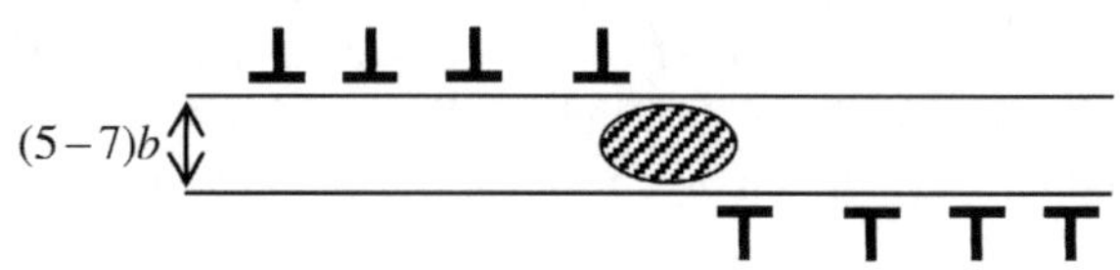

Figure 2. Crack nucleation scheme in the case of opposite dislocation pile-ups.

3. The mathematical formulation

The crystal plasticity based model allows describing the evolution of the internal structure of a material and the build-up of defects of various natures. The destruction research of components and constructions in the manufacture and operation process without the creation of appropriate mathematical models is impossible. The models must take into account the physical causes of the evolution of the material internal structure and damage accumulation at large deformations.

To describe the physical mechanisms of plastic deformation and fracture it is necessary to study the behavior of the material at a lower than macro scale levels. There are various physical mechanisms of damage accumulation (formation of embryonic cracks, their stabilization and growth the formation of the macroscopic crack) at each scale level. A complete description of deformation and fracture requires an approach that takes into account the multistage processes occurring in the material.

There are macro-, meso-I and mesolevel-II in the structure of the developed model. The macrolevel representative volume element is a polycrystalline aggregate of a large (at least 400) number of crystallites (grains, subgrains), each of which is considered as a mesolevel-I representative volume element. Parts of the grains and subgrains are considered at mesolevel-II. Different types of dislocations densities are evenly distributed within this level. At the macrolevel, the deformation of the material is described using an asymmetric indifferent strain rate [5]. Mesolevel-I is used to describe processes of plastic deformation in terms of shear in the slip systems of a crystallite and hardening. Description of damage accumulation and evolution of dislocation densities are given at the mesolevel-II. In addition, for an adequate description of the damage accumulation process is analyzed the dislocation interaction of various slip planes.

3.1 Macrolevel constitutive relations

The constitutive equation of the macrolevel is the Hooke's law in rate form (3):

$$\dot{\boldsymbol{\sigma}}+\boldsymbol{\sigma}\cdot\boldsymbol{\omega}-\boldsymbol{\omega}\cdot\boldsymbol{\sigma}=\boldsymbol{\pi}:(\mathbf{z}-\mathbf{z}^{in}) \tag{3}$$

where $\boldsymbol{\omega}=\dot{\mathbf{o}}\cdot\mathbf{o}^{\mathrm{T}}$ is the spin of the movable coordinate system defined by the quasi-solid motion, $\mathbf{z}=\hat{\nabla}\mathbf{v}^{T}-\boldsymbol{\omega}$ is the asymmetric indifferent strain rate. There are elastic and plastic components of the strain rate: $\mathbf{z}=\mathbf{z}^{e}+\mathbf{z}^{in}$. The plastic component is the sum of the shear strain rates on all slip systems can be written as equation (4).

$$\mathbf{z}^{in}=\sum_{k=1}^{n}\dot{\gamma}^{(k)}\mathbf{b}^{(k)}\mathbf{n}^{(k)} \tag{4}$$

Here, $\dot{\gamma}^{(k)}$ is the shear strain rate on the *k*-th slip system, $\mathbf{b}^{(k)}$ and $\mathbf{n}^{(k)}$ are the slip direction and slip plane normal of the *k*-th slip system, respectively.

3.2 Mesolevel-I constitutive relations

The physical mechanisms of deformation are considered in more details at the mesolevel-I. The total dislocation density can be decomposed additively into the densities of mobile and immobile dislocations: $\rho^{(k)}=\rho_{m}^{(k)}+\rho_{im}^{(k)}$. The immobile dislocations are inhibited by the obstacles of various nature (dislocation and non-dislocation barriers, grain boundaries, etc.) while the mobile dislocations are able to slide in the case of critical resolved shear stress. Furthermore, the mobile and immobile dislocation densities can be divided into the positive and the negative parts depending on the location of the extra plane as follows:

$$\rho_{m}^{(k)}=\rho_{m+}^{(k)}+\rho_{m-}^{(k)} \tag{5}$$

$$\rho_{im}^{(k)}=\rho_{im+}^{(k)}+\rho_{im-}^{(k)} \tag{6}$$

In the reference configuration, the initial densities of positive and negative dislocations are supposed to be equal, equation (7).

$$\rho_{m+}^{(k)}\big|_{t=0}=\rho_{m-}^{(k)}\big|_{t=0}=\frac{\rho_{m}^{(k)}}{2}\big|_{t=0} \tag{7}$$

Shear strain rate is necessary to evaluate the inelastic deformation at the macrolevel in equation (3). It is determined on each slip system by the Orowan equation with considering the separation for positive and negative dislocations density:

$$\dot{\gamma}^{(k)}=\dot{\gamma}_{+}^{(k)}+\dot{\gamma}_{-}^{(k)}=b\rho_{m+}^{(k)}v_{m+}^{(k)}+b\rho_{m-}^{(k)}v_{m-}^{(k)},$$

where b is the module of the Burgers vector, $v_{m\pm}^{(k)}$ is the dislocation motion velocity of mobile positive or negative dislocations.

The dislocation motion velocity depends on the temperature of the material under loading, actual and critical resolved shear stress on the slip system. Following [6] the dislocation motion velocity can be written as

$$v_{m\pm}^{(k)}=v_{0}^{(k)}\left(\frac{\tau^{(k)}}{\tau_{c\pm}^{(k)}}\right)^{m}\exp\left[-\frac{\Delta F}{\kappa\theta}\right]\mathrm{H}(\tau^{(k)}-\tau_{c\pm}^{(k)})$$

where ΔF is the enthalpy of dislocation movement activation, κ is the Boltzmann constant, θ is the absolute temperature, $\mathrm{H}(\cdot)$ is the Heaviside function. The resolved shear stress on a slip system is determined by the condition, $\tau^{k}=\mathbf{n}^{(k)}\cdot\boldsymbol{\sigma}\cdot\mathbf{b}^{(k)}$. The critical resolved shear stress depends on the total dislocation density and dislocation interactions on different slip systems and can be written in the following for:

$$\tau_c^{(k)} = \tau_{c_0}^{(k)} + Gb\left(\sum_{\beta} g^{(k,\beta)}\left(\rho_{m\pm}^{(\beta)} + \rho_{im\pm}^{(\beta)}\right)\right)^{1/2}$$

where $g^{(k,\beta)}$ is the dislocation interaction matrix [7].

3.3 Mesolevel-II constitutive relations

The model of the mesolevel-II is based on the kinetic equation for dislocation density and the relations of dislocation fracture theory. The main mechanism of inelastic deformation is the movement of edge dislocations, which interact with each other and with the lattice defects of various dimensions. Dislocation density based modeling is a universal approach to study the behavior of numerous dislocations and the various competing deformation mechanisms and their interaction. Such models allow obtaining the dependence of the dislocation density on the loading parameters and the parameters of the internal structure of a material. The advantage of these models is the ability to describe the structure, strength and plasticity of metals and alloys at the microstructural level.

The equations for describing the evolution of mobile dislocation densities can be written in the form (8) and similarly for immobile dislocation densities (9) [8].

$$\dot{\rho}_{m\pm}^{(k)} = \dot{\rho}_{nuc\pm}^{(k)} + \dot{\rho}_{mob\pm}^{(k)} - \dot{\rho}_{imm\pm}^{(k)} - \dot{\rho}_{ann(m\pm)}^{(k)} - \dot{\rho}_{ann(m\pm im\mp)}^{(k)}, \quad (8)$$

$$\dot{\rho}_{im\pm}^{(k)} = \dot{\rho}_{imm\pm}^{(k)} - \dot{\rho}_{mob\pm}^{(k)} - \dot{\rho}_{ann(m\mp im\pm)}^{(k)}. \quad (9)$$

The relations (8) and (9) take into account, that the densities of dislocations are dependent on such processes as annihilation ($\dot{\rho}_{ann(m\pm)}^{(k)}$, $\dot{\rho}_{ann(m\pm im\mp)}^{(k)}$), nucleation of new dislocations ($\dot{\rho}_{nuc\pm}^{(k)}$), mobilization ($\dot{\rho}_{mob\pm}^{(k)}$) and immobilization ($\dot{\rho}_{imm\pm}^{(k)}$).

Basic approximation, the criterion for the dislocation nucleation of microcracks, which is based on the Stroh's model described in section 2, was used. At high critical shear stress and immobile positive (or negative) dislocation densities exceeds the critical value density $\rho_{im+}^{(k)} \geq \rho^*$ (or $\rho_{im-}^{(k)} \geq \rho^*$). After that formed crack length $l_{crack} = \frac{(\rho_{im\pm}^{(k)})^2 b^5}{2}$.

4. Conclusion

There are the specific mechanisms of damage accumulation and fracture processes at different scale and structural levels. The properties of polycrystalline materials at the macrolevel are essentially determined by the state of the evolving meso- and microstructure of the material. In this article, we attempt to construct a multilevel mathematical model based on the crystal plasticity.

The descriptions of the basic models of dislocation crack nucleation, both the ones that take into account barriers and the ones, which are barrier-free, are given. The structure of the developed model is presented. At the macrolevel, an asymmetric indifferent strain rate is used to describe the material deformation. The mesolevel-I relations allow to describe the evolution of densities and the changes in dislocation movement velocities, shear rates, acting and critical resolved shear stresses for each slip system. The kinetic equations describing the change in the dislocation density on the slip system and the equations of dislocation fracture theory are used as a basis for the mesolevel-II model. The dislocation model developed to analyze the evolution of microstructure and deformation mechanisms includes equations for determining the densities of mobile and immobile dislocations, the density of dislocation sources and locks as well as equations for describing such processes as annihilation, etc.

The developed approach allows estimating the influence of the separated mechanisms of damage accumulation and deformation on the experimentally observed behavior features of materials.

Acknowledgments
The work was supported by the Russian Science Foundation (grant No. 17-19-01292).

References

[1] Ivanova V S, Gordienko L K and Seminov V N 1965 *The Role of Dislocations in Hardening and Fracture of Metals* (Moscow: Nauka) p 180
[2] Stroh A N 1954 *Proc. R. Soc. London,* A **223** 404–414.
[3] Vladimirov V I 1984 *Physical Nature of Metal Fracture* (Moscow: Metallurgy) p 280
[4] Finkel V M 1970 *Physics of Fracture* (Moscow: Metallurgy) p 376
[5] Trusov P V, Nechaeva E S and Shveykin A I 2013 *Physical Mesomechanics* **16** 15–31
[6] Meyers M A, Benson D J, Vohringer O, Kad BK, Xue Q and Fu H-H 202 *Mater. Sci. Engng.* **322** 194–216
[7] Khadyko M, Dumoulin S, Cailletaud G and Hopperstad O S 2016 *Int. J. Plasticity* **76** 51–74
[8] Austin R A and McDowell D L 2011 *Int. J. Plasticity* **27** 1–24

IOP Conf. Series: Materials Science and Engineering **286** (2017) 012019 doi:10.1088/1757-899X/286/1/012019

The technology of forming Ti-Nb layers on the surface of titanium alloy plates using a high-voltage electron beam output into a protective argon medium

I A Polyakov[1], E A Lozhkina[1], V V Samoylenko[1], M G Golkovski[2], V S Lozhkin[1], O G Lenivtseva[1], O E Matts[1] and L V Chuchkova[1]

[1] Novosibirsk State Technical University, 20, K. Marx Prospekt, Novosibirsk, 630073, Russia

[2] Budker Institute of Nuclear Physics, Siberian Branch of Russian Academy of Sciences, 11, Academician Lavrentiev Avenue, 630090, Novosibirsk, Russia

E-mail: status9@mail.ru

Abstract. The structure and properties of layers obtained by electron-beam melting of titanium-niobium powder mixtures in a protective argon medium are studied in the work. The method allows cladding traces 16.5 ... 17.5 mm wide with a maximum thickness of 3.5 ... 4.5 mm. The concentration of niobium in the formed surface layers is 6.6 and 13.8%. The cladded material has a predominantly lamellar structure. The production of layers in the protective medium of argon makes it possible to increase the microhardness level, as well as the corrosion resistance of titanium in boiling solutions of nitric and sulfuric acid.

1. Introduction

Titanium alloys have recently become more widely used in industry [1, 2]. Titanium possesses a high complex of mechanical and anticorrosive properties that draws the attention of researchers and engineers working in directions where the listed properties are necessary. For this reason, titanium and its alloys have proven themselves in the chemical and nuclear industries. However, in some cases, the use of metal is limited to particularly aggressive conditions of operation and, generally, titanium alloys are not chemically resistant in such mediums as boiling solutions of strong acids.

Addition of alloying elements with higher corrosion resistance under the similar conditions, but not forming chemical compounds and mechanical mixtures, improves the metal chemical resistance. These requirements are met by elements located next to the titanium in the periodic table: Zr, Hf, Nb, Ta, Mo. Zirconium and hafnium are analogues of titanium because these elements have completely isomorphic correspondence. However, a significant increase in chemical resistance is achieved at high alloying component concentrations: 50% and more [3]. Alloying with molybdenum makes it possible to protect titanium in boiling solutions of sulfuric and hydrochloric acid [4, 5]. Nevertheless, an increase in the molybdenum content leads to a decrease in the chemical resistance of the alloy in boiling nitric acid due to its re-passivation. One of the elements possessing high resistance both in boiling solutions of HCl, H_2SO_4, and in HNO_3 is tantalum. However, the addition of this component leads to a significant increase in the cost and weight of the alloy (density 16.6 g/cm^3). The cheaper analogue of tantalum is niobium. The element is less chemically resistant than tantalum, but has similar corrosive characteristics. In addition, the addition of niobium does not lead to a significant increase in the weight of the alloy (density 8.57 g/cm^3). For this reason, its use as the main alloying component for the purpose of complex increase in the corrosion resistance of titanium in various acids is the most rational.

Obtaining titanium products niobium-alloyed through all the volume is expedient in those cases when its introduction makes it possible to provide the necessary mechanical properties. From the standpoint of increasing the corrosion resistance, the presence of niobium throughout the material has practically no effect on corrosion resistance, since only the surface layer steps into the interaction. In this regard, it is sufficient to protect only the surface exposed to corrosion.

Previous studies on the creation of niobium-containing layers on the surface of titanium alloys have shown a high efficiency of the electron-beam cladding. The method allows applying homogeneous layers a few millimeters thick. Experiments on the formation of surface alloys, as a general rule, are conducted in air that leads to a slight saturation of the deposited material with gases of the air medium by a depth of ~ 30…60 μm [6], despite the presence of fluxes. In order to eliminate the influence of gases, electron-beam treatment is rationally to be carried out in a protective medium. For this reason, a new approach to the creation of niobium-containing layers in a protective medium of argon is implemented in the work.

2. Materials and methods

The layers were cladded on the basis of the industrial ELV-6 electron accelerator at the Institute of Nuclear Physics of the SB RAS. In order to implement the method in a protective medium, a sealed chamber was specially designed and created, in which a movable table was placed. Before treatment, the chamber was filled with protective gas (argon). The workpieces were titanium plates 12 mm thick and 100x25 mm in size with a previously applied trace of the powder layer. The trace was applied along the center of a large side plane. Since the electron-beam treatment was carried out in a protective medium, the flux was completely eliminated from powder portion, which was used to protect the melted metal when the method was performed in air [7]. The composition of the initial powder portion is shown in Table 1.

Table 1. Chemical composition of the powder portion and the cladded layer.

Specimen designation	Powder portion composition, wt. %		Cladded layer composition, wt. %	
	Nb	Ti	Nb	Ti
6.6 % Nb	27.2	72.8	6.6	93.4
13.8 % Nb	66.0	34.0	13.8	86.2

Electron-beam treatment was carried out at a beam current of 11 mA; electron energy 1.4 MeV; conveying speed 10 mm/s. The distance from the outlet to the surface of the workpiece was 136 mm. In the works carried out earlier, this distance was 90 mm [7]. This difference is explained by the availability of special technological equipment connecting the outlet of the accelerator with a protective chamber.

The structural study was carried out using the Carl Zeiss microscopes Axio Observer Z1m and EVO 50 XVP. The content of niobium in the formed layer was determined by micro-X-ray analysis using INCAX-Act (Oxford Instruments).

To determine the level of mechanical properties, the microhardness was measured. The microhardness was determined on the cross-sections using a WolpertGroup 402 MVD hardness tester.

The level of corrosion resistance of the cladded layers was estimated by the weight method. As an aggressive medium, a boiling solution of concentrated nitric acid (65 %) and dilute solution of boiling sulfuric acid (5%) were chosen. For the study, specimen with the size of 10x10x1 mm were prepared from a cladded layer. The plates of technically pure components of similar dimensions were used as materials of comparison. The total time of immersion in boiling solutions of nitric and sulfuric acid

IOP Conf. Series: Materials Science and Engineering **286** (2017) 012019 doi:10.1088/1757-899X/286/1/012019

was 120 hours and 50 minutes, respectively. The weight loss was monitored every 24 hours when tested in HNO_3 and 10 minutes when tested in H_2SO_4 solution. Changes in the mass of materials were recorded on the analytical balance AND GR-300 with an accuracy of 10^{-4} g.

3. Results and discussion

The cross-section of the layer with 6.6 % Nb is shown in Figure 1. The formed layer is characterized by a practically defect-free structure. A small amount of pores is detected in the fusion zone of the layer with a titanium base on the edge of the applied trace. The layer thickness in the central part is ~ 4.5 mm. The width of the trace is 16.5 mm. The cross section of the layer with 13.8 % Nb has a similar structure. A small difference is observed in the width and thickness of the formed material. The thickness of the surface alloy with 13.8 % Nb in the central part is 3.5 mm, and the width of the trace is ~ 17.5 mm.

Figure 1. The cross-section of the Ti-Nb layer cladded by an electron beam in a protective media of argon.

At the micro level, the structure of the cladded layers is mainly represented by lamellar morphology (Figure 2 a, b). In a specimen with 6.6 % Nb, this morphology is characterized by two types of structures: fine- and coarse-lamellar. The finely dispersed lamellar structure is identified only at high magnifications and forms the background of the cladded layer (Figure 2 a). The coarse-grained structure is fixed only in individual grains (Figure 2 b). The direction of growth of such plates coincides with the orientation of finely dispersed plates. For a surface alloy with 13.8 % Nb, a coarse-plate structure is not observed in the grain volume (Figure 3 a). However, there are small light areas that are poorly identified by chemical etching (Figure 3 b). Presumably, these areas are regions with a higher content of niobium. According to the works [8] when electron-beam cladding of niobium-containing mixtures on the surface of titanium workpieces, the formation of a structure with an apparent chemical heterogeneity is typical. Since only small local microvolumes are found in the layer with 13.8 % Nb, it can be assumed that a further increase in the concentration of niobium in the alloy will lead to an increase in aliquation.

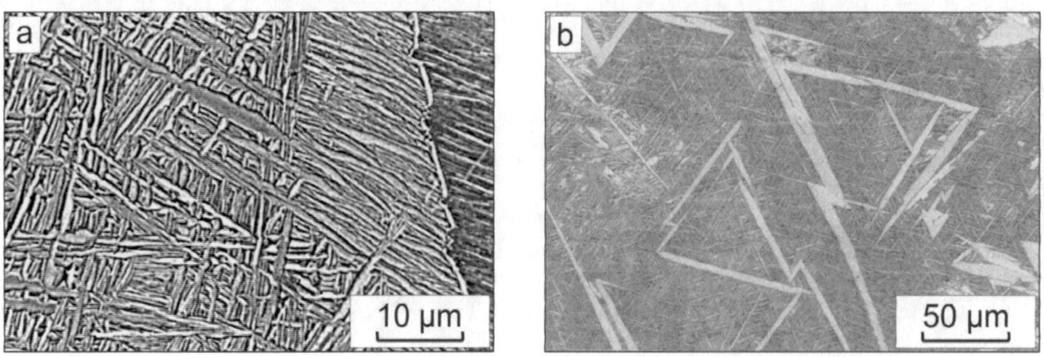

Figure 2. The structure of the layers with 6.6 % Nb.

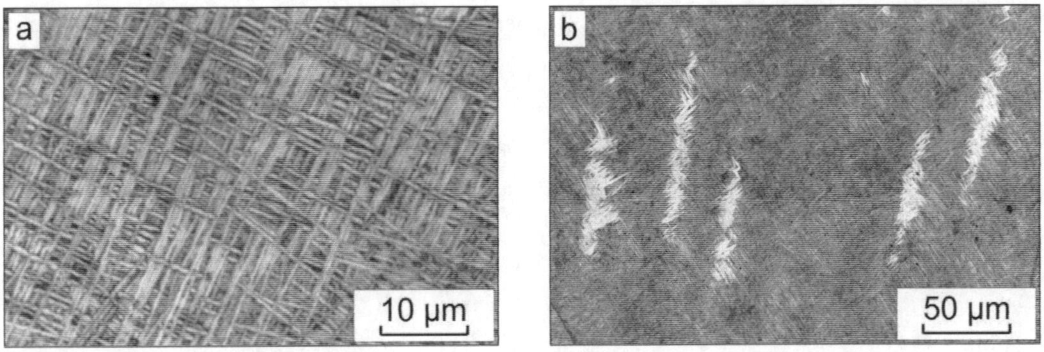

Figure 3. The structure of the layers with 13.8 % Nb.

To determine the level of mechanical properties, durometric tests were carried out. The results of the experiment are shown in Figure 3. The lowest value of the microhardness level was recorded for a cp-titanium base, which is ~ 185 HV. Electron-beam cladding of niobium in argon allows increasing this index by 1.4 ... 1.6 times. The HV value for the 6.6 % Nb layer was 255. The sample with the highest concentration of niobium in the experiment showed 295 HV.

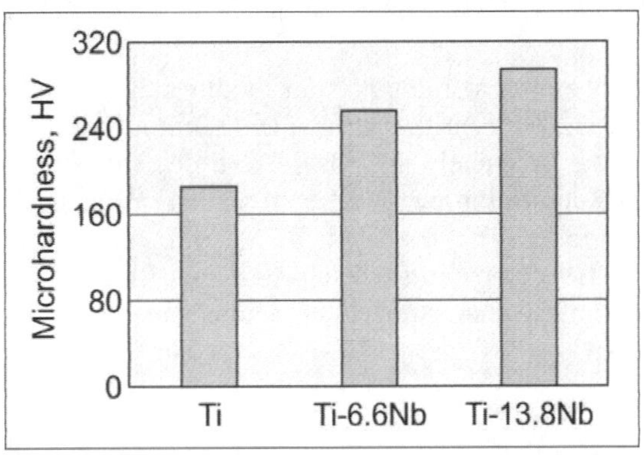

Figure 4. Microhardness of cp-titanium and layers with 6.6 % Nb and 13.8 % Nb.

Corrosion resistance of layers with different concentrations of niobium was estimated in boiling solutions of nitric and sulfuric acids. The results of the tests conducted in boiling concentrated HNO_3 are shown in Figure 5 a. The highest value of corrosion rate was recorded on a titanium specimen, and it was 0.158 mm/year. Surface alloying of titanium with niobium in an argon media promotes an increase in the level of corrosion resistance of the protected material. Forming a layer with 6.6 % Nb allows to reduce the corrosion rate to 0.037 mm/year, which is 4.3 times lower compared to titanium. A higher effect of surfacing is achieved when forming an alloy with 13.8 % Nb. The corrosion rate of this layer is 0.019 mm/year, and the anticorrosive effect of this type of protection reaches 8.3 times. Cp- niobium under the influence of aggressive solution for a year would lose a layer of metal equal to 1.2 microns.

Boiling solution of dilute (5 %) sulfuric acid has a more aggressive effect on the materials under study. This is evidenced by high values of corrosion rate (Figure 5 b). The lowest level of corrosion resistance in the experiment was shown by a sample made of cp-titanium. The corrosion rate of this material in the sulfuric acid solution was 64.3 mm/year. Electron-beam cladding of niobium-containing layers in the protective media of argon makes it possible to reduce the dissolution rate of the material by 2.3 ... 12.8 times. So for a sample with 6.6 % Nb, the corrosion rate is 28.4 mm/year. While for a surface alloy with 13.8 % Nb, dissolution of the material under the action of an aggressive medium proceeds at a rate of 5 mm/year. For samples prepared from cp- niobium, no changes in the sample mass were observed during the experiment.

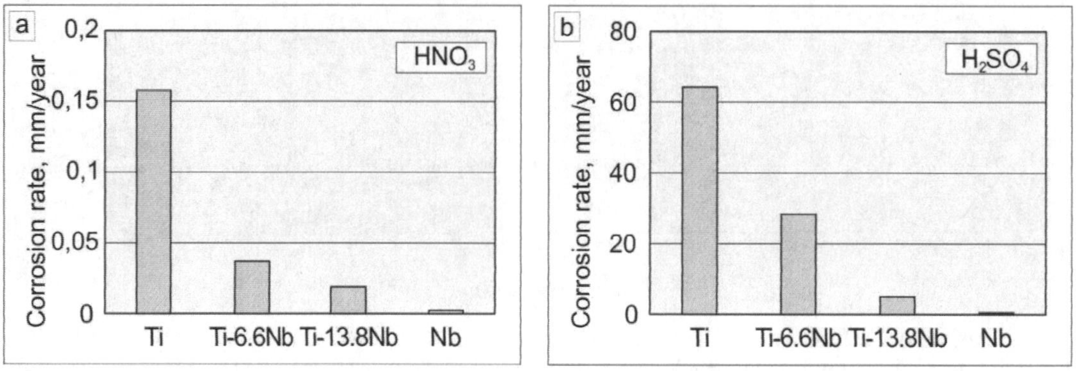

Figure 5. The corrosion rate of titanium, niobium and Ti-Nb layers in a boiling solution of 65 % nitric acid (a) and 5 % sulfuric acid solution (b).

4. Conclusion

Electron-beam cladding of niobium-containing powder mixtures allows the formation of layers with a maximum thickness of 3.5 ... 4.5 mm on the surface of titanium base. In the zone of fusion of the layers with the base, defects are practically not observed, except for single pores. Surface alloys are characterized by a lamellar structure throughout the cross-section of the cladded material. Fusing a surface with a previously applied powder layer helps to increase the microhardness level by 1.4 ... 1.6 times to 255 ... 295 HV compared to the cp- titanium. Electron-beam alloying with niobium makes it possible to obtain an alloy characterized by a higher resistance to corrosion in boiling acid solutions than the base material.

Acknowledgments

The reported study was funded by RFBR according to the research project No. 16-38-00733 mol_a.

References

[1] Leyens C and Peters M 2003 *Titanium and Titanium Alloys Fundamentals and Applications* (WILEY-VCH Verlag GmbH & Co. KGaA, Weinheim) p 532

[2] Matthew J and Donachie Jr 2000 *Titanium: A Technical Guide* (ASM International, Materials Park, Ohio) p 381

[3] Kieffer R Binder F and Bach H 1968 Beitrag zum physikalischen und korrosionschemischen Verhalten von IVa-Metallegierungen *Materials and Corrosion* **19**, 114-20 [in German].

[4] Cheng T-P Lee J-T and Tsai W-T 1991 Passivation of titanium in molybdate-containing sulphuric acid solution *Electrochimica Acta* **36** 2069-76

[5] Park P Y Akiyama E Habazaki H Kawashima A Asami K and Hashimoto K 1996 The corrosion behavior of sputter-deposited Mo-Ti alloys in concentrated hydrochloric acid *Corrosion Science* **38** 1649-67

[6] Zhuravina T V Bataev I A Ruktuev A A Alkhimov A P Lenivtseva O G and Butylenkova O A 2012 Non-vacuum electron beam cladding of titanium, tantalum and niobium powders on titanium substrate *Obrabotka metallov* **54** 90-95 [in Russian]

[7] Polyakov I A Krivezhenko D S Samoylenko V V Lenivtseva O G Ivanchik I S and Chuchkova L V 2016 Structure and Corrosion Resistance of Ti-Nb Layers Obtained by Non-vacuum Electron-Beam Cladding on CP-titanium Substrates *AIP Conference Proceedings* **1785**

IOP Conf. Series: Materials Science and Engineering **286** (2017) 012019 doi:10.1088/1757-899X/286/1/012019

040047
[8] Polyakov I A Samoylenko V V Lenivtseva O G and Golkovski M G 2015 The structure and corrosion resistance of the coatings obtained by non-vacuum electron beam cladding of the Ti-Nb powder mixture on a titanium substrates *Appled Mechanics ans Materials* **788** 267-73

IOP Conf. Series: Materials Science and Engineering **286** (2017) 012020 doi:10.1088/1757-899X/286/1/012020

Influence of dunite mineral additive on strength of cement

A A Vasilyeva[1], L V Moskvitina[1], S G Moskvitin[1], M P Lebedev[1] and G D Fedorova[2]

[1]V.P. Larionov Institute of Physical-Technical Problems of the North of the Siberian Branch of the Russian Academy of Sciences, 1 Oktyabrskaja str, Yakutsk, 677980, Russia
[2]M.K. Ammosov North-Eastern Federal University, 58 Belinsky str, Yakutsk, 677027, Russia

E-mail: kiir@mail.ru

Abstract. The work studies the applicability of dunite rocks from Inagli massif (South Yakutia) for the production of mixed (composite) cement. The paper reviews the implementation of dunite for manufacturing materials and products. The chemical and mineral compositions of Inagli massif dunite rocks are presented, which relegate the rocks to magnesia-silicate rocks of low-quality in terms of its application as refractory feedstock due to appreciable serpentinization of dunite. The work presents the results of dunite study in terms of its applicability as an additive to Portland cement. The authors have established that dunite does not feature hydraulicity and can be used as a filling additive to Portland cement in the amount of up to 40%. It was unveiled that the mixed grinding of Portland cement and dunite sand with specific surface area of 5500 cm^2/g yields the cement that complies with GOST 31108-2016 for CEM II and CEM V normal-cured cements with strength grades of 32.5 and 42.5. The work demonstrates the benefits of the studies of dunite as a filling additive for producing both Portland cement with mineral component and composite (mixed) cement.

1. Introduction

The elaboration of mixed cements using various mineral additives is an urgent problem in construction materials science, since it reduces the consumption of Portland cement clinker, which production consumes a lot of energy and releases significant amount of CO_2. The usage of mixed cements both reduces the cost of construction materials and ensures the environmental safety of the process. One of the main approaches to the enhancement of mixed cements is the optimization of the composition by various mineral additives.

In this connection, magnesia silicates that are widely used in production of refractory and ceramic materials, mineral wool and as an additive to heavy concrete [1–3] are of increased interest due to their structural peculiarities and chemical composition. Among strong points of dunites in production of articles and materials, along with increased material thermal resistance, are the stability of chemical composition, high dielectric properties and health safety (e.g. no silicosis danger, since dunite does not contain free silicon oxide). Dunite can be used as an additive to obtain composite cementing agent [4–5]. According to the researchers, mechanochemical activation of dunite in a mixture of Portland cement clinker and gypsum produces mixed calcium-magnesium silicates, which allows substituting 30–40% of Portland cement clinker without a decrease of the cement activity. The increased compressive and bending strength of dunite-based cement mixture is explained by the occurrence of magnesium hydrosilicates with fibrous structure [5]. These results were obtained by the scientists of Baikal Institute of Nature Management of SB RAS under the supervision of L.I. Khudokova, who were the first to introduce the term *dunite cement* (RF Patent No. 2288899 as of April 14, 2005). There is no other information on this topic both in national and international sources.

Taking into consideration that the authors of the aforementioned works have concluded the increased hydration activity of dunite cements, the present work is aimed at studying the applicability of Inagli massif dunite rocks (South Yakutia) as a mineral additive to Portland cement.

The goal is to be achieved by:

- fine grinding of the rock under study into mineral additive and its study as an additive to the cement, and
- studying the effect of mixed grinding of Portland cement with dunite sand on the strength of the cement.

2. Materials and methods

2.1. Dunite rock

The Inagli alkali-ultrabasic massif is located in the northern boundary area of the Aldan Shield, at the head of Inagli River, which is the right tributary of Aldan River, somewhat 30 km to the west of Aldan city. The central part of the massif with the area of 16 km^2 is composed of forsterite dunites on a narrow (from 5 to 25 meters) discontinuous margin of pyroxene-olivine rocks, which composition corresponds to peridotites and micaceous pyroxenites. Dunites are medium-to-small grain rocks of dark-grey color with steely shade. The composition of dunites of Inagli massif was mineralogically studied by L.V. Rasin in [6].

Table 1. Mineralogical composition of Inagli massif dunites [6].

Minerals	Weakly serpentized dunite (vol%)	Serpentized dunite (vol%)	Dunite from shatter zone (vol%)	Fresh dunite from massif core (vol%)
Olivine	73.5	60.5	61.5	85.5
Serpentine	21.5	33.0	30.0	12.0
Mica	2.5	3.5	3.0	0.5
Chrome-diopside	1.5	0.7	3.0	-
Magnitite	0.5	0.3	0.5	0.3
Chromospinelide	0.5	1.0	0.5	1.7
Calcite	-	1.0	1.5	-
Total	100	100	100	100

The rock sample under study is characterized by greenish light-brown color of porphyric structure with inclusions of dark grains. The sample was labeled as D1.

The chemical composition of the dunite sample was determined by energy-dispersive analysis using JCM-6000 electron microscope (JEOL, USA). The results are presented in table 2; the X-ray spectrum is depicted in figure 1.

Table 2. Chemical composition of dunite.

Name	Content of oxides (wt%)							Ratio	
Inagli massif dunite (Yakutia)	MgO	SiO_2	CaO	Al_2O_3	MnO	Fe_2O_3	Δm_{LOI}	MgO/ SiO_2	MgO/ Fe_2O_3
	45.61	34.59	1.06	0.41	0.39	8.07	10.88	1.34	5.65

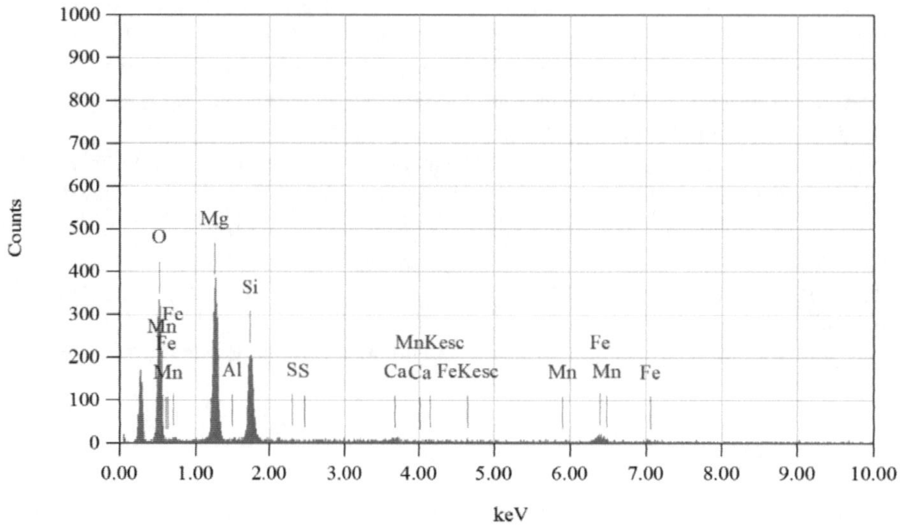

Figure 1. X-ray spectrum of Inagli massif dunite elements.

The mineral composition of the samples was determined by:

- XRD analysis using XRD-7000 X-ray diffractometer (Shimadzu, Japan) with Cu anode (the scan step was 0.05 °/min, time of intensity measurement in scan points was 0.5 s, tube voltage was 40 kV, current was 30 mA), and
- thermal analysis (TGA, TGA-DTA and TGA-DSC) using STA 449 F3 Jupiter® simultaneous thermal analyzer (NETZSCH, Germany) at a temperature of up to 1000 °C in air.

The X-ray pattern is presented in figure 2; the thermogram is depicted in figure 3.

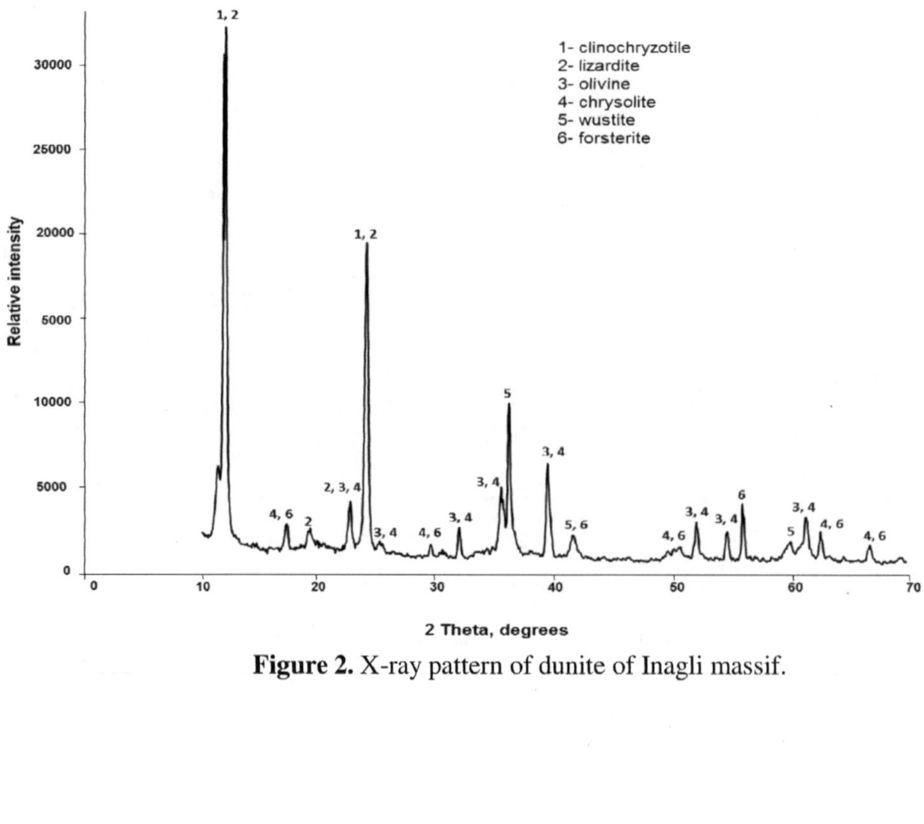

Figure 2. X-ray pattern of dunite of Inagli massif.

IOP Conf. Series: Materials Science and Engineering **286** (2017) 012020 doi:10.1088/1757-899X/286/1/012020

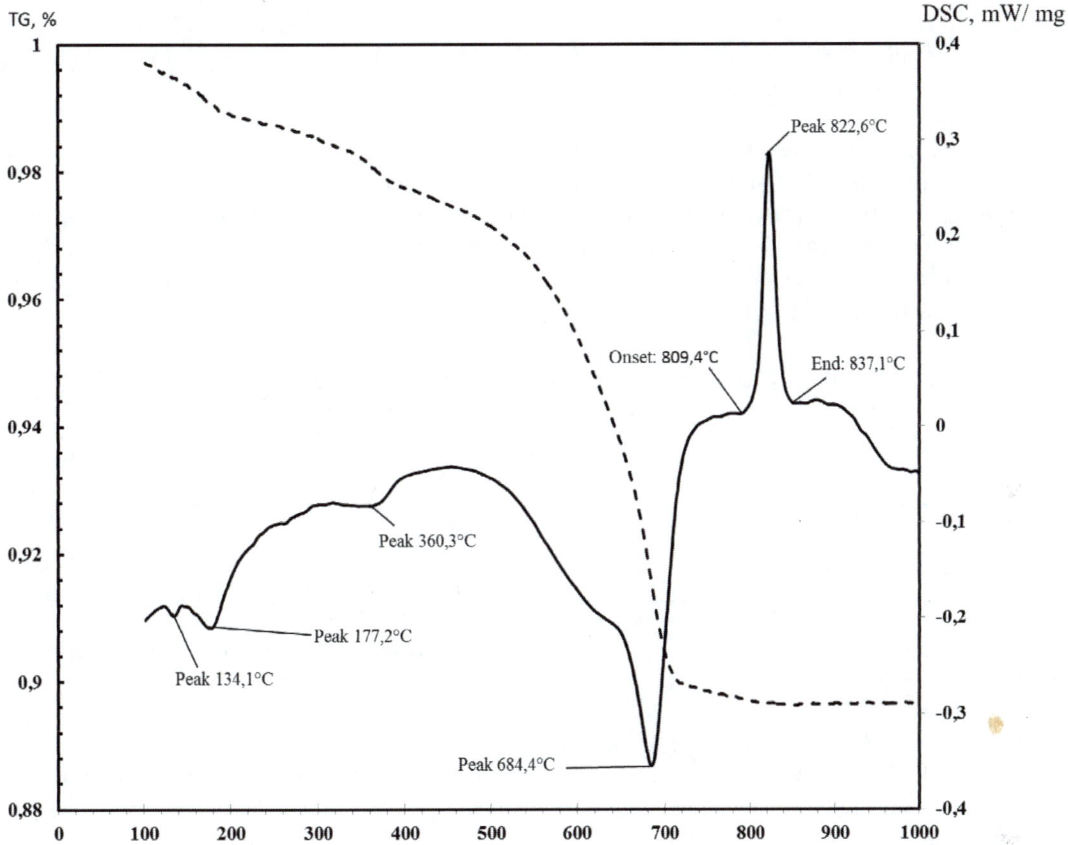

Figure 3. Thermogram of Inagli massif dunite.

The analysis of the X-ray patterns of Inagli massif dunite elucidates the main rock-forming minerals to be olivine $(Mg,Fe)_2SiO_4$, chrysolite Mg_2SiO_4, chrysolite $(Mg,Fe)_2SiO_4$ and wustite FeO. The presence of such serpentinite minerals as clinochrysotile $Mg_3Si_2O_5\ (OH)_4$ ($3MgO*2SiO_2*2H_2O$) and lizardite $Mg_3Si_2O_5\ (OH)_4$ testifies a prominent alteration and serpentinization of the dunite rock. This is confirmed by the large value of loss on ignition of 10.88 wt% and low silicate ratio of 1.34, which corresponds to forsterite-1.34. The thermogram vindicates the XRD analysis results on the presence of serpentinite minerals. For instance, the thermogram features endothermic effects that at 81.4 °C correspond to the removal of free water; at 134.1 °C, 177.2 °C and 360.3 °C, step-wise dehydration of clinochrysotile and lizardite; at 684.4 °C, with the inflection at somewhat 630 °C, the removal of constitutional, i.e. chemically bound, water. The exothermic effect with the maximum at 822.6 °C corresponds to the crystallization of a new phase—forsterite—from dehydrated serpentinite minerals. The elevated magnesium content of the studied samples can be due to intense serpentinization of rocks under hydrothermal and hydrothermal-contact alteration of ultrabasic rocks. As a result of such alteration, the serpentine minerals are formed that are collectively referred to as serpentines, antigorite $(Mg_3[Si_2O_5](OH)_4$, lizardite $(Mg_3[Si_2O_5](OH)_4$, chrysotile $(Mg_3[Si_2O_5](OH)_4$, that contain magnesium. The results confirm the correspondence of the sample chemical composition to ultrabasic rocks, dunites, from the central core of Inagli massif.

2.2. Portland cement

The experiments used PC400-D0 Portland cement (Yakutskcement, Russia) that complies with GOST 10178-85. The mineral composition is 58.70% C_3S, 16.38% C_2S, 6.44% C_3A, 14.35% C_4AF; specific surface area is 3140 cm^2/g.

IOP Conf. Series: Materials Science and Engineering **286** (2017) 012020 doi:10.1088/1757-899X/286/1/012020

2.3. Preparation of dunite additive and its characteristics

Dunite additive was produced as follows:

- crushing of rock in a laboratory jaw crusher;
- screening of material through a 5-mm sieve;
- milling in a laboratory drum mill during 1, 2 and 3 hours.

The true density of grains of the rock under study was determined by pycnometer method. The specific surface area and mean particle size of finely milled additive was determined on PSKh-2 device (Russia) by the air permeability method.

The characteristics of the prepared additives are given in table 3.

Table 3. Characteristics of finely milled mineral (dunite) additive.

Item no.	Additive label	Rock true density (g/cm^3)	Milling time (hours)	Specific surface area (cm^2/kg)	Mean particle size (μm)
1	D_1		1	3750	5.0
2	D_2	2.69	2	5510	3.9
3	D_3		3	7280	3.03

Further works involved additive D_3 to ensure the specific surface area of the cement with the additive of 4000–4500 cm^2/g.

2.4. Preparation of mixed cement and testing methods

To evaluate the impact of the mineral additive (dunite) content on cement properties, two samples of cements were prepared with different content of the mineral additive (20% and 40%) in the cementing agent by mechanical mixing of the initial Portland cement with mineral additive D_3. The samples were mixed in a laboratory vibration mixer. The reference samples were prepared from Portland cement without any additive.

The cements under study were also prepared by mixed grinding of Portland cement PC 400-D0 and dunite sand in Aktivator 2S ball mill (Aktivator, Russia).

The cement properties were determined as per GOST 310 except for the samples that were intended for strength testing. These samples after extraction from the mold were placed into a chamber of standard curing (t_{air} = 20±2 °C, W_{rel} = 95±5%).

To determine the compression strength, the samples with the dimensions of 2x2x2 cm were mold from cement paste of normal density that were wet-cured at the temperature of 20±2 °C.

The kinetics of cement paste curing was studied on the samples from cement paste with water-to-cement ratio of 0.4 and dimensions of 2x2x2 cm that were cured in the chamber of standard curing and tested after 3, 7, 14 and 28 days of curing.

3. Results and discussion

The physical and mechanical properties of the cements are presented in table 4.

Table 4. Physical and mechanical properties of cements.

Cement characteristics	Testing method	Requirements as per GOST 10178-85	Real values		
			PC 400-D0	80%PC + 20% D_3	60%PC + 40% D_3
1 Specific surface area (cm^2/g) (mean grain diameter (μm))	GOST 310.2-76*		3140 (6.15)	3840 (5.20)	4100 (5.00)
2 Cement paste thickness (%)	GOST 310.3-76*	-	26.75	26.50	27.00
3 Curing time:	GOST 310.3-76*				
beginning, h-min		not earlier than 0-45	2-50	3-05	2-50
ending, h-min		not later than 10-00	3-30	3-40	3-50
4 Uniformity of cement volume change	GOST 310.3-76*		-	-	-
5 Compression strength (MPa)	GOST 310.4-81*				
7 days		-	29.9*	13.2*	9.4*
28 days		at least 39.2	37.4*	33.2*	21.1*
6 Bending strength (MPa)	GOST 310.4-81*				
7 days			4.6*	4.0*	3.2*
28 days			6.7*	5.9*	4.6*
7 Cement grade		400			

*Samples were cured in the chamber of standard curing

The specific surface area of initial Portland cement and cements under study with 20% and 40% of mineral additive amounted to 3140, 3840 and 4100 cm^2/g, correspondingly. The cements with mineral additive have elevated specific surface area as compared to initial Portland cement. In spite of this effect, the water demand of cements with mineral additives remains virtually unchanged. The normal thickness of the cement paste lies within ±0.25%.

The introduction of the mineral additive in the amount of up to 40% of the cement mass also has no significant effect on the time of cement curing. In terms of curing time, the curing beginning time of the cements (not earlier than 75 min) complies with the requirements of GOST 10178-85 *Portland cement and Portland Blastfurnace cement. Product specifications* and GOST 31108-2003 *Standard cements. Product specifications*.

The activity of initial Portland cement amounted to 37.4 MPa, which is lower than required by GOST 10178-85 (39.2 MPa) by 4.6%. This can be explained by the curing of the samples in the chamber of standard curing instead of wet-curing. The activity of the Portland cements with the content of mineral additives of 20% and 40% of the cement mass amounted to 33.2 and 21.1 MPa, respectively, which is less than that of initial Portland cement; i.e. the growth of the specific surface area has no effect on the strength. This indicates the inertness of the dunite additive.

Figure 4 demonstrates the growth of the strength of cement samples from cement paste with normal thickness having the dimensions of 2x2x2 cm that were wet-cured.

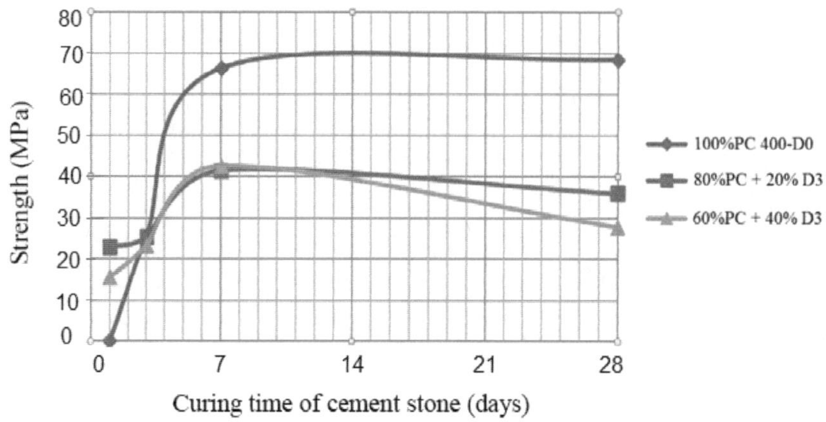

Figure 4. Strengthening of samples from cement paste of normal thickness.

Obviously, the strength of specimens with the mineral additives is nearly two times lower than that of specimens from initial cement PC 400-D0; this also indicates that the mineral additive with the specific surface area of 7280 cm^2/g is an inert additive (filler). The experiment has established as well that the cements with mineral additive fairly intensely strengthen during first 7 days, while further they weaken. Probably, this is connected with the change to the uniformity of the volume during curing in the presence of magnesium oxide. This effect is to be studied further.

To elucidate the causes of the determined strength characteristics, we decided to study the structure of the cement stones. The specimens were selected from among the 7-day samples after testing. The images taken by 3700 scanning electron microscope (JEOL, USA) are presented in figure 5. The comparison of the images of cement stones at 500x magnification (a, d, h) demonstrates that the cements with mineral additives feature appreciable cracking in the cement stone as compared to initial cement. In addition, the opening of the cracks increases with raising content of the additive. This complies with the received strength characteristics. The images with 5000x and 10,000x magnification show the grains of the mineral additive and the contact zone between the additive and cement stone. The cracks discovered in the contact zone have possibly occurred during the tests and reflect weak strength of the contact zone.

The determined strength properties of the cements denote that the mineral additive from the rock under investigation is an inert additive even after its mechanical activation by milling down to the specific surface area of 7280 cm^2/g. The received result does not agree with that in [5], which demonstrated the increased binding of dunite with Portland cement clinker into mixed hydrosilicates. Probably, during mixed grinding of clinker, mineral additive and gypsum, hard dunite rock acts as a grinding agent and increases the specific surface area of the clinker fraction of the cement, which leads to the activation of the cement mixture. This can probably increase the strength in the case of simultaneous milling with dunite sand.

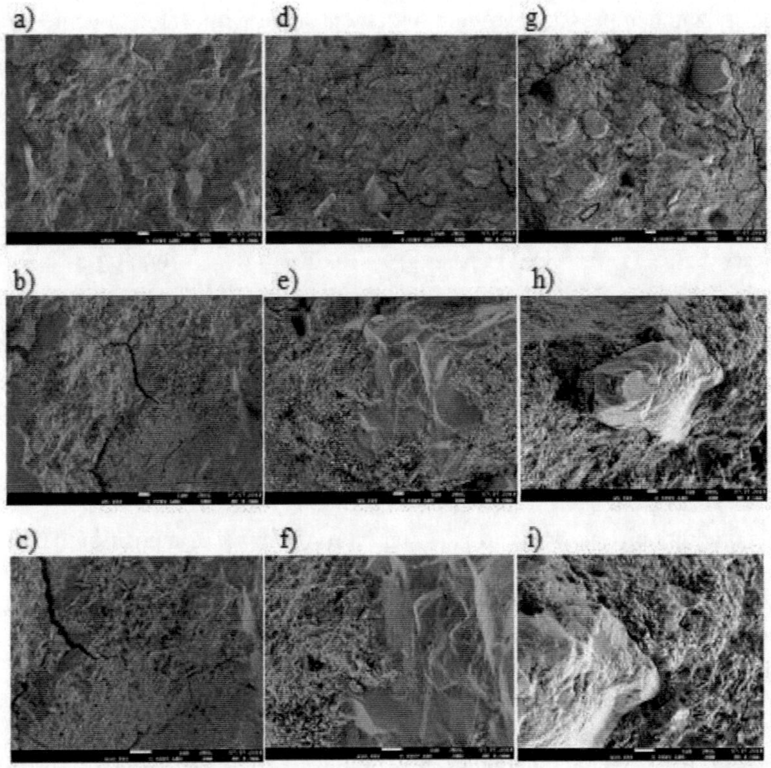

Figure 5. SEM images of cement stones structure at 500x (a, d, g), 5000x (b, e, h) and 10,000x (c, f, i) magnification: 100%PC 400-D0 (a, b, c), 80%PC 400-D0 + 20%D3 (d, e, f) and 60%PC 400-D0 + 40%D3 (g, h, i).

The characteristics of cements under study prepared by different methods are presented in Table 5; the results of density and strength testing of cement stones for different curing time are in table 6.

Table 5. Specific surface area and mean particle diameter of cements.

Cement label	Cement composition	Method of cement preparation with additive	Specific surface area (cm^2/kg)	Mean particle size (μm)
D0	100%PC00-D	-	3140	6.2
D20	80%PC + 20% D_3	Mechanical mixing	3840	5.2
D40	60%PC+40% D_3		4100	5.0
Dds20	80%PC + 20% DS	Mixed grinding	5570	3.5
Dds40	60%PC + 40% DS		5540	3.7

Evidently from table 6, the cements prepared by mixed grinding of cement and dunite sand demonstrated higher strength (41.7 and 35.1 MPa, respectively for 80%PC + 20% DS and 60%PC + 40% DS specimens) as compared to cements of corresponding composition prepared by mechanical mixing of cement with mineral additive. The received results also demonstrate that dunite rock is inert. Nevertheless, preliminary testing has shown that the mixed grinding of cement with dunite sand has better efficiency against finely milled mineral additive for cement.

Table 6. Density and compression strength of cement stone with water-to-cement ratio of 0.4.

Cement label	Cement composition	Density (kg/m^3) /compression strength of cement stones in MPa (% of R_{28}) after curing time (days)			
		3	7	14	28
D0	PC 400-D0	1990/26.0	2010/32.2	2082/43.0	2018/51.0
D20	80%PC + 20% D_3	1930/18.2	1860/18.4	1854/18.2	1910/23.4
D40	60%PC+40% D_3	1870/13.6	1910/12.9	1909/22.4	187320.9
Dds20	80%PC + 20% DS	2040/28.7	1910/36.6	1937/41.6	1923/41.7
Dds40	60%PC + 40% DS	1870/17.6	1890/24.8	1870/34.3	1948/35.1

Apparently, the increased strength vs. the introduction of milled mineral additive is conditioned by high specific surface area of the cement.

4. Conclusion

The studied mineral additive from dunite rock in terms of chemical composition belongs to ultrabasic rocks—dunites of Inagli massif central core; in terms of chemical and mineral composition it belongs to magnesia-silicate rocks of low quality regarding its application as refractory feedstock due to appreciable serpentinization of dunite.

The study has established that the investigated dunite rock does not feature hydraulicity and belongs to inert mineral additives.

The work elucidated that the mixed grinding of Portland cement and dunite sand down to specific surface area of 5500 cm^2/g allows producing cement that complies with GOST 31108-2016 for normal-cured CEM II and CEM V cements with strength grades of 32.5 and 42.5.

The work has demonstrated the benefits of the studies of dunite as a filling additive for producing both Portland cement with mineral component and composite (mixed) cement.

References

[1] Khoroshavin L B, Kosolapova E P 1968 Refractory concretes from dunite *Collected papers of the Eastern Refractories Institute* **7** 127-133

[2] Belogurova O A, Grishin N N and Mamontov V P 2001 Evaluation of dunites of Pados-tundra deposit as refractory raw material *Building and technical materials from natural and technogenic raw materials of the Kola Peninsula* (Apatity: Publishing House of the Kola Science Centre of the Russian Academy of Sciences) pp 45-53

[3] Bondarenko A V, Balamygin D I 2008 Modern refractory materials based on dunite of Solovyevogorskoe deposit *Refractories and technical ceramics* **3** 5-9

[4] Khudyakova L I, Konstantinova K K and Narkhinova B A 2002 Acquisition of heat-resistant composite binder materials *Building Materials* **11** 44-45

[5] Khudyakova L I, Voiloshnikov O V, Kotova I I et al. 2010 Waste of mining enterprises as raw materials for production of building materials *Vestnik of the Far East Branch of the Russian Academy of Sciences* **1** 81-84

[6] Razin L V, Khvostova V P and Novikov V A 1965 Metals of platinum group in rock-forming and accessory minerals of ultrabasic rocks *Geochemistry* **2** 159-174

IOP Conf. Series: Materials Science and Engineering **286** (2017) 012021 doi:10.1088/1757-899X/286/1/012021

Influence of Material Structure Crystallography on its Formability in Sheet Metal Forming Processes

F V Grechnikov[1], S V Surudin[2], Ya A Erisov[2], A O Kuzin[2] and I N Bobrovskiy [3]

[1]Samara Scientific Center of the Russian Academy of Sciences, Studencheskii pereulok 3a, Samara, Russia
[2] Department of Metal Forming, Samara National Research University, Moskovskoe shosse 34, Samara, Russia
[3] Department of Togliatti State University, Belorusskaya 14, Togliatti, Russia

E-mail: alexandrkuzin88@gmail.com

Abstract. The finite element simulation was used to analyze the effect of crystallographic texture on material formability in typical sheet metal forming processes: stretch forming, hole expansion, drawing and bending. It is defined that the best formability (maximum cupping depth, flanging ratio, drawing depth, minimum bending radius) is provided by a rotated cube orientation; the worst – by Goss orientation. Some texture components provide better formability than the isotropic case (rotated cube), some - worse (cube, goss) or close to isotropy (copper, brass or S-orientation).

1. Introduction

It is well known that such widespread in the aerospace industry semi-finished products like sheets, bars, extrusions, tubes, etc. possess a pronounced anisotropy of properties. Ignoring the anisotropy in technological calculations not only reduces the expected limit strains of the blanks, but also leads to a number of other undesirable phenomena: increased metal consumption, distortion of shape and size, decrease of finished part performance [1-6]. On the other hand, rational anisotropy is a serious factor for the intensification of the metal forming processes and improvement of the parts performance in certain directions [6-11].

The optimal anisotropy of properties for a particular metal forming process can be determined by means of computer simulation, the modern development level of which allows transferring a significant part of the work on the evaluation and analysis of the stress-strain state to the area of numerical experiment, receiving a great amount of information; implementing a comprehensive study not only of the forming processes, but also of the material behavior depending on its structure; considering and comparing a great number of alternatives [12-14].

The plasticity criterion developed in [15] allows taking into account the crystallographic texture and formulating requirements to it. The peculiarity of the mentioned above criterion is that it explicitly includes the parameters of the crystallographic orientation and the crystal lattice constants. On the basis of this criterion, a material model was developed [16], which made it possible to perform an analysis of the metal forming processes taking into account the crystallographic orientation of the blanks structure [17-18].

IOP Conf. Series: Materials Science and Engineering **286** (2017) 012021 doi:10.1088/1757-899X/286/1/012021

Using this material model, the influence of the crystallographic orientation of the structure on material formability in various sheet metal forming operations (stretch forming, drawing, bending, flanging) has been studied.

2. Numerical Simulation

Figure 1 shows the finite element models of the sheet metal forming processes under consideration. The dimensions and geometry of the models correspond to the standard tests for stretch forming (ISO 20422), hole expansion (ISO 16630), earing (EN 1669) and variable-radius bending [19]. In all cases, the blank thickness is 1 mm. Finite element models are made using 4-node shell elements with 5 integration points over the thickness. In order to reduce the number of elements, quater of the volume was modeled (for bending – half). Between the tool and the blank contact pairs were prescribed, the friction on which obeys the Coulomb law (the friction coefficient is assumed to be 0.12). The tool was taken absolutely rigid.

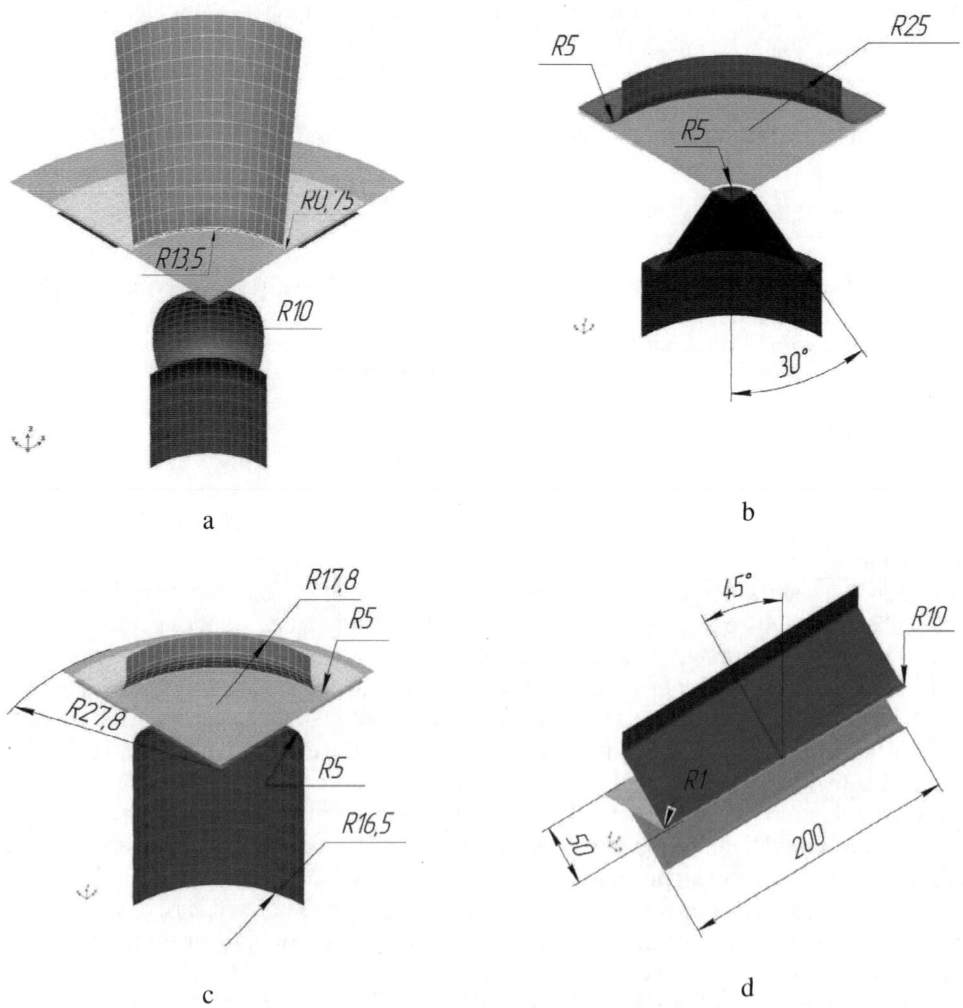

Figure 1. Finite element models: a – stretch forming; b – hole expansion; c - drawing; d - bending with variable radius

The model of orthotropic elasto-plastic material considering crystallographic texture is used to describe the blank material behavior - high-strength aluminum-lithium alloy V-1461 [15]:

$$\sigma_i = \sqrt{\frac{\eta_{12}+\eta_{31}}{2}\cdot\sigma_{12}^2 + \frac{\eta_{12}+\eta_{23}}{2}\cdot\sigma_{22}^2 - \eta_{12}\sigma_{11}\sigma_{22} + (5-2\eta_{12})\sigma_{12}^2} \quad (1)$$

Here σ_i is the equivalent stress; σ_{ij} is stress tensor. The generalized anisotropy parameters η_{ij} are defined by:

$$\eta_{ij} = 1 - \frac{15(A'-1)}{3+2A'}\left(\Delta_i + \Delta_j - \Delta_k - \frac{1}{5}\right) \quad (2)$$

where A' is the anisotropy parameter of the crystal lattice; Δ_i are the orientation factors of crystallographic orientation:

$$\Delta_i = \frac{h_i^2k_i^2 + k_i^2l_i^2 + l_i^2h_i^2}{(h_i^2+k_i^2+l_i^2)^2} \quad (3)$$

h_i, k_i, l_i are Miller indices defining the i-th direction in the crystal with respect to the coordinate system associated with the blank. Rolling direction is set along the axis X (for bending test – along bending line).

Material hardening during plastic deformation obeys the Swift's law [20]:

$$\sigma_i = k(\varepsilon_0 + \varepsilon_i)^n \quad (4)$$

where ε_i is the equivalent strain; ε_0 is the initial strain of plastic flow; n is the hardening exponent; k is the hardening coefficient. The strain ε_0, at which the material changes from elastic to plastic state, has been found by equaling the Eq. (4) and Hooke's law ($\sigma=E\varepsilon$, here E is the Young's modulus). Properties of V-1461 aluminum alloy are given in table 1.

Table 1. The properties of alloy V-1461

Parameter	Value
Young Modulus, E (GPa)	79.5
Puasson coefficient	0.33
Density, ρ (kg/m^3)	2.6
Hardening coefficient, k (GPa)	0.7091
Hardening exponent, n	0.01627
Initial strain of plastic flow, ε_0	0.102

In order to assess the influence of structure crystallography on the formability it was modeled anisotropic material, the texture of which is represented by only one ideal crystallographic orientation. It was considered the most characteristic for the rolled material deformation orientations (copper {112} <111>, brass {110} <112>, S {123} <634>, rotated cube {100} <011>) and recrystallization orientations (cube {100} <001> and Goss {110} <001>) [21]. The orientation parameters are given in table 2. In the calculations it was assumed that for aluminum-lithium alloys $A' = 2.05$ [22].

To determine the formability of the material during the forming process, the following fracture criterion was used [17]:

$$\psi = \frac{\varepsilon_i}{\varepsilon_i^{lim}} < 1 \quad (5)$$

IOP Conf. Series: Materials Science and Engineering **286** (2017) 012021 doi:10.1088/1757-899X/286/1/012021

Table 2. Crystallographic orientations and their parameters.

Orientation		Generalized anisotropy parameters		
Name	Designation	η_{12}	η_{23}	η_{31}
Copper	{112}<111>	0.705	1.073	0.705
Brass	{110}<112>	0.705	0.705	1.073
S	{123}<634>	0.758	0.896	0.882
Rotated cube	{100}<011>	0.335	1.444	1.444
Cube	{100}<001>	1.444	1.444	1.444
Goss	{110}<001>	1.444	0.335	1.444
Isotropy		1	1	1

where ε_i^{lim} is the limit strain, which depends on the stress-strain state and the material texture. So ε_i^{lim} for stretch forming and hole expansion was calculated by the next formula:

$$\varepsilon_i^{lim} = \frac{K}{(\eta_{12}+\eta_{31})-\eta_{12}m}n \qquad (6)$$

for drawing:

$$\varepsilon_i^{lim} = \frac{K}{(\eta_{12}+\eta_{23})m-\eta_{12}}n \qquad (7)$$

where $K = \sqrt{2}\sqrt{(\eta_{12}+\eta_{31})-\eta_{12}m+(\eta_{12}+\eta_{23})m^2}$, and $m = \sigma_{11}/\sigma_{22}$ is the stress state index.

3. Results and discussion

Tables 3 and figure 2 show the parameters of the stretch forming, hole expansion, drawing and bending processes at the moment of fracture initiation ($\psi = 0.9$). It is evident that the maximum formability of the material (maximum cupping depth, flanging ratio, drawing depth, minimum bending radius) is provided by an ideal crystallographic orientation {100} <011>; the minimum – by orientation {110} <001>. Some texture components provide better formability in comparison with the isotropic case ({100} <011>), some worse ({100} <001>, {110} <001>) or close to isotropy ({112} <111) >, {110} <112>, {123} <634>).

IOP Conf. Series: Materials Science and Engineering **286** (2017) 012021 doi:10.1088/1757-899X/286/1/012021

Table 3. Limiting parameters of the sheet metal forming process depending on the crystallographic orientation

Orientation	Process/ parameter		
	Stretch forming	Hole expansion	Drawing
	Cupping depth (mm)	Flanging ratio	Drawing depth (mm)
{112}<111>	5.3	1.244	9.4
{110}<112>	5.0	1.260	9.8
{123}<634>	5.1	1.208	9.55
{100}<011>	6.05	1.420	10.35
{100}<001>	4.35	1.108	6.55
{110}<001>	4.6	1.059	5.3
Isotropic	5.05	1.154	9.55

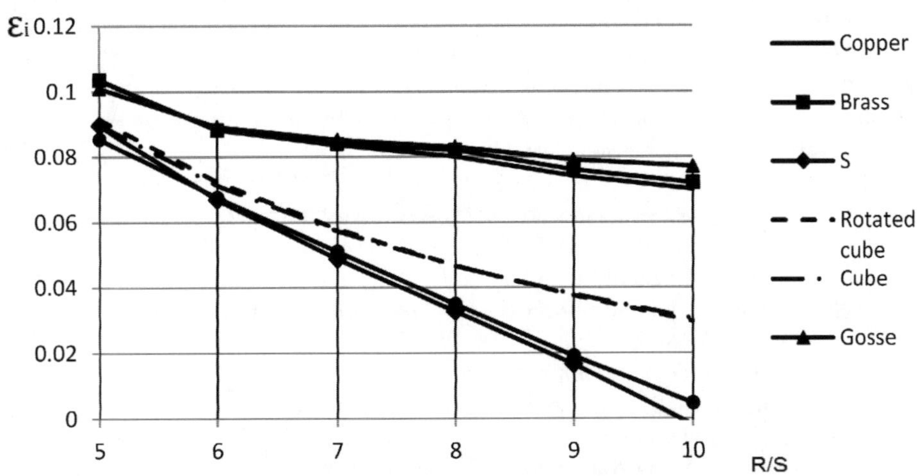

Figure 2. Dependence of the equivalent strain on the relative bending radius

Thus, when developing the thermomechanical conditions of production sheets from aluminum alloys, including aluminum-lithium, it is necessary to assign consistent rolling and intermediate heat treatment (annealing) conditions, since only a combination of various ideal crystallographic orientations of deformation and recrystallization types can provide an increase in formability in sheet forming technological processes. Upon that, the composition of the texture components will be individual for each specific forming process.

4. Conclusion

The proposed calculation procedure considers the crystal lattice constants and the parameters of crystallographic orientation of material. The main practical significance of this procedure is possibility to predict the effect of crystallographic texture of rolled sheets on limiting strains and formability of material in different metal forming process. In addition, the proposed procedure allows designing the composition of crystallographic texture depending on the requirements of the metal forming processes.

Acknowledgements

The reported study was funded by RFBR according to the research project no. 16-48-630828.

References

[1] Hutchinson W B, Oscarsson A and Karlsson 1989 *Materials Science and Technology* **5** 1118-1127
[2] Engler O and Hirsch J 2002 *Materials Science and Engineering* **336** 249-262
[3] Demyanenko E G A 2014 *Type Pats-Russian Aeronautics* **57** 204-211
[4] Banabic D, Bunge H J, Pohlandt K and Tekkaya 2000 A E *Formability of Metallic Materials: Plastic Anisotropy, Formability Testing, Forming Limits* (Springer, Berlin, Germany) p 334
[5] Tóth L S, Hirsch J and Van Houtte P 1996 *International Journal of Mechanical Sciences* **38** 1117-1126
[6] Barlat F 1987 *Materials Science and Engineering* **91** 55-72
[7] Grechnikov F V , Antipov V V, Erisov Y A and Grechnikova A F 2015 *Russian Journal of Non-Ferrous Metals* **56** 39-43
[8] Peters M, Gysler A and Lotjering G 1984 *Metallurgical Transactions* **15A** 1597-1605
[9] Venkateswara Rao K T, Bucci R J, Jata K V and Ritchie R O 1991 *Materials Science and Engineering* **A141** 39-48
[10] Wu X J, Wallace W, Raizenne M D and Koul A K 1994 *Metallurgical and Materials Transactions* **25(3)** 575-588
[11] Banabic D, Barlat F, Cazacu O and Kuwabara T 2010 *International Journal of Material Forming* **3** 165-189
[12] Raabe D, Roters F, Barlat F and Chen L Q 2004 *Continuum Scale Simulation of Engineering Materials* (WILEY, Berlin, Germany) p 885
[13] Rybin Yu I, Rudskoy A I and Zolotov A M 2004 *Mathematical Modeling and Design of Metal Forming Technological Processes* (Science, Moscow, Russia) p. 644
[14] Owen D R J and Hinton E 1980 *Finite Elements in Plasticity: Theory and Practice* (Pineridge Press Limited, London, England) p. 450
[15] Erisov Y A, Grechnikov F V and Surudin S V 2016 *Structural Engineering and Mechanics* **58(4)** 677-687
[16] Grechnikov F V and Erisov Y A 2016 *Key Engineering Materials* **684** 134-142
[17] Erisov Y and Grechnikov F 2015 *Proceeding of 2015 International Conference on Mechanical Engineering, Automation and Control Systems MEACS 2015* **7414959**
[18] Grechnikov F V, Erisov Y A and Alexandrov S E 2016 *CEUR Workshop Proceedings* **1638** 569-577
[19] Averkiev A Yu 1985 *Methods for Evaluating of Sheet Metal Formability* (Mechanical Engineering, Moscow, Russia) p. 176

[20] Gronostajski Z 2000 *Journal of Materials Processing Technology* **106** 40-44
[21] Choi S H, Cho J H, Barlat F, Chung K, Kwon J W and Oh K H 1999 *Metallurgical and Materials Transactions* **30(A)** 377-386
[22] Landolt – Bornstein 1966 *Numerical Data and Functional Relationships in Science and Technology. New Series. Group III: Crystal and Solid State Physics. Volume I: Elastic, Piezoelectric and Electrooptic Constants of Crystals* (Springer, Berlin, Germany)

IOP Conf. Series: Materials Science and Engineering **286** (2017) 012022 doi:10.1088/1757-899X/286/1/012022

Microstructure and properties of 3Y-TZP ceramic fabricated using PEG temporary binder

S Veselov, N Cherkasova, R Kuzmin, E Drobyaz, A Felofyanova and Y Kalugina

Novosibirsk State Technical University, 20, K. Marks ave., Novosibirsk, 630073, Russia

E-mail: ninacherkasova60@gmail.com

Abstract. 3Y-TZP ceramics prepared with uniaxial pressing technology followed by free sintering were investigated. Polyethylene glycols (PEG) with different molecular weight were used for the granules preparation. Green and sintered samples with different granulated powders were analyzed. The forming uniaxial pressure was 100 MPa. Sintering of samples was performed at 1520 °C during 5 hours of soaking. Microstructural investigation was carried out by scanning electron microscope. Obtained relative green and sintered densities for all sample series are equal (50 % and 95 %, corresponding). Fracture surfaces of green samples are uniform and do not contain defects as large pores. Microstructural investigation of sintered samples revealed large cracks with a size in the range between 20 and 40 μm. The average grain size is about 0.7 μm. The average microhardness for all series is about 1150±25 $HV_{0.5}$. The minimum strength was obtained for a samples series prepared using PEG 1500 and PEG 6000 mixture. All other series show a flexure strength on one level (880±50 MPa). It is determined, that usage of polyethylene glycols with 400-6000 molecular weight as temporary organic additives is an effective for the 3Y-TZP ceramic production.

1. Introduction

The dry pressing technology followed by free sintering have a wide area of application in production of modern dense high-strength ceramics for different purposes [1]. As usually, the high strength of sintered ceramics is achieved by means of using finely dispersed powders as a raw material. At the same time submicron powder pressing leads to nonuniform distribution of pressure and formation of macrodefects in the compacts. This is due to the low technological properties and tendency to aggregating of submicron powders.

The negative effects of compacting can be reduced by granulation of raw powders (preparation of conglomerates with controlled shape, size and properties). Strength and the ability of granules to deform plastically affect their rearrangement in the press-form on initial pressing stage, as well as on destroying during final stage. This leads to the uniform distribution of pressure and elimination of large defects in compacts. The usage of various binders and plasticizers as organic additives is an effective method of controlling the granulated powder behavior. Typically, high molecular weight polymers are used as binders, and low molecular weight polymers are used as plasticizers.

The main parameters that determine the choice of polymer additives are dissolubility in water or alcohol media, the glass transition temperature (T_g) and the thermomechanical behavior [2]. The molecular weight (M_w) of the polymer significantly effects on the described parameters. It is known that the increase of molecular weight results in increasing of the glass transition temperature and melting temperature as well as plateau of viscous-flow state changing. Besides, in case when T_g is higher that the pressing temperature the polymer ductility is limited and its granule strength is higher than the pressure in the press-form. For example, the granules obtained by using only polyvinyl alcohol (PVA) deform during compaction but practically do not destroy.

At the stage of sintering, the processes promoting the reduction of the residual porosity take place. The intensity of these processes depends on the activity of the raw powders. In this case, large defects (for example, surfaces and junctions of unbroken granules) partially retained and lead to a decrease of mechanical properties of ceramics [3, 4]. Y-TZP shows a limited ability to reduce porosity during ceramics sintering [5]. Thereby, temporary organic additives choice should be made in the way that if fully allows eliminating of macrodefects formation at the pressing stage. The purpose of this research is to estimate the usefulness of polyethylene glycols (PEG) with different molecular weight application for the manufacturing of 3Y-TZP ceramics with high mechanical properties by uniaxial pressing followed by free sintering technology.

2. Materials and experimental procedure

The submicron powder 3Y-TZP (PSZ-5.5YS, Stanford Materials) with an average particle size d_{50}=0.5–1 μm was used as a raw material for preparation of the experimental materials. Dispersing of water suspension was carried out by using ultrasonic treatment (plant power 2 kW). Deflocculation agent Dolapix CE 64 was used for preventing particles agglomeration.

The suspension granulation was performed with a Mobile Minor 0.8 (GEA Niro) spray dryer. Before the granulation stage 2 wt. % of organic additives was introduced into the suspension. PEG M_w=400 (T_g≈–50 °C), 20 wt. % water solution of PEG 1500 (M_w=1500, T_g≈–28 °C), 20 wt. % water solution of PEG 6000 (M_w=6000, T_g≈–17 °C) and their mixtures were used as organic additives [6]. Granules size was 15–30 μm for all experimental series. The press-powder moisture was 0.5–0.7 % depending on the composition of the used organics. To exclude moisture difference influence, all press-powders were saturated with water to moisture about 1 %.

Bar-shaped samples of 5x5x50 mm size were prepared by unaxial pressing on the Instron 3369 test machine. The forming uniaxial pressure was 100 MPa. Sintering of samples was performed in the LHT 02/17 (Nabertherm) laboratory furnace at 1520 °C during 5 hours of soaking [7].

Green and sintered samples were analyzed. Investigation of the samples after pressing was carried out based on geometric density measurement data and SEM images of fracture surface. The density of the sintered samples was determined using Archimedes' method. Microhardness and flexure strength were investigated to estimate the mechanical properties of the obtained ceramics. Microhardness was measured by the Vickers method on 402MVD semiautomatic instrument (Wolpert Group) under loading of 500 g. Three-point flexure strength tests were performed according to standard EN 843-1-2008 at the same equipment that was used at the pressing stage. Structural and fractographic studies of the samples were carried out with scanning electron microscope (SEM) Carl Zeiss EVO 50. Investigations were conducted at NSTU Materials Research Center.

3. Results and discussion

For research, four sample series of zirconia ceramics were obtained. The granulated powders with temporary organic additives of different molecular weight were used in the experimental series preparation (Table 1). The used polymers have the same molecular structure but different chain length. As a result, when two different binders were added simultaneously it was supposed that a homogeneous mixture of polymers with intermediate molecular weight is formed in suspension. Furthermore, in case of mixing two types of polymers the result physical and mechanical properties of binder are of an intermediate state between the initial components.

Green samples obtained from different granulated powders were analyzed. The density of all sample series are equal (Table 1). It can be assumed, that all press-powder granules have similar physical and mechanical properties and equal pressing behavior. The obtained density also suggest a low concentration of such macrodefect as pores in the samples.

Table 1. Green and sintered ceramic properties

Series N	Organic binder, (wt. %)	Relative green density, % of theoretic density	Relative sintered density, % of theoretic density	Flexure strength, MPa
1	PEG 1500 (2)	50 ± 1	96 ± 1	820 ± 50
2	PEG 6000 (2)	50 ± 1	95 ± 1	910 ± 50
3	PEG 6000 (1) + PEG 400 (1)	50 ± 1	95 ± 1	880 ± 50
4	PEG 1500 (1) + PEG 6000 (1)	50 ± 1	95 ± 1	530 ± 50

The fracture surfaces of the green samples are shown in figure 1. The analysis indicates that the fracture surfaces are homogeneous and do not contain such defects as large pores (Figure 1a). At the same time, a small content of granule boundaries is visible in figure 1b. It also should be noted, that large defects such as diagonal cracks are absent. The formation of these cracks is associated with the overpress effect. Similar defects occur in powder materials with high plasticity. Such materials do not have enough strength and the formation of overpress defect is caused by the elastic after-effect of the steel die [7, 8].

Figure 1. Fracture surface of the green samples (1st series).

It is assumed that granules with low strength were formed using all considered organic binders. Using considered PEG the obtained granules strength is sufficient to avoid overpress defect. At the same time during the pressing process at the initial stage of the load application, the granules are destroyed without redistribution. As the result the quantity of such defects as granule surfaces and junctions in the green ceramic decreased, but a uniform pressure distribution in the compact is not achieved.

Microstructural investigations of sintered samples showed, that large cracks with a size in the range between 20 and 40 μm were observed (Figure 2a). This defect size is the equal to the diameter of the initial granules. It is assumed that due to the uneven pressure distribution the concentration of defects in green ceramic (the surface of undestroyed granule) is quite high. The average grain size is about 0.7 μm. Figure 2b also shows that the maximum grain size was up to 3.8 μm in the local areas.

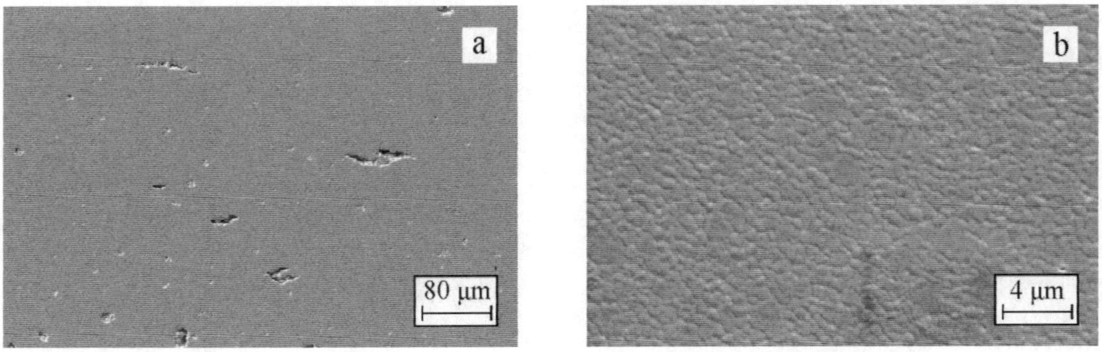

Figure 2. Typical defects (a) and microstructure (b) of the 1st samples series.

Although the fine-grained microstructure was obtained, the presence of large defects in sintered ceramic are affected on mechanical properties level. The average microhardness for all series is about 1150 ± 25 $HV_{0,5}$. The minimum strength was obtained from the samples series prepared using PEG 1500 and PEG 6000 binder mixture. All other series show a flexure strength on one level (880 ±50 MPa). This level corresponds to the properties of modern high strength zirconia ceramic obtained by such expensive technologies as cold and hot isostatic pressing. In particular, authors of [9, 10] showed that ceramics, obtained by cold isostatic pressing, have strength at the level of 600-1100 MPa. Moreover, according to [3] the usage of PEG as an organic binder for zirconia ceramic is more valid in comparison with polyvinyl alcohol when the samples strength was about 700 MPa.

The fracture surfaces of the sintered samples are shown in figure 3. It was stated that for all experimental series the origins of fracture were undestroyed granules. Figure 3a shows interface between undestroyed granule and the rest structure. It was assumed that the presence of almost spherical shape granules in sintered ceramic (Figure 3b) results from non-uniform pressure distribution. According to microstructure analysis, low flexure strength of the 4th experimental series samples were related with higher volume fracture these defects in comparison with other samples.

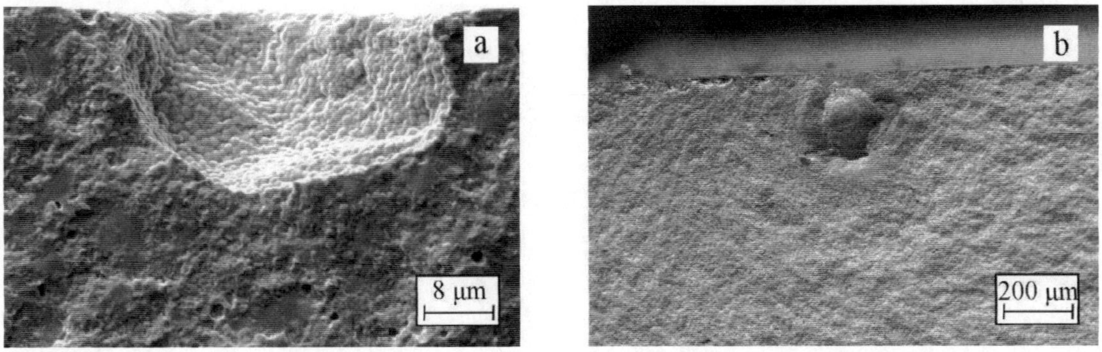

Figure 3. Typical fracture origin: an interface between undestroyed granule and the rest structure of the 1st samples series (a), undestroyed granule of the sample of the 4th series (b).

4. Conclusions

Consequently, usage of polyethylene glycols with 400-6000 molecular weight as temporary organic additives is an effective method for the production of 3Y-TZP ceramic by the axial pressing followed by free sintering technology. All studied samples series had a relative density about 95 % of theoretical value and a microhardness about 1150 ± 25 $HV_{0.5}$.

The low strength of polymers leads to destroying of the most powders granules during the pressing stage. As a result, the volume fraction of such defects as surfaces and junctions of undestroyed granules was reduced and sintered ceramics strength increased. The highest flexure strength was obtained for ceramic samples with the injection of 2 wt. % PEG 6000 and was about 910 MPa.

IOP Conf. Series: Materials Science and Engineering **286** (2017) 012022 doi:10.1088/1757-899X/286/1/012022

At the same time, we expect that the low strength of temporary organics excludes the stages of granules redistribution and pressure equalizing in the green sample. Therefore, in areas with low specific pressure macrodefects are forming. Microstructural studies have shown that the major structural defect of all sintered ceramic samples is large cracks with a size from 20 to 40 μm.

Acknowledgments

This research was performed at Novosibirsk State Technical University under the financial support of the Ministry of Education and Science of the Russian Federation (contract No 02.G25.31.0144 on 01.12.2015).

References

[1] Carter C B and Norton M G 2007 *Ceramic materials: science and engineering* (Springer Science & Business Media)
[2] Taktak R, Baklouti S and Bouaziz J 2011 *Mater. Charact.* **62(9)** 912-6
[3] Veselov S V, Stukacheva N S, Kuzmin R I, Cherkasova N Yu, Perepelkin M Yu and Myz A L 2016 *Science bulletin of NSTU* **4(65)** 207-17
[4] Adolfsson E and Shen Z 2012 *J. Eur. Ceram. Soc.* **32(11)** 2653-9
[5] Hwang S L and Chen I W 1990 *J. Am. Ceram. Soc.* **73** 3269-77
[6] Faucher J A, Koleske J V, Santee Jr E R, Stratta J J and Wilson III C W 1966 *J. Appl. Phys.* **37** 3962-4
[7] Belousova N; Veselov S; Kuzmin R and Melnikova E 2014 *Applied Mechanics & Materials* **698** 513-8
[8] Carneim R D and Messing G L 2001 *Powder Technol.* **115(2)** 131-8
[9] Li J, Liao H and Hermansson L 1996 *Biomaterials* **17(18)** 1787-90
[10] Liang X,Qiu Y, Zhou S, Hu X, Yu G, Deng X 2008 *Journal of University of Science and Technology Beijing* **15(6)** 764-8

IOP Conf. Series: Materials Science and Engineering **286** (2017) 012023 doi:10.1088/1757-899X/286/1/012023

Dislocation based multilevel model for elastic-plastic deformation of polycrystalline materials

E A Chechulina and P V Trusov

Department of Mathematical Modeling of Systems and Processes, Perm National Research Polytechnic University, 29 Komsomolsky prospekt, Perm, 614990, Russia

E-mail: zhenya-chechulina@yandex.ru

Abstract. The main aim of the work is to study the Portevin–Le Chatelier effect. The occurrence of the effect is closely connected to the phenomenon of dynamic strain ageing, i.e. the additional pinning of mobile dislocations by foreign atoms diffusing into the dislocation core during their arrest at obstacles, e.g. forest dislocation. The description of interaction of dislocation with impurities plays a great role, that`s why it is necessary to develop a several submodels: dislocation submodel for the interaction between dislocations and dislocation submodel for the interaction between dislocations and impurities. The multilevel approach based on using the internal variables, i.e. parameters describing the evolution of meso- and microstructure of the material, was applied to construct the model. The approach to model the dislocation structure is based on introduction of homogeneous dislocations densities on each slip system and obtaining evolutionary equations describing the mechanisms of their generation and interaction. The deformation mechanisms are evaluated and evolution of structural parameters during the process of deformation is analyzed. The dislocation submodel for the interaction between dislocations and impurities will be described in the next papers.

1. Introduction

The Portevin–Le Chatelier (PLC) effect manifests itself as an unstable plastic flow on the macrolevel. It is observed in a wide class of plastic materials under certain regimes of strain rate and temperature. The PLC effect is mostly studied through tensile testing either at constantly applied stress rate ('soft' loading) or at constantly applied strain rate ('hard' loading). The typical shapes of the stress–strain curves displayed in those two cases are, respectively, of staircase type, with bursts of plastic strain at almost constant stress, and serrated, with sudden stress drops at almost constant strain.

One of the main problems we have in the process of plastic deformation is establishing a necessary condition for loss of material stability and establishing the boundaries of the PLC effect range. A huge number of publications is devoted at the present time [1, 2] to studying of the general regularities and the character of the distribution of inhomogeneous plastic deformation at meso- and macro-scales, because the serrated yielding is detrimental and leads to loss of formability and ductility.

Numerous experiments of discontinuous plasticity have demonstrated that the plastic behavior of materials under constant loading depends on inhomogeneities properties of the material and the cooperative motion of large number of dislocations at different scale levels, from nano- to macroscale. Experimental studies of the internal structure evolution, dislocation substructures and point defects, which influence physical and mechanical processes of the materials, are very difficult and expensive. Therefore, in the last decades, there is an increasingly more need for constructing mathematical models of different scale levels.

The main mechanism responsible for the manifestation of the PLC effect is dynamic strain aging (DSA), it results from the interaction between solute atoms and mobile dislocation, whereas the mobile dislocation is temporarily arrested in the slip path by obstacles, e.g. forest dislocation, solute atom aggregations, etc.

The majority of researchers (A. Benallal, A. Bertram, T. Böhlke, L.P. Kubin, Y. Estrin, R. Larsson, S.-Y. Yang, D. Yu, X. Chen, S. Zhang and others) use macro phenomenological models of inelastic deformation to describe the effects associated with dynamic strain aging (DSA) [1]. The original formulation of the phenomenological viscoplastic model of strain aging proposed by Kubin and Estrin (1985) is based on an implicit mathematical description of homogeneous material exhibiting negative strain rate sensitivity [3]. Macro phenomenological models based on the macroscopic description of the deformation establish the relationships between the parameters at the macrolevel, but they don't consider the microstructure of the material.

The crystal plasticity based models (J. Alcala, H. Askari, R.A. Austin, D.L. McDowell, J. Balik, P. Lukac, M.S. Bharathi and others) allow to simultaneously analyze deformation processes at the macrolevel and to describe explicitly inelastic deformation mechanisms, material internal structure and its evolution at lower scale levels. A brief review of models, which are based on crystal plasticity, devoted to the description of the materials deformation, associated with the diffusion processes, in wide ranges of strain rates and temperatures is given in [2]. One of the important issues that arise in construction of this class of models: 'closing' set of relationships, a large number of parameters and their identification, is the problem of numerical implementation.

The article discusses physical mechanisms of serrated yielding. Plastic deformation is associated with a non-uniform motion of elementary carriers of plastic flow, i.e. dislocations (in time and space). Thus, for a correct description of plastic deformation, its heterogeneity and taking into account all its attendant real physical mechanisms, there is need to study the material behavior not only on the macrolevel, but on a lower meso- and microstructural levels, that is why a number of scientists attempted to construct dislocation dynamics models (A. Alankar, A. Arsenlis, K.M.Davoudi, L. Kiely, D. Li, H. Zbib, C. Reuber, M. Zecevic, M. Knezevic, M. Lebedkin and others).

This paper considers the multilevel mathematical model [4] based on the crystal plasticity theories, proposed to describe the Portevin–Le Chatelier effect. The description of interaction of dislocation with impurities plays a great role. The approach to model the dislocation structure is based on introduction of homogeneous dislocations densities on each slip system and obtaining evolutionary equations describing the mechanisms of their generation and interaction. The dislocation submodel for describing the mechanisms of the interaction between dislocations and impurities is in development stage and it isn`t considered in the present paper.

2. PLC constitutive model

Using the two-level models for describing the PLC effect isn't correct, because it is impossible to take into account the changes of the internal stresses of the material (the critical shear stress on slip systems) in the process of inelastic deformation caused by dynamic strain aging in the frame of two-level models. To describe the phenomenon of the Portevin–Le Chatelier effect we produced three-level model based on the crystal plasticity theories by applying dislocation submodel.

The generic structure of direct three-level model has been presented in [5]. In this model, the hierarchy of scale levels is defined as follows: macrolevel (representative volume of polycrystalline material) – mesolevel-1 (the level of the crystallite (grain, subgrain)), mesolevel-2 (grain, subgrain, fragment). Mesolevel-1 is used to describe the processes of hardening and of plastic deformation in terms of shear on the slip systems of a crystallite. Mesolevel-2 is applied to describe complex behaviour of the defect structure evolution and the evolution of dislocations densities. The general problem is facilitated by the definition of three subtasks: 1) determining the stress-strain state, 2) determining of the temperature, and 3) the diffusion. These subtasks can be realize using a dislocation submodel.

The system of relation of material constitutive model with internal variables at the macrolevel includes a set of three groups of equations (state (constitutive relations), evolutionary and closing):

$$\mathbf{\Sigma}^r = \mathbf{F}_r(\mathbf{P}_\alpha, \mathbf{J}^e_\gamma); \;\; \mathbf{J}^{ir}_\delta = \mathbf{R}_{r\delta}(\mathbf{P}_\alpha, \mathbf{J}^i_\beta); \;\; \mathbf{J}^{er}_\gamma = \mathbf{C}_{r\gamma}(\mathbf{P}_\alpha, \mathbf{J}^i_\delta), \tag{1}$$

where $\mathbf{\Sigma}, \mathbf{\Sigma}^r$ is a measure of the stress and its objective rate of change; $\mathbf{F}_r$, $\mathbf{R}_{r\delta}$, $\mathbf{C}_{r\gamma}$are tensor functions or simple (for example, differential) operators over tensor-valued arguments; $\mathbf{P}_\alpha$ - parameters of thermomechanical and non-thermomechanical nature; the superscript «r» represents a derivative, that does not depend on the coordinate system choice; $\mathbf{J}_\gamma^e$ are the explicit internal variables which enter the structure of the constitutive relations; $\mathbf{J}_\beta^i$ are the implicit internal variables, characterizing structure evolution and related to deeper scale levels [6].

Strain rate tensor is additively decomposed into elastic and plastic parts, the tensor is applied at the macrolevel and mesolevel-1. The strain rates calculation at nodes is performed using finite element method (FEM), the description of the evolution of dislocations densities, barriers and shear strain rate on the k-th slip system are considered at the mesolevel-2. Asymmetric measure of the deformed state and the modified Hooke's law are used at all levels [7].

The main mode of plastic deformation is the slip of edge dislocations. At the mesolevel-1, the response (stress) of the crystallite is determined from the velocity gradient, which is transmitted from the macrolevel. The velocities of dislocation movement on the slip system are transmitted from the mesolevel-2 to the mesolevel-1, the inelastic part of the velocity gradient and the components of the Cauchy stress tensor are established using Hooke's law and compute using the FEM.

3. Dislocation model (mesolevel-2)

The dislocations velocity may decrease due to their interaction with each other and with other defects in nonideal crystals. Dislocation motion can be hindered by dislocation interaction. The description of dislocation evolution takes into account the basic mechanisms: generation of dislocations by Frank-Read sources, annihilation of dislocations, and the formation of dislocation barriers (immobilization), detaching barriers by thermally activated mechanism or the respective critical shear stress (mobilization). The total dislocation density can be decomposed additively into the densities of mobile and immobile dislocations: $\rho_m^{(k)} = \rho_m^{(k)} + \rho_m^{(k)}$. It is proposed that, during the motion glide edge dislocations annihilate mutually with dislocations of opposite sign, the mobile and immobile dislocation densities can be divided into positive and negative parts depending on the location of the extra plane:

$$\begin{aligned} \rho_m^{(k)} &= \rho_{m+}^{(k)} + \rho_{m-}^{(k)}, \\ \rho_{im}^{(k)} &= \rho_{im+}^{(k)} + \rho_{im-}^{(k)}. \end{aligned} \quad (2)$$

In the reference configuration, the initial densities of positive and negative dislocations are supposed to be equal: $\rho_{m+}^{(k)}|_{t=0} = \rho_{m-}^{(k)}|_{t=0} = \dfrac{\rho_{m0}^{(k)}}{2}|_{t=0}$.

The positive (negative) mobile dislocations velocity depends on the temperature of the material under loading, actual and critical resolved shear stress on the slip system and is defined as follows [8]:

$$v_{m\pm}^{(k)} = v_0^{(k)} \exp\left(-\Delta G (\kappa\theta)^{-1} \left(1 - \left(\left|\tau^{(k)}\right| \left(\tau_{\text{crit}}^{(k)}\right)^{-1} \right)^p \right)^q \right), \quad (3)$$

where ΔG is the activation energy of dislocation movement, κ is Boltzmann constant, $\tau^{(k)}, \tau_c^{(k)}$ are the resolved shear stress and critical resolved shear stress, θ is the absolute temperature, p, q are dimensionless parameters. The critical resolved shear stress for positive and the negative dislocations in first-order match together.

The critical resolved shear stress depends on the total dislocation density and dislocation interactions on different slip systems:

$$\dot{\tau}_{\text{crit}}^{(i)} = |\mathbf{b}| \sum_{j=1,24} C_{ij} \left(\rho_m^{(i)} + \rho_{im}^{(i)} \right)^{1/2}, \quad (4)$$

where C_{ij} is the dislocation interaction matrix, $\mathbf{b}$ is the Burgers vector.

IOP Conf. Series: Materials Science and Engineering **286** (2017) 012023 doi:10.1088/1757-899X/286/1/012023

The equations describing the evolution of mobile and immobile dislocation densities are presented in the following way:

$$\begin{aligned}
&\rho_m^{(k)} = \rho_{m+}^{(k)} + \rho_{m-}^{(k)},\\
&\rho_{im}^{(k)} = \rho_{im+}^{(k)} + \rho_{im-}^{(k)}.\\
&\dot{\rho}_{m\pm}^{(k)} = \dot{\rho}_{gen\pm}^{(k)} + \dot{\rho}_{mob\pm}^{(k)} - \dot{\rho}_{imm\pm}^{(k)} - \dot{\rho}_{ann\,m\pm}^{(k)} - \dot{\rho}_{ann(m\pm im\mp)}^{(k)},\\
&\dot{\rho}_{im\pm}^{(k)} = \dot{\rho}_{imm\pm}^{(k)} - \dot{\rho}_{mob\pm}^{(k)} - \dot{\rho}_{ann(m\mp im\pm)}^{(k)}.
\end{aligned} \tag{5}$$

Here, the subscripts *gen*, *ann*, *imm* and *mob* refer to the generation dislocations, annihilation, trapping (locking by the obstacles: grain boundaries, forest dislocations, second-phase inclusions) and mobilization of dislocations, respectively.

In equation (5) the change rate of mobile dislocations density, both positive and negative, consists of 5 terms: dislocations generation rate in the grain by Frank-Read sources $\dot{\rho}_{gen\pm}^{(k)}$; an increase of the mobile dislocations density by detachment dislocations from barriers $\dot{\rho}_{mob\pm}^{(k)}$; a decrease of the mobile dislocations density by locking the obstacles (of dislocation origin or non-dislocation origin)on the slip systems $\dot{\rho}_{imm\pm}^{(k)}$; a decrease of the mobile dislocations density at dislocations annihilation of opposite signs $\dot{\rho}_{ann\,m\pm}^{(k)}$; a decrease of the mobile dislocations density at the annihilation of mobile and immobile dislocations of opposite signs $\dot{\rho}_{ann(m\pm im\mp)}^{(k)}$. An increase of the immobile dislocations density occurs due to inverse processes.

The generation rate depends on the density of active sources of dislocations and may be written as follows from [9]:

$$\dot{\rho}_{nuc}^{(k)} = \rho_s^{(k)} \frac{v^{(k)}}{h_{ann}}, \tag{6}$$

where $\rho_s^{(k)}$ is the density of active sources of dislocations; $h_{ann} \sim (\rho_m)^{-1/2}$ is the distance between newly formed dislocations, when the dislocations of opposite signs approach each other in a smaller distance, they annihilate.

During the process of plastic deformation, the dislocation density increases, and the number of active sources increases too. The authors [10] assume that the density of active sources is proportional to the density of dislocations. Then the generation rate of dislocations is taken in the following way:

$$\dot{\rho}_{gen}^{(k)} = \rho_s^{(k)} \frac{v^{(k)}}{h_{ann}} = v^{(k)} (\rho_m)^{3/2}. \tag{7}$$

During the motion, dislocations will tend to interact. When they are in the same plane on a critical distance h_{ann} ,they repel each other if they have the same sign, and they annihilate reciprocally if they have opposite signs (producing a perfect crystal). The rate of dislocation annihilation is taken in the following way [8]:

$$\begin{aligned}
&\dot{\rho}_{ann\,m+}^{(k)} = \dot{\rho}_{ann\,m-}^{(k)} = h_{ann} \rho_{m+}^{(k)} \rho_{m-}^{(k)} (v_{m+}^{(k)} + v_{m-}^{(k)}),\\
&\dot{\rho}_{ann(m\pm im\mp)}^{(k)} = h_{ann} \rho_{im\pm}^{(k)} \rho_{m\mp}^{(k)} v_{m\mp}^{(k)}.
\end{aligned} \tag{8}$$

The critical shear stress is determined by the current state of the defect densities on the slip system. The average dislocations velocities are determined by the shear stress, temperature, and defects density. The average dislocations velocities determined at the mesolevel-2 are used on the mesolevel-1, where shear strain rate is set according to the Orovan's equation.

4. Results

Numerical experiments have been carried out to determine the change in the density of dislocations, changes in critical stresses on slip systems over time under uniaxial loadings along the [100] direction.

Under such loading there are eight equal active slip systems, critical stresses on these slip systems changes in the same way.

Figure 1 shows the dependence diagram of critical shear stress – inelastic strain with under a constant number of dislocation sources and in the case when the sources density increases.

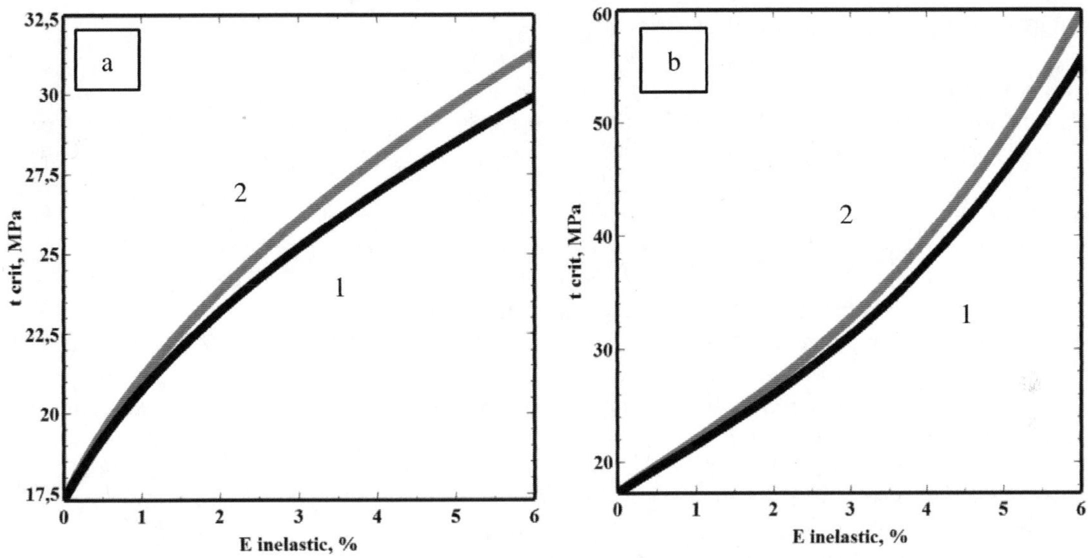

Figure 1. The critical shear stress – inelastic strain on active (1) and inactive (2) slip system with: a) constant number of dislocation sources, b) the increase of sources density.

Figure 2 shows the dependence diagram of critical shear stress – mobile dislocations density under a constant number of dislocation sources and in the case when the sources density increases.

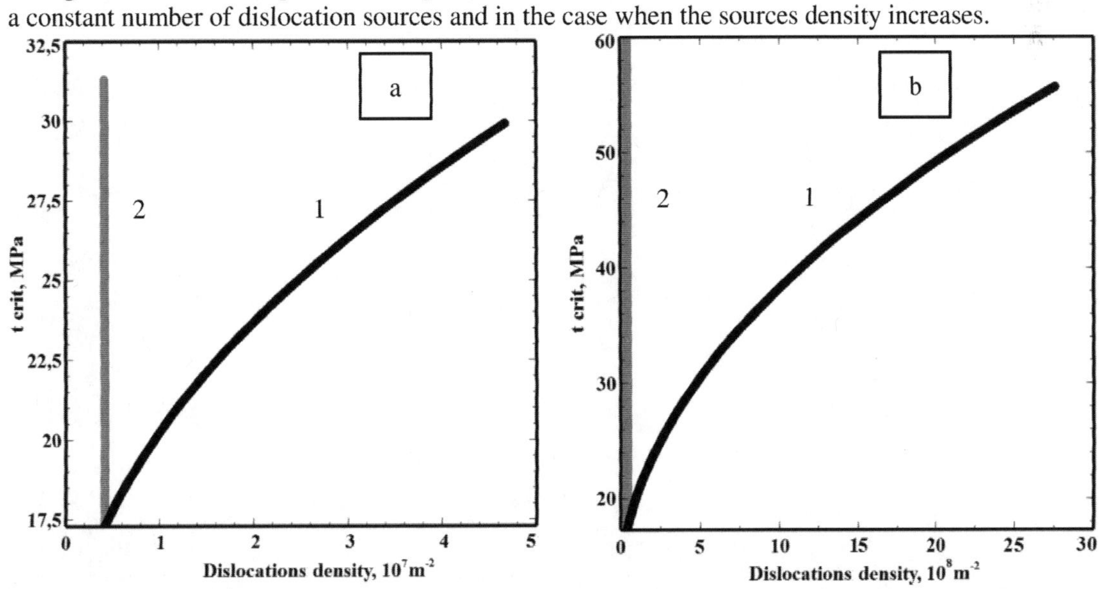

Figure 2. The critical shear stress – mobile dislocations density on active (1) and inactive (2) slip system with: a) constant number of dislocation sources, b) the increase of sources density.

5. Conclusion

In this paper, we consider multilevel constitutive model to describe the serrated yielding, taking into account the dislocation submodel. Only two first terms in the evolution equations for the dislocation densities are considered. The numerical experiments have been carried out to determine the change rate of mobile dislocations density and the critical shear stress on slip system over time. The dependence diagrams of critical shear stress – mobile dislocations density, critical shear stress – inelastic strain on active and inactive slip system under uniaxial loadings are obtained. The results of numerical experiments of the developed model graphically demonstrated some features of the deformation behaviour and the evolution of the dislocation structure.

Acknowledgments
The work was supported by the Russian Federation Ministry of Education and Science (the basic part of PNRPU state order, state. register. No. 9.7434.2017/8.9.), RFBR (grants Nos. 16-31-00215 mol_a).

References

[1] Trusov P V and Chechulina E A 2014 *PNRPU Mechanics Bulletin* **3** 186-232
[2] Trusov P V and Chechulina E A 2017 *PNRPU Mechanics Bulletin* **1** 134-163
[3] Kubin L P and Estrin Y 1990 *Acta metall. mater* **38** 697–708
[4] Trusov P V, Shveykin A I, Nechaeva E S and Volegov P S 2012 *Physical Mesomechanics* **15 3–4** 33–56
[5] Chechulina E A 2016 *Bulletin of Tambov University. Series of Natural and Technical Sciences. Physics* **21** no 3 pp 1419-22
[6] Trusov P V and Shveykin A I 2011 *Perm: Publ. of Perm State Technical University* p 419
[7] Trusov P V, Nechaeva E S and Shveykin A I 2013 *Physical Mesomechanics* **16 2** 15-31
[8] Alankar A, Field D P and Zbib H M 2012 *Philosophical Magazine* **92 24** 3084–3100
[9] Davoudi K M and Vlassak J J 2014 *Materials Science*pp 1-23
[10] Kocks U F, Argon A S and Ashby M F 1975 *Progress in materials science* (*Pergamon Press)* **19** p 291

Effect of copper on the structure and antifriction properties of cast hypoeutectoid steel

N Stepanova, T Zimogliadova, A Ognev, D Krivizhenko, Yu Maliutina and O Zimogliadova

Novosibirsk State Technical University, 20, K. Marks ave., Novosibirsk, 630073, Russia

E-mail: stepanova@corp.nstu.ru

Abstract. The structure, mechanical and tribotechnical properties of medium-carbon steel doped with 0...9 wt. % of copper with the 1% increment were investigated. During this study, the structural changes in the alloy with an increase of the copper amount were analyzed. Increase of the copper amount in the structure of medium-carbon steel leads to the gradually refinement of the ferrite grains and perlite components. In the samples doped with 9 wt. % of copper the tertiary cementite was found at the boundaries between ferrite grains and the colonies of lamellar perlite. The ε-Cu phase was revealed using X-ray diffraction of the cast steel with ~ 3 wt. % of Cu. Further increase of the copper amount promotes the increase of ε-Cu volume fraction. The change in hardness and wear resistance with an increase in the copper amount in medium-carbon steels was detected. The samples doped with ~ 6 wt. % of copper showed the maximum wear resistance at the non-fixed particles condition that corresponds to the hardness of material 240 HB.

1. Introduction

In some cases, copper-doped steel is used as tribotechnical material and is considered as an alternative material to expensive bronzes [1, 2]. The replacement of bronze on steel is the most economically rational decision at the manufacture of parts of large-sized heavy-loaded sliding friction units. The operating conditions of these units assume the presence of high specific loads. Thus, the materials for these applications must exhibit a high complex of strength and tribological properties.

Previously, the studies of such iron-carbon alloys as cast iron or hypereutectoidal steel have been done [3, 4]. The effect of copper on the process of graphitization, as well as on the tribotechnical properties of cast irons, was described in [5]. When carrying out structural studies of alloyed cast iron by light metallography and transmission electron microscopy, a variety of shapes and sizes of the ε-phase particles was detected. Classification of ε-Cu particles by size and also the determination of the conditions for the formation of this phase were carried out in previous studies while the investigation of the cast iron doped with copper [3-5]. It is obvious that the influence of these particles makes a significant contribution to the complex of strength and tribological properties of the material. The strength characteristics of medium-carbon steels are significantly higher than those of cast irons. For this reason, the use of steels alloyed with copper for the manufacture of elements of sliding friction units is more promising than the use of cast irons. However, the influence of copper on the tribotechnical properties of steels is poorly represented in the literature. The purpose of this study is to investigate the special aspects of influencing copper on the structure and tribotechnical properties of steels.

2. Materials and methods

The series of castings based on steel 45 (containing 0.45 wt. % of carbon) with a copper additions of 0…9 wt. % with the increment of 1% were made. Steel 45 and electrotechnical copper were used as a charge. The charge was melted in an induction furnace with acidic lining. The capacity of the furnace crucible was 40 kg. The weight of the casting of each chemical composition was 30 kg. Thus, 10 castings were performed. Elemental analysis of the steels obtained during the experiments was performed on the ARL-3460 optical-emission spectrometer. The results of the analysis are presented in Table 1.

Table 1. Element composition of copper-alloyed steel.

№ of sample	C	Mn	Si	P	S	Ni	Cr	Cu	Al	Fe
0	0.58	0.39	0.84	0.02	0.02	0.02	0.02	0.03	0.16	bal.
1	0.5	0.31	1.44	0.01	0.02	0.02	0.02	1.08	0.15	bal.
2	0.38	0.19	0.36	0.01	0.02	0.02	0.02	2.05	0.15	bal.
3	0.39	0.25	0.73	0.02	0.02	0.02	0.02	3.11	0.16	bal.
4	0.46	0.3	1.05	0.02	0.01	0.02	0.01	4.37	0.09	bal.
5	0.38	0.26	0.98	0.02	0.02	0.02	0.02	5.36	0.06	bal.
6	0.51	0.3	0.35	0.02	0.02	0.02	0.02	6.06	0.11	bal.
7	0.38	0.2	0.3	0.02	0.02	0.02	0.01	7.26	0.1	bal.
8	0.39	0.17	0.5	0.02	0.02	0.02	0.01	7.88	0.07	bal.
9	0.38	0.21	0.39	0.02	0.02	0.02	0.02	8.7	0.07	bal.

Metallographic studies of the materials were carried out on a Carl Zeiss microscope Axio Observer Z1m. Structural studies at elevated magnifications were performed on a FEI Tecnai G2 20 TWIN transmission electron microscope. Phase analysis of the obtained alloys was studied using X-ray diffractometer ARL X'TRA. The source of X-ray radiation was a copper tube (voltage 40 kV, current 40 mA). The analysis of the materials was performed in the geometry of the reflected radiation. Monochromatization of the primary and reflected radiation of the source was not applied. The average value of the beam wavelength λ was 0.15406 nm. For the X-ray investigations the energy-dispersed Si(Li) detector was used.

The diffraction patterns were recorded in a step mode $\Delta 2\theta = 0.02$ and 0.05°, dwell time was 12 s. A pinhole spot emitter with diameter 1.5 mm was used. The hardness of the obtained castings was measured by the Brinell method in accordance with Russian standard 9012, load on a steel ball with a diameter of 10 mm was 3000 kg. Tests on the wear resistance of quenched and low-tempered steels under friction conditions with non-fixed abrasive particles were carried out in accordance with Russian standard 23.208-79. The friction path in the wear process was 942 m. River sand with a grain size of ~ 200 μm was used as an abrasive material. The rotational speed of the roller was 60 rpm. The pressing force of the roller was 44 N, cast steel 45 was used as the standard for determining the level of wear resistance.

3. Results and discussion

Non-doped steel 45 in the cast state had a ferrite-pearlite structure. The amount of ferrite was ~ 30 %. The average grain size of ferrite was ~ 50 μm.

With an increase in the copper amount the size of structural components decreases. The average size of ferritic grains in the alloy containing 5 wt. % copper was ~ 30 μm. With a further increase of copper amount to 9 wt. % the average size of the ferritic grains doesn't decrease (Figure 1). The volume fraction of ferrite doesn't change with increasing copper amount.

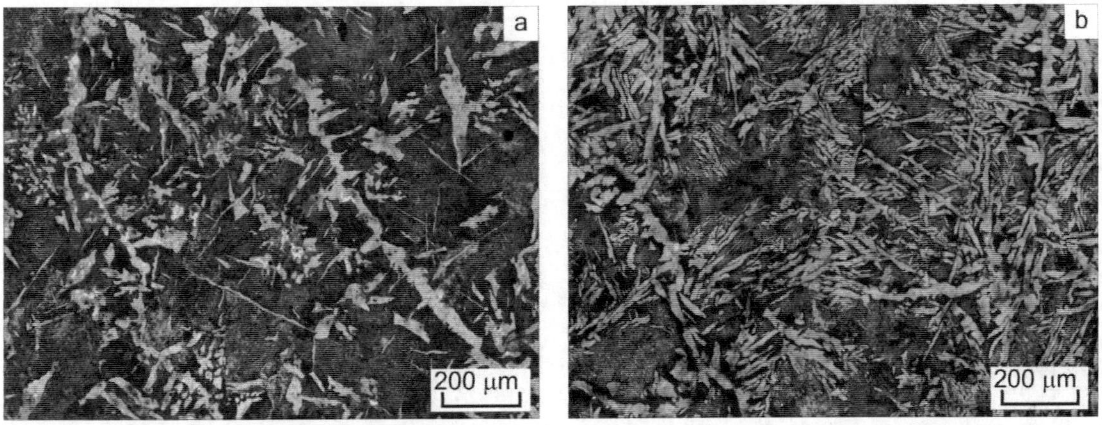

Figure 1. Effect of copper on the size of ferritic grains in steel 45 (a – 0.03 % Cu, b – 8.7 % Cu).

An increase in the copper amount is accompanied by an increase of the degree of dispersion of perlite components. The interlaminate distance becomes smaller, the thickness of the cementite plates decreases. The reason for this is probably due to a change in the rate of the alloys cooling. In its turn, the increase of the cooling rate is associated with a change of the thermal conductivity with an increase of copper content in the alloy [6]. This is due to the fact that the thermal conductivity of copper is 5 times higher than the thermal conductivity of iron [7]. Another confirmation of this statement is the rim of the tertiary cementite which located at the interfaces of the ferrite grains and the lamellar perlite colonies in the samples with 9 wt. % of copper (4 in Figure 2a). The precipitation of this structural component is due to a decrease of the solubility of carbon in α-iron at the cooling stage of castings below 727 ° C.

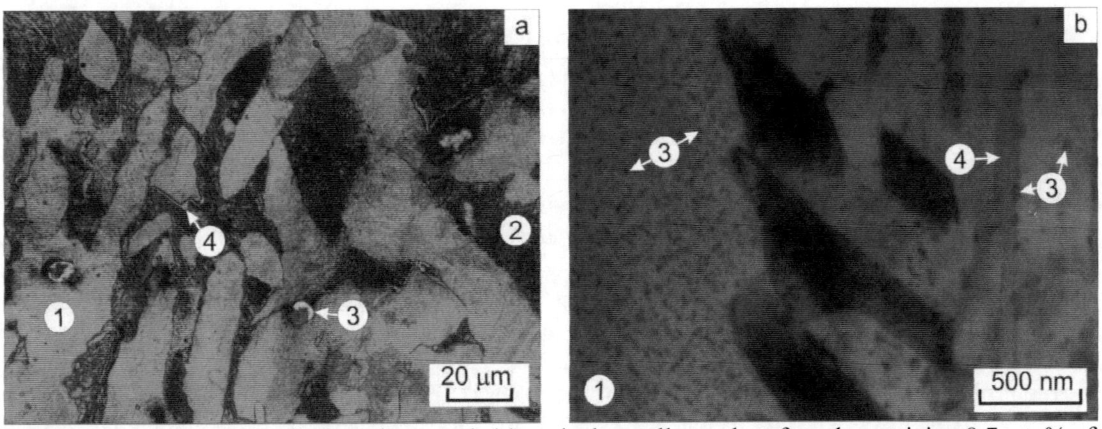

Figure 2. Structure of medium-carbon steel, (a) optical metallography of steel containing 8.7 wt. % of copper, (b) TEM of steel, containing 3.1 wt. % of copper (1 – ferrite, 2 – perlite, 3 – ε-Cu, 4 – Fe_3C).

With the method of transmition electron microsccopy (TEM) ε-Cu particles were revealed inside the ferritic grains (on the left in Figure 2b) and inside ferritic interlayers of perlite (on the right in Figure 2b). Some ε-Cu particles are located on the surface of the cementite plates.

Copper-based inclusions were observed metallographically in alloys containing not less then 4 wt. % of copper. In castings with 5 wt. % of copper inclusions have a size of 2.5...3 μm and are predominantly compact. In steel with 9 wt. % of copper the size of the copper-containing particles reaches 8 ... 10 μm (Figure 3). In the alloys with 7 ... 9 wt. % of copper the part of inclusions are located on the grain boundaries and has an elongated shape.

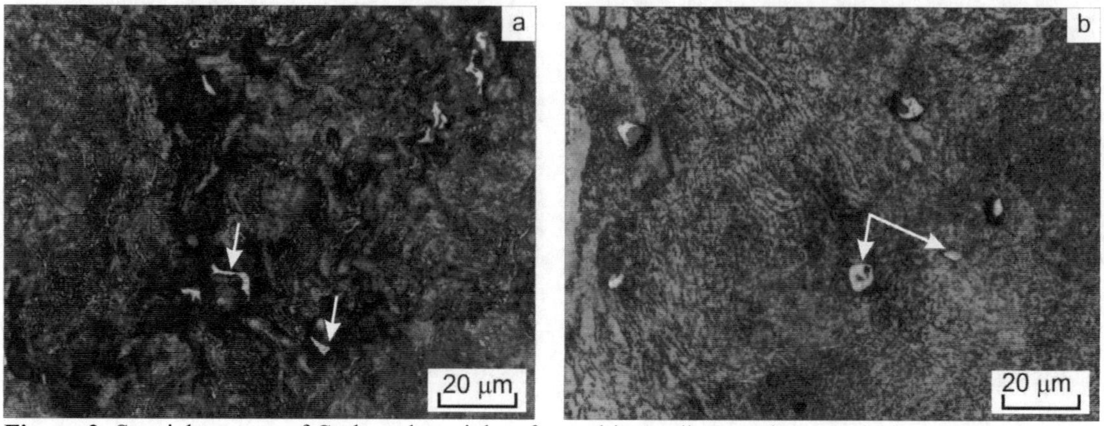

Figure 3. Special aspects of Cu-based particles, formed in medium-carbon steels (a – 7.88 % Cu, b – 8.7 % Cu).

The X-ray diffraction patterns of cast steel 45 with different amount of copper are shown in Figure 4. X-ray diffraction patterns of non-alloyed steel 45 peaks corresponding to α-iron were detected. A complex analysis of X-ray diffraction and microstructural studies allows to conclude that fixed peaks correspond to ferrite phase. On the pattern corresponding to steel with 3 % copper both the peaks of α-iron and the peaks of ε-Cu were found. X-ray diffraction patterns corresponding to samples with 6 and 9 wt. % of copper show an increase of ε-Cu peaks the intensity.

The effect of copper addition on the hardness of medium-carbon steel castings is illustrated in Figure 5. An increase of the copper amount to 6 wt. % promotes raise the hardness level of cast non-heat treated steels from 180 to 230 HB. Further growth of copper amount to 9 wt. % is accompanied by a decline of the hardness level to 200...210 HB (Figure 5).

The dependence of the mass loss of cast steels on the copper amount is shown in Figure 5. The mass loss of cast alloys containing 3.11, 6.06 and 8.7 wt. % of copper is approximately the same. In the castings with copper amount of 6.06 wt. %, the level of wear resistance reaches a maximum value, which is associated with a significant growth of the hardness level of material.

4. Conclusions

An increase of the copper amount from 0 to ~ 5 wt. % in the structure of medium-carbon steel is accompanied by a gradual decrease in the size of the ferritic grain from ~ 50 μm to ~ 30 μm. The change in the volume fraction of ferrite was not found. The degree of dispersion of perlite increases with increasing copper amount. The structure of the medium-carbon steel alloy with 9 wt. % of copper addition has a rim of tertiary cementite at the interfaces of ferrite grains and lamellar perlite colonies.

The X-ray patterns of the cast steel with ~ 3 wt. % of copper have peaks belonging to ε-Cu phase. With a further increase of the copper content the volume fraction of ε-Cu increases. Using TEM methods ε-Cu particles with an average size of 20 nm were revealed in ferritic grains and ferritic perlite layers in the samples with copper addition not more than ~ 1 wt. %.

The hardness of the material increases from 180 to 240 HB with an increase in the copper amount to ~ 6 wt. %, further increase of copper to ~ 9 wt. % leads to the declining of hardness level to 225 HB. The wear resistance of the material in a friction condition with non-fixed abrasive particles varies in full accordance with the change of the hardness level.

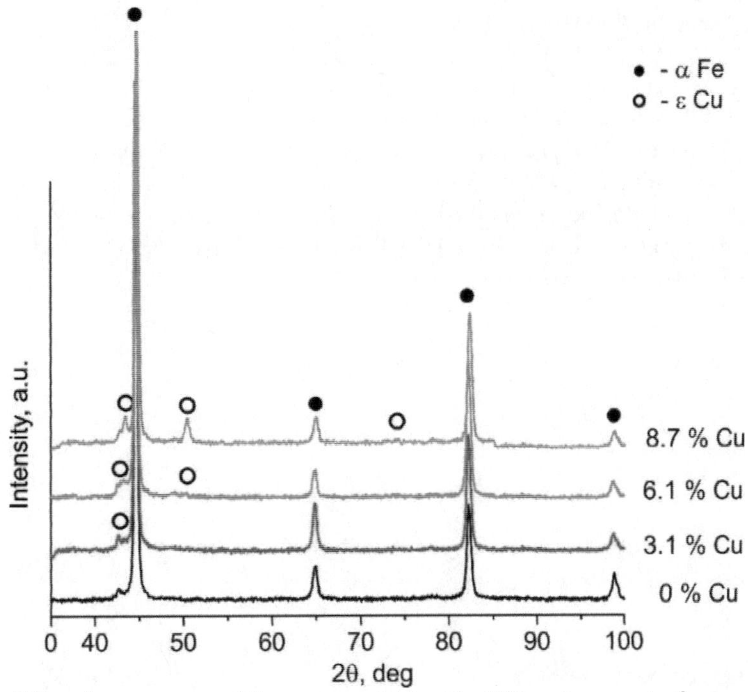

Figure 4. X-ray diffraction patterns of the cast steel 45 with different amount of copper.

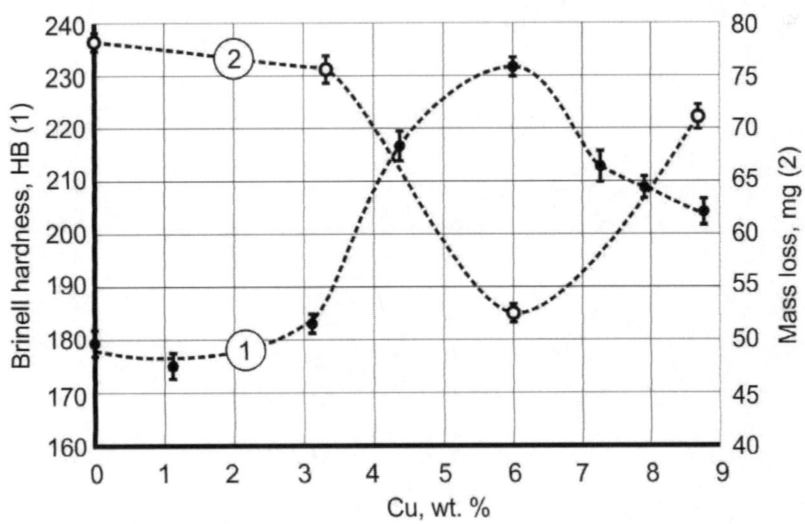

Figure 5. Dependence of hardness on the copper amount in the castings of medium-carbon steels (1) Effect of copper on weight loss of steels during wear testing in a condition of non-fixed abrasive particles (2).

Acknowledgments
This study was supported by Russian Science Foundation (project No. 15-19-00230).

References

[1] Sil'man G I, Kaminin V V and Goncharov V V 2007 *Met. Sci. Heat Treat.* **9** 16-22

IOP Conf. Series: Materials Science and Engineering **286** (2017) 012024 doi:10.1088/1757-899X/286/1/012024

[2] Golovin E D, Kuznetsov V A and Kumar V 2012 *Obrabotka metallov (Metal Working and Material Science) (in Russian)* **54** 81-4
[3] Stepanova N V, Bataev A A, Sitnikov A A and Oskolkova T N 2015 *Obrabotka metallov (Metal Working and Material Science) (in Russian)* **69** 72-9
[4] Sil'man G I 2009 *Met. Sci. Heat Treat.* 51 19-24
[5] Bataev I A, Stepanova N V, Bataev A A, Nikulina A A and Razumakov A A 2016 *Physics of Metals and Metallography* **117** 901-5
[6] Yakovlev A Yu and Volchok I P 2008 *Met. Sci. Heat Treat.* **1** 44-6
[7] Kutateladze S S and Borishanskii V M 1958 *Reference book of Heat Conductance* (Moskow: State Power Publishers) p 416

IOP Conf. Series: Materials Science and Engineering **286** (2017) 012025 doi:10.1088/1757-899X/286/1/012025

Structure and Properties of Coatings Formed by Detonation Spraying of Titanium Powder

S V Panin[1,2,3], I V Vlasov[2,3], R V Stankevich[3], D V Dudina[4,5], V Yu Ulyanitsky[4] and I S Batraev[4]

[1]Jilin University, 4026 Yatai St, Nanguan Qu, Changchun Shi, 130022, China
[2]Institute of Strength Physics and Materials Science SB RAS, 2/4, pr. Akademicheskii, Tomsk, 634055, Russia
[3]National Research Tomsk Polytechnic University, 30, Lenin Avenue, Tomsk, 634050, Russia
[4]Lavrentyev Institute of Hydrodynamics SB RAS, 15, pr. ac. Lavrentyeva, Novosibirsk, 630090, Russia
[5]Novosibirsk State Technical University, 20, pr. K. Marksa, Novosibirsk, 630073, Russia

E-mail: svp@ispms.tsc.ru

Abstract. Structure and properties of the coatings formed by detonation spraying of titanium powder have been studied at varying the deposition parameters (nature of carrier gas, spraying distance, O_2/C_2H_2 ratio, and the volume of explosive mixture). It is shown that when air is used as the carrier gas the primary strengthening by the titanium oxide phases is realized. When nitrogen is introduced into the explosive mixture, a more complicated mechanism of hardening is realized due to the formation of titanium oxides, carbides, oxynitrides and carbonitrides. The aspects of employing the revealed "rational" modes of the detonation spraying for formation of protective composite coatings based on titanium possessing a complex of improved physical and mechanical properties are discussed.

1. Introduction

At present, several coating deposition methods are known and widely used in the industry. One of the efficient technologies for restoration, repair and strengthening of machine parts is cold gas-dynamic spraying. A key problem solved at thermal spraying is formation of protective coatings containing (new) high-strength phases [1], as well as deposition of multicomponent coatings with a uniform structure reinforced with finely dispersed inclusions (precipitations) including nanosized ones that ensures high performance properties [2, 3].

Currently, detonation spraying and high-velocity oxy-fuel spraying (HVOF) are the most efficient methods of depositing coatings that can protect machine parts from erosion, corrosion and different types of wear [4]. It advantages over the others by realizing dispersed hardening due to formation of ceramic and cermet phases which are synthesized under high temperatures during the detonation. At the same time, low porosity of the coatings is provided due to the high rate of spraying [5]. The method also might be applied for periodic restoration of worn out parts of machines [6].

High efficiency of coatings formed by detonation spraying with the use of titanium powder is worth mentioning. High temperatures and pressures to take place under the titanium deposition process give rise to formation of high-strength intermetallic, carbide and ceramic compounds. Titanium and its alloys based coatings possess high wear and corrosion resistance [7] which promotes their use in medical applications [8] as well as in aircraft industry [3].

In this paper the structure and mechanical properties of layers (coatings) formed on titanium substrates by detonation spraying of titanium powder were investigated. It is of scientific and practical interest to study the structure and mechanical properties of sprayed coatings as well as the effect of various detonation spraying parameters on the strength properties of the formed layers. Previously, the structure of such coatings has been studied in [5]. The aim of the present work is to assess the effect of the formed structure on the mechanical micro- and macroscale properties of the coatings.

2. Materials and Methods

The deposition of composite coatings onto titanium substrates was carried out using a computer controlled detonation spraying (CCDS2000) facility (table 1). Due to the interaction of titanium with the components of the spraying atmosphere, titanium oxides, nitrides, carbides, oxynitrides and carbonitrides phases were formed. Titanium (99 % purity, average particle size 15 μm, referred to as "PTOM-2", Russia) was used as a feedstock powder. Parameters of the detonation spraying for the formation of coatings are given in table 1. Microhardness of the coatings was measured on the polished surface parallel to the coating/substrate interface. Local microanalysis of the coatings was carried out by energy dispersive spectroscopy capacity INCA (Oxford, Instruments), equipped to the scanning electron microscope LEO EVO 50 (Zeiss, Germany). To establish the correlation between coating structure and mechanical properties the three-point bending tests were carried out. To this end, an electromechanical testing machine Instron 5582 was utilized. This loading scheme is motivated by the fact that under applying the external force in this way a surface layer (in this case – deposited coating) plays a major role in ensuring resistance to deformation and failure. Thus, the three-point bending testing in combination with photographing the specimen lateral surface allows one to gain more complete information on the deformation behavior as well as to estimate the cracking stress σ_{cr} [9]. The stress under the 3-point bending was calculated with the help of the formula (1) [10]:

$$\sigma_{cr} = \frac{M_{max}}{W} \quad (1)$$

where $M_{\max}$ – largest value of bending moment, W – the resistance momentum for rectangular-section specimens. The shear stress was calculated by the formula (2) [11]:

$$\tau = \frac{h_c \times \sigma_{delam}}{L_c} \quad (2)$$

where L_c – the distance between the cracks in the coating, σ_{delam} – the stress at which the coating starts to delaminate (delamination stress), h_c – coating thickness. The interface fracture toughness ("coating shear resistance") was determined by the formula (3) [11]:

$$K_c = \tau \times (\pi \times L_c)^2 \quad (3)$$

Specimens for mechanical testing were cut by electro-erosion machine to have the size 20×10×2.5 mm. The span (distance between the supports) was made 19 mm. Since the titanium substrate possesses high ductility, it was nearly impossible to fail the specimen under the used loading scheme. For this reason, the evaluation of mechanical properties was carried out at the bending deflection of 2.5 mm.

3. Results and discussion

For the specimen's No. 1 and 2 the spraying distance was 100 mm and filling of the barrel was 30 %. They differed by the explosive mixture ratio O_2/C_2H_2 only (table 1). Analysis of micrographs (Figure 3a,b) shows that the coating structure represents a set of densely "packed" splats that is typical for thermal spraying techniques.

This ensures low porosity and high microhardness – at a level not exceeding 3.7 MPa (table 1). The data of scanning electron microscopy, as well as the microanalysis for the coating No. 1, allows revealing two main phases being particles of titanium and its oxides. Coating No. 2 has a similar structure; however, a larger number of splat-shaped inclusions of smaller sizes are observed.

Table 1. Parameters of spraying and mechanical properties of the specimens.

Specimen No.	Composition of explosive mixture, O_2/C_2H_2	Spraying distance (mm)	Explosive charge (%)	Carrier gas	H_μ (GPa)	σ_{cr} (MPa)	K_c (MPa·m$^{1/2}$)
1	1.1	100	30	air	3.70±0.03	98.4	10.4
2	2.5	100	30	air	3.84±0.07	140.5	12.7
3	1.1	10	40	nitrogen	2.68±0.07	87.9	4.8
4	0.7	10	40	nitrogen	2.45±0.11	44.9	0.9
5	0.7	10	50	nitrogen	2.72±0.09	94.2	7.6
6	0.7	100	50	nitrogen	2.02±0.01	10	-
7	1.1+33% N_2	10	60	nitrogen	2.74±0.09	372.1	11.6
8	1.1+33% N_2	100	60	nitrogen	2.29±0.02	33.5	6.8

The microanalysis data have shown the oxygen content (specimen No. 2) in the oxide rich regions make up to 45 %, which is by ~ 10 % higher in contrast with the specimen No. 1. The key mechanical characteristics of coating No. 2 are improved relative to those of coating No. 1: the fracture initiation stress is increased by 43 % and interface fracture toughness – by 22 %, (table 1). According to the microstructure data these changes should be associated with the formation of a high strength titanium oxide based matrix.

For the specimen's No. 3 and 4 reduction of oxygen in the explosive mixture resulting in decreasing of the content of the oxidizing component in the spraying atmosphere should primarily be accompanied by diminishing the amount of oxides (Figure 1,c,d). On the other hand, the use of nitrogen as the carrier gas can be accompanied by the formation of nitride phases. The structure of the coating No. 4 is denser while its porosity is much lower (in contrast with specimen No. 3). The higher content of C_2H_2 in the explosive mixture has led to the increase in the carbon content in the coating as is evidenced from the microanalysis data (table 1).

For specimens No. 5 and No. 6, the spraying distance was 10 and 100 mm, respectively. The coatings of both types have a homogeneous fine-dispersed structure: titanium particles are surrounded by nitride, carbonitride and titanium carbide matrix (Figure 1,e,f). The increase in the deposition distance by a factor of 10 promoted the increase in the reaction time in the detonation jet and resulted in the formation of a larger number of ceramic compounds of titanium, nitrogen, and carbon. From comparison of the mechanical properties of the specimens, the following conclusions can be drawn. The thickness of the coating in specimen No. 6 was 4 times greater than that in the rest of the specimens), which significantly affected the crack resistance measured during bending and scratching (σ_{cr} and K_c). Nevertheless, the bending cracking resistance of the specimen No. 5 is higher than that of specimen No. 6. The reason might be attributed to more a uniform structure of the specimen No. 5. At the same time, the bending delamination stress of coating No. 5 (table 1) is noticeably higher than the values observed in coatings No. 3 and No. 4.

For the specimen's No. 7 and 8 the nitrogen was added to the O2/C2H2 mixture (at the ratio of 1:1) in an amount of 33 % by volume (Figure 1,g,h). It was assumed that introduction of the nitrogen into the explosive mixture would favor formation of a greater amount of nitride/carbonitride phases. It is revealed that the fracture initiation stress σcr increases by a factor of ~ 10 for the specimen No. 7 in comparison with No. 8. At the same time the interface fracture toughness Kc is increased by 40 % (table 1). A shorter deposition distance has led to increasing the oxygen content that results in the formation of larger amount of the oxide phases. The introduction of nitrogen into the explosive mixture has increased the amount of titanium nitrides and carbonitrides. The latter should be responsible for improving mechanical properties. Generally speaking, the specimen No. 7 possesses the maximum cracking resistance among all the studied compositions.

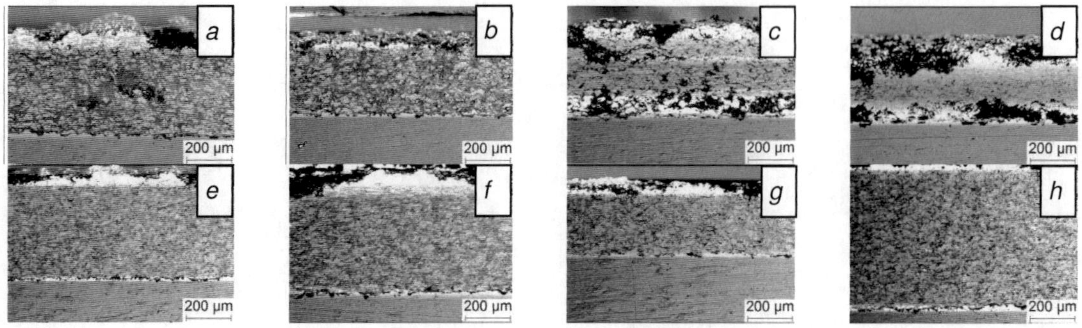

Figure 1. SEM-micrographs of the cross section for the specimens No. 1 (a); No. 2 (b); No. 3 (c); No. 4 (d); No. 5 (e); No. 6 (f); No. 7 (g); No. 8 (h).

4. Conclusion

The structure and mechanical properties of the coatings deposited by reactive detonation spraying of a titanium powder were studied. The composition of the individual phases of coatings as well as the effect of spraying regimes on the mechanical properties of the formed coatings were investigated. For the three characteristic modes of the coating formation, the following recommendations were formulated:

1. In case of the spraying with air as a carrier gas, no nitrogen added into the explosive mixture, an increase in the O_2/C_2H_2 from 1.1 and 2.5 leads to an increase in the cracking resistance under 3-point bending due to formation of a hierarchically organized coating structure with clearly exhibited phase boundaries. Deposition of titanium at O_2/C_2H_2=2.5 allows obtaining coatings with a high fracture toughness and a high microhardness;
2. In the event of spraying with nitrogen as a carrier gas, no nitrogen added into the explosive mixture, a short spraying distance (10 mm) and a low oxygen content in the explosive mixture (O_2/C_2H_2 = 0.7) makes it possible to form a heterogeneous structure of the coating. This ensures mechanical properties comparable to those of the coatings, in which the oxide phases were predominantly formed. The coatings show a moderate crack resistance and a high microhardness (H_{100} = 2.72 GPa);
3. In case of the spraying with nitrogen as a carrier gas and nitrogen added to the explosive mixture, the formation of a complex heterogeneous structure makes it possible to achieve a high crack resistance under 3-point bending, a high interface fracture toughness under the scratch test and a high microhardness (H_{100} = 2.74 GPa).

Acknowledgments

The work was partially supported by Cheung Kong Scholar of Jilin University, and performed with a partial support by Grant No. 11.2 of the Russian Academy of Sciences (Department of Power Engineering, Mechanical Engineering, Mechanics, and Control Processes). The authors are grateful to "Nanotech" Shared Use Center of ISPMS SB RAS for the assistance in running fractographic investigations. The experimental calculations are carried out at Tomsk Polytechnic University within the framework of Tomsk Polytechnic University Competitiveness Enhancement Program grant.

References

[1] Deevi S and Sikka V 1997 *Journal of Materials Science* **32** 3315-3325
[2] Fauchais P, Montavon G, Lima R and Marple B 2011 *J. Phys. D: Appl. Phys.* **44** 093001
[3] Kovaleva M, Tyurin Yu, Kolisnichenko O, Prozorova M and Arseenko M 2013 *Journal of Thermal Spray Technology* **22(4)** 518–524
[4] Smurov I and Ulianitsky V 2011 *Surface Engineering* **27(8)** 557-559
[5] Ulianitsky V, Dudina D, Batraev I, Kovalenko A, et. al. B 2015 *Surf. Coat. Tech.* **261** 174–180

IOP Conf. Series: Materials Science and Engineering **286** (2017) 012025 doi:10.1088/1757-899X/286/1/012025

[6] Senderowski C, Chodala M and Bojar Z 2015 *Materials* **8(3)** 1108-1123
[7] Zhu J, Huang J, Wang H, et. al. 2008 *Materials Letters* **62** 2009–2012
[8] Kawakita J, Kuroda S, Fukushima T, et. al. 2006 *Surf. Coat. Tech.* **201** 1250–1255
[9] Isakov M, Matikainen V, Koivuluoto H and May M 2017 *Surf. Coat. Tech.* **324** 264–280
[10] Gere J and Goodno B 2012 *Mechanics of Materials* (Canada: Cengage Learning) p 1045
[11] Zhou Y, Tonomori T, Yoshida A, et. al. 2002 *Surf. Coat. Tech.* **157** 118–127

Multilevel model of polycrystalline materials: grain boundary sliding description

E Sharifullina, A Shveykin and P Trusov

Department of Mathematical Modeling of Systems and Processes, Perm National Research Polytechnic University, 29 Komsomolsky prospekt, Perm, 614990, Russia

E-mail: elvira16_90@mail.ru

Abstract. Material behavior description in a wide range of thermomechanical effects is one of the topical areas in mathematical modeling. Inclusion of grain boundary sliding as an important mechanism of polycrystalline material deformation at elevated temperatures and predominant deformation mechanism of metals and alloys in structural superplasticity allows to simulate various deformation regimes and their transitions (including superplasticity regime with switch-on and switch-off regimes). The paper is devoted to description of grain boundary sliding in structure of two-level model, based on crystal plasticity, and relations for determination the contribution of this mechanism to inelastic deformation. Some results are presented concerning computational experiments of polycrystalline representative volume deformation using developed model.

1. Introduction

There are different approaches within the framework of multilevel and multiscale modeling of material inelastic deformation, based on crystal plasticity, in relation formulation for polycrystal inelastic behavior in the case of large deformation gradients. These models have two main aspects [1]: crystallite behavior description (crystal plasticity model is the basis of this level) and description of transition from crystallite level to macroscopic level. The last aspect is one of the important tasks, which solution could be achieved by different ways.

The most popular Taylor – Bishop – Hill models [2] (or Full Constraints (FC) models) are classical statistical models, based on relatively independent consideration mesolevel elements (crystallites, grains, subgrains); integration of mesolevel elements into macrolevel element (representative macrovolume or representative volume of macrolevel) is carried out by using the hypothesis about plastic strain homogeneity (plastic strain rate homogeneity) assumed a priori in all crystallites. It is noted in [1], that more common assumption – generalization of Voigt (Taylor) hypothesis about equality the velocity gradient in each crystallite of polycrystal to macroscopic velocity gradient ($\hat{\nabla}\mathbf{v}^{\mathrm{T}} = \hat{\nabla}\mathbf{V}^{\mathrm{T}}$) could be performed. In a number of papers it is proposed Relaxed Constraints (RC) models, which don't strictly enforce condition about velocity gradient equality and «relax» it for some components. In these models the internal structure of a polycrystalline material is modeled as lamellar structure consisting of elongated grains. The assumption about homogeneity of velocity gradient is used not for each crystallite but for stacks of two crystallites, other relations are formulated for these stacks too. Different models (Pancake, Lath, Lamel and ALamel models) exist depending on hypothesis about possible directions of relative crystallite displacements along a common boundary, shears are defined using these displacements, which act as additional ones with respect to shears from Taylor model. Other type of models are viscoplastic self-consistent (VPSC) [3, 4, 5], which consider crystallites in a representative macrovolume as inhomogeneous inclusions interacting with some homogeneous effective medium with characteristics weighted averaged over all the crystallites from this representative volume. Cluster model is another common type of models [6, 7] (above-mentioned models, considering interactions between crystallite pairs, can be classified as cluster models, in which the cluster consists of two crystallites). In [7] it is pointed out that introduction of additional (intermediate) scale level between macrolevel and mesolevel (level of individual grains) comprising of cluster of several crystallites provides the possibility to come close to physics of their interaction.

The description of statistical model modification is performed in this paper including grain boundaries along with grains to take into account the mechanism of grain boundary sliding (GBS). We focus on description of GBS mechanism in the structure of two-level model, relationships between levels and equations for determination the contribution of this mechanism to inelastic deformation.

2. Grain boundary sliding description within two-level model

The structure of basic two-level model and its general relations are presented in detail, for example, in [8]. The developed model unlike the last one includes on macro- and mesolevel not only stress state measure (and its objective rate) and strain rate measure, but temperature and grain structure state (in view of GBS description in wide range of influence changes included superplastic deformation regime) as process characteristics. Besides, it is used a new method of motion decomposition, proposed by the authors of paper [9]: Cartesian orthogonal coordinate system with quasi-solid motion (moving coordinate system) [9, 10] is introduced for each crystallite and is associated with its symmetry elements – crystallographic direction and plane, containing this direction.

In accordance with modifications, noted above, the constitutive mesolevel model for each crystallite assumes the following form (crystallite index is omitted):

$$\mathbf{\kappa}^{cor} \equiv \dot{\mathbf{\kappa}} - \bar{\mathbf{\omega}}\cdot\mathbf{\kappa} + \mathbf{\kappa}\cdot\bar{\mathbf{\omega}} = \mathbf{\bar{\pi}} : (\mathbf{z} - \mathbf{z}_s^{in} - \mathbf{z}^{th}) = \mathbf{\bar{\pi}} : (\hat{\nabla}\mathbf{v}^{T} - \bar{\mathbf{\omega}} - \mathbf{z}_s^{in} - \mathbf{z}^{th}), \tag{1}$$

$$\theta = \Theta,\ \dot{\theta} = \dot{\Theta}, \tag{2}$$

$$\hat{\nabla}\mathbf{v}^{T} = \hat{\nabla}\mathbf{V}^{T} - \mathbf{Z}_{gb}^{in}, \tag{3}$$

$$\mathbf{z}_s^{in} = \sum_{i=1}^{K} \dot{\gamma}^{(i)}\mathbf{b}^{(i)}\mathbf{n}^{(i)}, \tag{4}$$

$$\dot{\gamma}^{(i)} = \mathrm{f}_1(\tau^{(i)}, \tau_c^{(i)}, \theta),\ i = 1,..,K, \tag{5}$$

$$\dot{\tau}_c^{(i)} = \mathrm{f}_2(\dot{\gamma}^{(i)}, \gamma^{(i)}, \theta),\ i = 1,..,K, \tag{6}$$

$$\mathrm{q} = \mathrm{f}_3(\mathbf{\kappa}, \mathbf{z}_s^{in}), \tag{7}$$

$$\mathbf{z}^{th} = \mathbf{\alpha}\dot{\theta}, \tag{8}$$

$$\bar{\mathbf{\omega}} = \mathbf{f}_4\left(\hat{\nabla}\mathbf{v}^{T}, \dot{\gamma}^{(i)}, \mathbf{\mu}\right),\ \bar{\mathbf{\omega}} = \dot{\mathbf{r}}\cdot\mathbf{r}^{T}, \tag{9}$$

$$\mathrm{r}\cdot\mathbf{g}\cdot\mathrm{r} = 1,\ \mathbf{g} = \mathbf{f}^{-T}\cdot\mathbf{g}_0\cdot\mathbf{f}^{-1}, \tag{10}$$

where (1) is constitutive relation in terms of actual configuration (hypoelastic law), $\mathbf{\kappa}$ is Kirchhoff stress tensor of mesolevel, $\mathbf{\kappa} = \overset{o}{\rho}(\hat{\rho}\mathbf{\sigma})^{-1}$, $\mathbf{\sigma}$ is Cauchy stress tensor, $\overset{o}{\rho}$ is material density of crystallite in reference configuration, $\hat{\rho}$ is material density of crystallite in actual configuration, $\mathbf{\kappa}^{cor} = \dot{\mathbf{\kappa}} - \bar{\mathbf{\omega}}\cdot\mathbf{\kappa} + \mathbf{\kappa}\cdot\bar{\mathbf{\omega}}$ is objective rate (corotational derivative) of stress state measure on mesolevel,

$\bar{\boldsymbol{\omega}}$ is spin tensor, which defines rotation rate of crystallite moving coordinate system, $\boldsymbol{\pi}$ is elastic modulus tensor of crystallite (its components are constant in the basis of moving coordinate system), $\hat{\nabla}\mathbf{v}^{\mathrm{T}} = \dot{\mathbf{f}} \cdot \mathbf{f}^{-1}$ is transposed velocity gradient, which is transmitted from macrolevel with extended Voigt hypothesis (3) taking into account the implementation of some forcing due to GBS mechanism $\mathbf{Z}^{\mathrm{in}}_{\mathrm{gb}}$, $\mathbf{z} = \bar{\nabla}\mathbf{v}_r^{\ \mathrm{T}} = \hat{\nabla}\mathbf{v}^{\mathrm{T}} - \bar{\boldsymbol{\omega}}$ is strain rate measure, determined as transposed relative velocity gradient $\mathbf{v}_r$ (relative to moving coordinate system of mesolevel), $\bar{\nabla}$ and $\hat{\nabla}$ are nabla–operators in moving and actual Lagrangian coordinate systems respectively, $\mathbf{z}^{\mathrm{in}}_{\mathrm{s}}$ and $\mathbf{z}^{\mathrm{th}}$ are inelastic and thermal components of transposed relative velocity gradient. Inelastic component $\mathbf{z}^{\mathrm{in}}_{\mathrm{s}}$ is defined by relation (4), in which $\mathbf{b}^{(i)}$ and $\mathbf{n}^{(i)}$ are unit vectors of slip direction and normal to slip planes in actual configuration, intragranular shear rates $\dot{\gamma}^{(i)}$ (relation (5)) in each crystallite describe the contribution to inelastic deformations due to intragranular dislocation sliding in terms of crystallographic slip planes for this material (K is number of slip systems) using Arrhenius–type relation. Strengthening law (6) is used to describe the evolution of critical shear stresses $\tau_c^{(i)}$, rates of critical shear stresses $\dot{\tau}_c^{(i)}$ are generally a function of accumulated shears and shear rates on slip systems as well as temperature [8]. Relation (7) includes internal heat source capacity q, which is determined due to energy dissipation on inelastic strains in crystallite and transmitted to macrolevel. Thermal component $\mathbf{z}^{\mathrm{th}}$ is defined by relations (8), in which $\boldsymbol{\alpha}$ is thermal expansion tensor, $\dot{\theta}$ is temperature rate (from relation (2)). To determine $\bar{\boldsymbol{\omega}}$ (relation (9)) the rotation model from [9] is used. Orientation change of moving coordinate system for each crystallite during deformation process is described with orthogonal tensor $\mathbf{r}$ established by integrating the rotation rate $\bar{\boldsymbol{\omega}} = \dot{\mathbf{r}} \cdot \mathbf{r}^{\mathrm{T}}$. The first relation in (10) determines surface of crystallite (its shape is assumed to be ellipsoidal), $\mathbf{g}$ is second rank tensor defining both crystallite geometry (shape and dimensions using the lengths of three semi-axes) and its spatial orientation; the second relation in (10) characterizes form change using applied kinematics (strain gradient $\mathbf{f}$), change of tensor form can be also formulated taking into account realization of fragmentation and breaking (more detail in [11]).

System of equations on macroscale level assumes the following form:

$$\mathbf{K}^{\mathrm{cor}} \equiv \dot{\mathbf{K}} - \bar{\boldsymbol{\Omega}} \cdot \mathbf{K} + \mathbf{K} \cdot \bar{\boldsymbol{\Omega}} = \bar{\boldsymbol{\Pi}} : (\mathbf{Z} - \mathbf{Z}^{\mathrm{in}}_{\mathrm{gb}} - \mathbf{Z}^{\mathrm{in}}_{\mathrm{s}} - \mathbf{Z}^{\mathrm{th}}) = \bar{\boldsymbol{\Pi}} : (\hat{\nabla}\mathbf{V}^{\mathrm{T}} - \bar{\boldsymbol{\Omega}} - \mathbf{Z}^{\mathrm{in}}_{\mathrm{gb}} - \mathbf{Z}^{\mathrm{in}}_{\mathrm{s}} - \mathbf{Z}^{\mathrm{th}}), \quad (11)$$

$$\bar{\boldsymbol{\Omega}} = \langle \bar{\boldsymbol{\omega}} \rangle, \quad (12)$$

$$\bar{\boldsymbol{\Pi}} = \langle \bar{\boldsymbol{\pi}} \rangle, \quad (13)$$

$$\mathbf{Z}^{\mathrm{in}}_{\mathrm{s}} = \langle \mathbf{z}^{\mathrm{in}}_{\mathrm{s}} \rangle + \bar{\boldsymbol{\Pi}}^{-1} : \langle \bar{\boldsymbol{\pi}}' : \mathbf{z}^{\mathrm{in}'}_{\mathrm{s}} \rangle + \bar{\boldsymbol{\Pi}}^{-1} : (\langle \boldsymbol{\kappa}' \cdot \bar{\boldsymbol{\omega}}' \rangle - \langle \bar{\boldsymbol{\omega}}' \cdot \boldsymbol{\kappa}' \rangle), \quad (14)$$

$$\mathbf{Z}^{\mathrm{th}} = \langle \mathbf{z}^{\mathrm{th}} \rangle + \bar{\boldsymbol{\Pi}}^{-1} : \langle \bar{\boldsymbol{\pi}}' : \mathbf{z}^{\mathrm{th}'} \rangle, \quad (15)$$

$$\mathrm{k}\dot{\Theta} = \langle \mathrm{q} \rangle, \quad (16)$$

where constitutive relation (11) is also formulated in actual configuration, $\mathbf{K}$ is Kirchhoff stress tensor of macrolevel, $\mathbf{K}^{\mathrm{cor}} = \dot{\mathbf{K}} - \bar{\boldsymbol{\Omega}} \cdot \mathbf{K} + \mathbf{K} \cdot \bar{\boldsymbol{\Omega}}$ is objective rate (corotational derivative) of stress state measure on macrolevel. Spin tensor $\bar{\boldsymbol{\Omega}}$ of moving coordinate system (relation (12)) and elastic modulus tensor $\bar{\boldsymbol{\Pi}}$ (relation (13)) are transmitted from mesolevel to macrolevel using the averaging operation over a representative macrovolume. Tensors $\mathbf{Z}^{\mathrm{in}}_{\mathrm{s}}$ (relation (14)) and $\mathbf{Z}^{\mathrm{th}}$ (relation (15)) are defined from mesolevel model with consistency conditions [12], the following notations are used: $\langle \ \rangle$ as in relations (12) and (13) is for averaging operation over a representative macrovolume, line quantities $\mathbf{a}'$ are for deviation of tensor characteristic $\mathbf{a}$ in crystallite (to reduce notation the crystallite index is omitted) from the mean value ($\mathbf{a}' = \mathbf{a} - \langle \mathbf{a} \rangle$), averaging operation has a property: $\langle \mathbf{a}' \rangle = 0$ for any characteristic **a.** Temperature rate on macrolevel $\dot{\Theta}$ in relation (16) is defined due to averaging of internal heat source capacity q over a representative macrovolume, k is heat capacity (generally temperature and temperature rate are defined from boundary value problem for temperature). We will have more to say below about inclusion of GBS mechanism in structure of performed two-level statistical model and description of inelastic component $\mathbf{Z}_{\mathrm{gb}}^{\mathrm{in}}$ due to realization of this mechanism.

The following modification of model is proposed. Two structural levels are distinguished by mechanisms: one of them is associated with description of intragranular dislocation sliding (this structural level refers to the scale of mesolevel), the other – with description of GBS (this structural level refers to the scale of macrolevel). The level of GBS description is introduced as a separate structural level, and the model is modified into a three-level one: macrolevel – structural level of GBS description (it is considered as a macrolevel submodel) – mesolevel. Relative displacements of crystallites (grains, subgrains) along the common boundary during GBS are observed in bicrystals and polycrystals [13]. In polycrystalline materials GBS is limited by neighboring crystallites, which is modeled in grain boundary hardening law [12]. It is assumed that grain boundary of each crystallite is a set of flat facets with zero thickness, each pair of neighboring crystallites in a representative macrovolume has a common boundary facet, and relative displacements of crystallites occur along a common facet by moving grain boundary dislocations (however the last one is not modeled in explicit form). Basis of vectors $\mathbf{n}_{gb}$, $\mathbf{b}_{gb}^{(1)}$ and $\mathbf{b}_{gb}^{(2)}$ in three-dimensional space is introduced to determine GBS for each intergranular facet: $\mathbf{n}_{gb}$ – normal to the facet plane, $\mathbf{b}_{gb}^{(1)}$ and $\mathbf{b}_{gb}^{(2)}$ – sliding directions in the facet plane. The transition from displacements to deformation characteristics is carried out for model construction: both for intraglanular dislocation sliding and GBS displacement or set of displacements (velocities) is approximated by shear (shear rate) on a certain volume.

Now we formulate the relation for shear rate due to GBS using analogy with Orowan equation for intragranular shear rate [14]. The following model description of real kinematics during GBS is adopted: the relative velocity of two crystallites along the common boundary is approximated by shear rate of the entire representative macrovolume, this procedure is repeated for each pair of neighboring crystallites. Representative macrovolume contains, in general case, differently oriented boundaries with various directions of displacement vectors of grain boundary dislocations, that makes it impossible to define slip systems for GBS by analogy with slip systems for intragranular dislocation sliding. Each boundary facet is considered as a separate slip plane with its orientation and directions of displacement vectors in it, therefore it can be assumed that in this slip plane (boundary facet) N_{gb} grain boundary dislocations move (i.e. in this case each slip system of GBS is represented by one slip plane). Then the value of shear rate in a representative macrovolume as a result of displacement of N_{gb} grain boundary dislocations in direction i in boundary facet with using relation for $\bar{v}_{gb}$ [15, 16] can be written:

$$\dot{\gamma}_{gb}^{(i)} = \frac{1}{\mathrm{h}} \mathrm{K}_V^{(i)} \mathrm{v}_{gb0}^{(i)} \left(\frac{\tau_{gb}^{(i)}}{\tau_{cgb}^{(i)}} \right)^{\frac{1}{n}} \exp\left(-\frac{\Delta F}{k\theta} \right), \; i = 1, 2, ..., 4K_{gb}, \sum_i\!\!\!\!\!\!\diagdown \;, \tag{17}$$

where h is crystallite size–height (the average crystallite size in a representative macrovolume), coefficient $\mathrm{K}_V^{(i)} = \mathrm{h}\mathrm{l}_{gb}\mathrm{d}_{gb}(\mathrm{HLD})^{-1}$ is ratio of crystallite volume to representative volume of macrolevel, l_{gb} is boundary length, $\mathrm{v}_{gb0}^{(i)} = \mathrm{b}_{gb}^{(i)} \rho_{gb}^{(i)} \nu_0 \Delta \mathrm{l}_{gb} A$ is initial velocity of grain boundary dislocations in boundary facet, $\mathrm{b}_{gb}^{(i)}$ is displacement vector module of grain boundary dislocations, $\rho_{gb}^{(i)} = \mathrm{N}_{gb}\mathrm{d}_{gb}(\mathrm{l}_{gb}\mathrm{d}_{gb})^{-1}$ is density of grain boundary dislocations (dimension is 1/m) as a ratio of total length of grain boundary dislocations in boundary facet to the boundary facet area [17], the absence of a model description of breaking processes (changes in crystallite size during deformation) allows to combine the quantities into one parameter $(\mathrm{h})^{-1}\mathrm{K}_V^{(i)}\mathrm{v}_{gb0}^{(i)} \equiv \dot{\gamma}_{gb0}^{(i)}$, $\dot{\gamma}_{gb0}^{(i)}$ is initial shear rate in boundary facet, n is rate sensitivity parameter, ΔF is magnitude of the energy barrier (activation energy of GBS), k is Boltzmann constant, $4K_{gb}$ is total number of sliding directions for all boundary facets in a representative macrovolume.

The inelastic component of transposed relative velocity gradient $\mathbf{Z}_{gb}^{in}$ due to GBS realization is determined from the shear rates (17) as follows:

$$\mathbf{Z}_{gb}^{in} = \sum_{i=1}^{4K_{gb}} \dot{\gamma}_{gb}^{(i)} \mathbf{b}_{gb}^{(i)} \mathbf{n}_{gb}^{(i)}. \tag{18}$$

3. Results of modelling

The results for simple shear of polycrystalline representative macrovolume (343 crystallites) in isothermal case using parameters of material model close to copper are performed below. Intragranular hardening law (6) with parameters is from [18]. Relation for critical stress evolution of grain boundary shears $\tau_{cgb}^{(i)}$ is taken as in the work [11], submodel for GBS description includes following parameters: $\dot{\gamma}_{gb0}\exp(-\Delta F(k\theta)^{-1}) = 10^{-4}\,s^{-1}$, $n = 0.02$, $\tau_{cgb}^{(i)}(0) = 80$ MPa, $h = 3000$ MPa, $g = 0.01$, $g_i^k - 1000$MPa $\forall k,i$, $c(S_t)^{-1}\exp(-U_d(k\theta)^{-1}) = 0.0005s^{-1}$.

Figure 1 and figure 2 show macrolevel stress intensity – strain intensity diagrams when process of simple shear of polycrystal using two-level model (without GBS) and three-level model (with GBS) is considered.

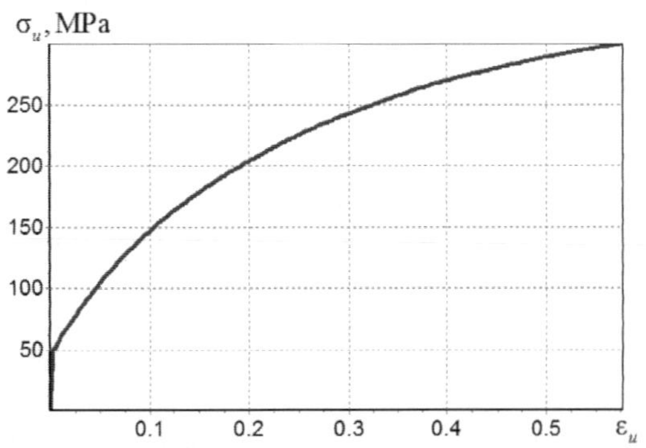

Figure 1. Stress intensity – strain intensity diagrams, obtained in calculations with two-level model without GBS.

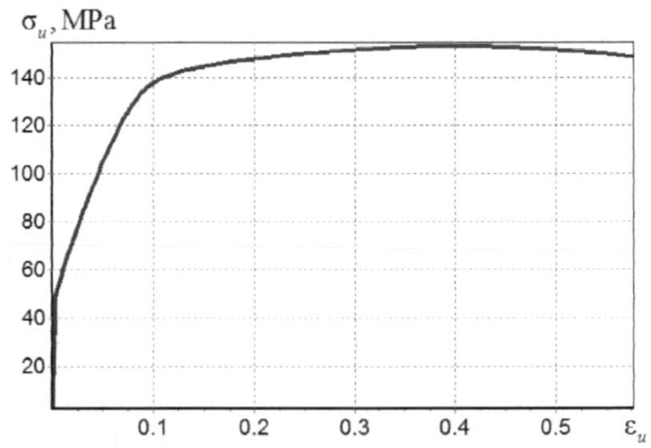

Figure 2. Stress intensity – strain intensity diagrams, obtained in calculations with three-level model with GBS.

The obtained curve in figure 2 has lower flow stress and flattened form compared with curve in figure 1. It means that after reaching a certain amount of strain intensity (in this case at about 0.1) GBS mechanism switches on and becomes dominant.The presented results demonstrate model capabilities in description of GBS and transitions to deformation regimes with reduced flow stresses, particularly, regime of structural superplasticity, when GBS is important mechanism.

4. Conclusion

Modification of statistical model – three-level constitutive model of polycrystal, which takes into account intraglanular dislocation sliding and GBS as main mechanisms of inelastic deformation, is developed. The model also includes temperature, characteristics of grain structure and relationship between the levels. Special attention is given to description of GBS within two-level model and relations, which determine the contribution of this mechanism to inelastic deformation. The results of numerical experiments on the modeling of a simple shear of polycrystal copper sample are presented.

They demonstrate significant role of GBS and characterize the model as promising from the point of view of its capabilities for describing different deformation regimes (depending on the temperature-rate conditions and the state of grain structure) and transitions between them.

Acknowledgements
The work was supported by the Russian Federation Ministry of Education and Science (basic part of the state task PNRPU, project №9.7434.2017/8.9), RFBR (grants Nos. 15-08-06866-a, 17-41-590694-r_a).

References

[1] Van Houtte P, Delannay L and Samajdar I 1999 *Textures and Microstructures* **31** 109
[2] Taylor G 1938 *J. Inst. Metals* **62** 307
[3] Lebensohn R 1999 *Modelling Simul. Mater. Sci. Eng.* 7 739
[4] Molinari A, Canova G and Ahzi S 1987 *Acta metall* **35** 2983
[5] Tom´e C 1999 *Modelling Simul. Mater. Sci. Eng.* **7** 723
[6] Eisenlohr P, Tjahjanto D, Hochrainer T, Roters F and Raabe D 2009 *International Journal of Materials Research* **100** 500
[7] Xie Q, Van Bael A, Sidor J, Moerman J and Van Houtte P 2014 *Acta Materialia* **69** 175
[8] Trusov P, Shveykin A, Nechaeva E and Volegov P 2012 *Physical Mesomechanics* **15** 155
[9] Trusov P and Shveykin A 2016 *Physical Mesomechanics* **19** 25
[10] Pozdeev A, Trusov P and Nyashin Yu 1986 *Large elastoplastic deformations: theory, algorithms, applications* (Moscow: Nauka) p 232
[11] Shveykin A and Sharifullina E 2015 *Nanomechanics Science and Technology: An International Journal* **6** 281
[12] Trusov P, Sharifullina E and Shveykin A 2015 *IOP Conf. Series: Materials Science and Engineering* **71** 012081
[13] Gleiter H and Chalmers B 1975 *High-angle grain boundaries* (Moscow: Mir) p 375
[14] Orowan E 1940 *Proc. Phys. Soc.* **62** 8
[15] Asaro R and Needleman A 1985 *Acta metal* **33** 923
[16] Kocks U, Argon A and Ashby M 1975 *Prog. Mater. Sci.* **19** p 291
[17] Orlov A, Perevezentsev V and Rybin V 1980 *Grain Boundaries in Metals* (Moscow: Metallurgiya) p 156
[18] Anand L 2004 *Comput. Methods Appl. Mech. Engrg.* **193** 5359

IOP Conf. Series: Materials Science and Engineering **286** (2017) 012027 doi:10.1088/1757-899X/286/1/012027

Study of the effect of heat treatment on the structure of surface alloys formed on cp-titanium plates by electron beam cladding in the air atmosphere

V V Samoylenko[1], I A Lozhkina[1], I A Polyakov[1], M G Golkovski[2], V S Lozhkin[1], V A Bataev[1], O G Lenivtseva[1] and O E Matts[1]

[1] Novosibirsk State Technical University, 20, K. Marx Prospekt, Novosibirsk, 630073, Russia
[2] Budker Institute of Nuclear Physics, Siberian Branch of Russian Academy of Sciences, 11, Academician Lavrentiev Avenue, 630090, Novosibirsk, Russia

E-mail: samoylenko.vitaliy@mail.ru

Abstract. The effect of different heat treatment on the structure of Ti-Ta-Zr layers formed by electron beam cladding in the air atmosphere on titanium substrates was investigated. It was established that annealing of surface alloys at the temperature of 900 °C promoted the formation of an equilibrium coarse-plate structure regardless of the concentration of the alloying elements. No complete annealing occurred at temperatures below 800 °C. Quenching of the layer with 26 % Ta, 15 % Zr from 900 °C made it possible to obtain a more homogeneous structure of the material. Significant changes in the morphology of alloys were also observed for layers with a lower concentration of tantalum after quenching from 600...900 °C.

1. Introduction

The current level of the chemical industry development puts forward new requirements for the materials operating under conditions of constant chemical aggression of the various types of corrosive media. Materials of this type should have a high complex of mechanical and anticorrosion properties. Titanium is one of the materials that can provide the listed properties. Titanium is characterized by high specific strength, ductility and corrosion resistance while the metal has low density [1-3]. The disadvantage of the metal is its low resistance to hot acids. The impact of these media contributes to the dissolution of the surface at high speed, which does not allow using it under similar conditions. Since only areas contacting with the medium are subjected to corrosion, the formation of protective layers with a high level of corrosion resistance on the surface of the metal will solve this problem. In addition, this approach will preserve the complex of mechanical properties of the titanium substrate.

The investigations carried out earlier have shown that an increase in the resistance of titanium to corrosion in boiling acid solutions is possible due to combining tantalum and zirconium powders over electron beam cladding on the surface of metal substrate [4, 5]. This solution makes it possible to increase titanium resistance to corrosion, both in boiling nitric acid and hydrochloric acid solutions. It should be noted that in the above papers the structure of the cladded layer is presented by a finely dispersed plates. From the standpoint of corrosion resistance, this structure is unfavorable, since it is possible to form galvanic pairs between the plates and the space between it. In addition, each boundary of the plate will serve as an additional stress concentrator in the emerging film, due to a violation of coherence at the junction between the phases.

For this reason, the formation of a single-phase structure is one of the ways to increase the corrosion resistance of the alloy without changing the chemical composition. Another possible way to increase durability is to form a coarser structure of the plates. Such a solution will reduce the number of interface boundaries that will lead to the formation of a film with fewer defects on the alloy surface. Quenching allows the formation of single-phase structures for alloys containing β-stabilizers (tantalum). Reducing the dispersion of the plates is achieved by performing such a heat treatment as annealing.

In this regard, the structure of surface layers obtained by combined tantalum and zirconium electron-beam cladding on titanium alloys after various types of heat treatment was studied in this paper.

2. Materials and methods

The preparation of surface-alloyed titanium substrate was carried out at the ELV-6 electron accelerator at the Institute of Nuclear Physics of the SB RAS. In the course of the experiment, cladding of two compositions with a different ratio of tantalum to zirconium in the initial powder mixture (Table 1) was realized. The total weight of the powder mixture in both cases was 22.5 g. The plates of commercially pure titanium with a thickness of 12 mm and a size of 100x50 mm were used as the substrate. Before melting, the alloy powder mixture was applied evenly to the substrate surface. Electron-beam treatment was carried out according to the operating conditions given in [6]. After cladding the layers, the formed workpiece was machined to remove the flux residues.

Table 1. Composition of thc initial powder portion.

Sample number	Powders mass (g)			
	Ta	Zr	CaF_2	LiF
No. 1	4.9	9.6	6.0	2.0
No. 2	9.8	6.0	5.0	1.7

The obtained materials were annealed in the temperature range 600 ... 900 °C during 1 hour. Heat treatment at the temperature above 900 °C was not conducted because of the propensity of technically pure titanium to a strong grain growth in the β-region. Quenching of the samples was carried out in water from similar temperatures.

The structure of the materials before and after heat treatment was studied using a Carl Zeiss Axio Observer Z1m optical microscope and a Carl Zeiss EVO 50 XVP scanning electron microscope. The content of the alloying elements in the cladded layers was determined using INCA X-ACT.

3. Results and discussion

The chemical composition of the samples after cladding is presented in Table 2. Melting of the powder mixture No. 1 (Table 1) with an electron beam makes it possible to form a layer with a concentration of 12 % Ta and 23 % Zr. The change in the ratio of alloying components in the initial powder mixture to the higher tantalum concentrations promotes the formation of a layer in which the tantalum content is 26 %, and zirconium content is 15 %.

Table 2. Chemical composition of the cladded layers.

Sample number	Elements composition (wt. %)		
	Ta	Zr	Ti
No. 1	12.0	23.0	65.0

IOP Conf. Series: Materials Science and Engineering **286** (2017) 012027 doi:10.1088/1757-899X/286/1/012027

No. 2	26.0	15.0	59.0

The morphology of surface alloys without thermal treatment is characterized by a dendritic structure (Figure 1 a). At large magnifications, a plate-like structure is revealed (Figure 1 b). Annealing of these layers at the temperature of 600 °C does not lead to the changes in the structure of the material. Increasing the heat treatment temperature to 700 °C promotes the appearance of the first indications of a change in the structure of alloys. For a layer with 26 % Ta, 15 % Zr, it is expressed in the local growth of α-plates (Figure 1 c). Presumably, the plates grow in regions enriched in tantalum, since this surface alloying method is characterized by the presence of dendritic liquation [6, 7]. The layer with 12 % Ta, 23 % Zr at the temperature of 700 °C has no significant changes in the structure of the alloy. However, when the annealing temperature increases to 800 °C, the first indications of the plate growth appear on the sample (Figure 1 d), and the structure of the alloy has a similar morphology with a 26 % Ta layer, 15 % Zr after annealing at 700 °C. This difference in the beginning of the diffusion processes is explained by the difference in the chemical composition of the cladded layers. So for an alloy with a higher concentration of tantalum, phase transformations will begin at a lower temperature due to stabilization of the high-temperature β-phase of titanium with tantalum.

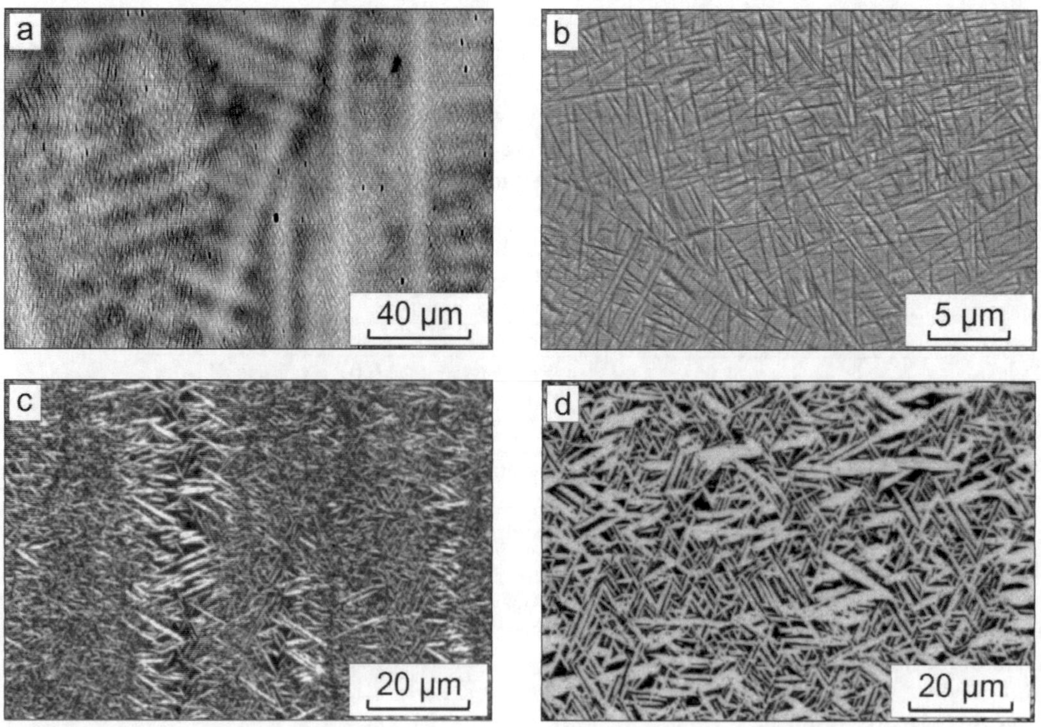

Figure 1. The structure of the layers after cladding (a, b), annealing at 700 ºC (c) and 800 ºC (d).

Annealing of the workpiece with 26 % Ta, 15 % Zr at 800 °C promotes an increase in the proportion of growing plates (Figure 2 a). There is no complete annealing of the material, since small interlayers with a finely dispersed lamellar structure are observed between regions of different chemical composition (Figure 2 b). Heating the layers to 900 °C with subsequent slow cooling promotes the formation of the layers with a completely annealed structure. This is evidenced by the formation of a typical α+β structure having a characteristic lamellar packets structure of the α-phase with interlayers of the β-phase (Figure 2 c, d). In addition, with this annealing, the dimensions of the plates are much larger than that of the alloy without heat treatment.

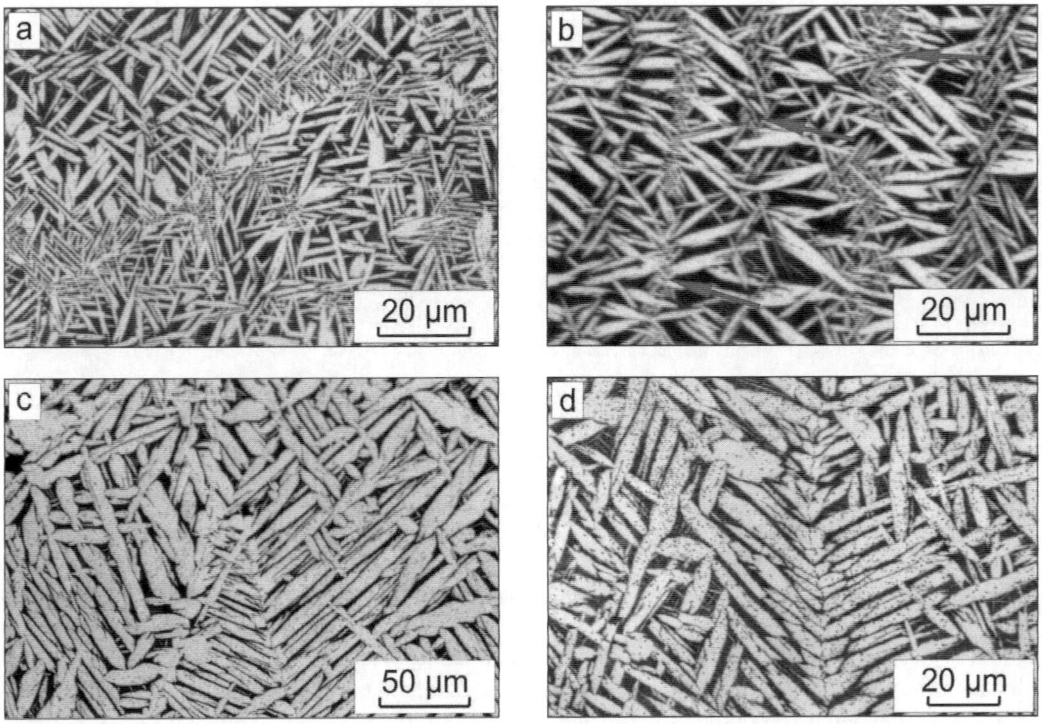

Figure 2. The structure of the layers after cladding (a, b), annealing at 800 °C (c) and 900 °C (d).

The structure of surface alloys after quenching is shown in Figure 3. Heating the layers to the temperature of 600, 700 and 800 °C followed by quenching in water does not lead to a significant change in the structure of the cladded materials compared to the material without heat treatment. The morphology of the alloys is represented by a dendritic structure against which a plate-like structure is revealed during etching (Figure 3 a, b). An increase in the heating temperature to 900 °C does not lead to the significant changes in the structure of the layer with 12 % Ta, 23 % Zr, but it contributes to the formation of a more homogeneous structure on the 26 % Ta alloy, 15 % Zr (Figure 4 a). Nevertheless, prolonged etching of the material still helps to identify the lamellar structure. It should be noted that the plates are etched significantly worse than the surrounding space (Figure 4 b). This is attested to by a large number of holes, chaotically located in the body of grain. This fact suggests that quenching in water from 900 °C of the layers with 26 % Ta, 15 % Zr can positively affect the corrosion resistance of this material.

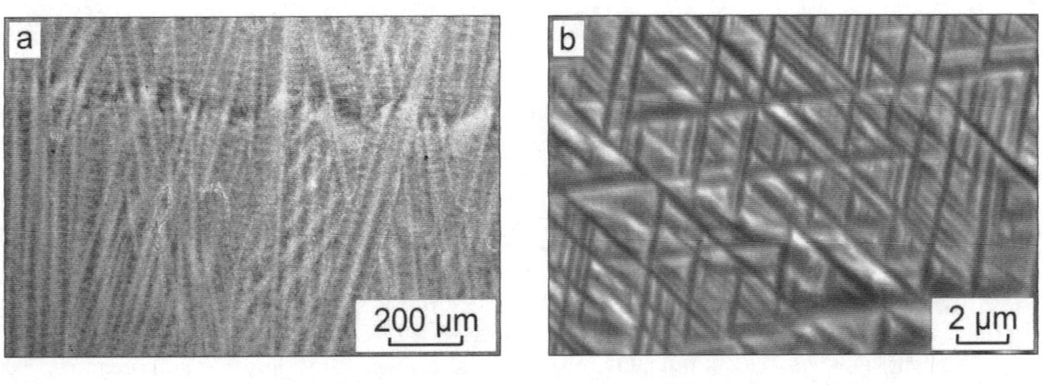

Figure 3. The structure of the layers after cladding.

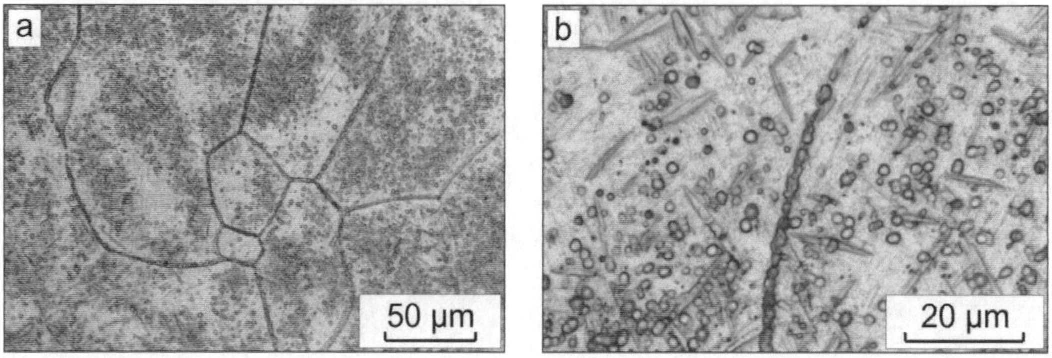

Figure 4. The structure of the layers after annealing at 800 °C (a) and 900 °C (b).

4. Conclusion

Temperatures for the heat treatment of surface alloys formed by the method of electron-beam cladding of tantalum and zirconium powders were determined. It was established that annealing at the temperature of 900 °C promotes the formation of an equilibrium structure having the coarse-plate morphology, compared to the material without thermal treatment. Quenching of the cladded layers from temperatures of 600 ... 900 °C does not allow obtaining a single-phase structure. The only exception is the layer with 26 % Ta, 15 % Zr after quenching from 900 °C, which is characterized by the formation of a more homogeneous structure. Long-term etching of this alloy promotes the identification of lamellar morphology. However, the structural components are etched out worse than the surrounding space, which can indirectly indicate an increase in corrosion resistance compared to layers without heat treatment.

Acknowledgments

The reported study was funded by RFBR according to the research project No. 16-33-00900 mol_a.

References

[1] Leyens C and Peters M 2003 *Titanium and Titanium Alloys Fundamentals and Applications* (WILEY-VCH Verlag GmbH & Co. KGaA, Weinheim) p 532

[2] Matthew J and Donachie Jr 2000 *Titanium: A Technical Guide* (ASM International, Materials Park, Ohio) p 381

[3] Lütjering G and Williams J C 2007 *Titanium (Engineering Materials and Processes)* (Springer, Berlin) p 442

[4] Samoylenko V V, Ogneva T S, Polyakov I A, Ivanchik I S, Lenivtseva O G and Matts O E 2016 Structure and Properties of Surface Alloyed Layers Formed by Non-vacuum Electron Beam Cladding of Ta and Zr Powders on Commercially Pure Titanium Plates *AIP Conference Proceedings* **1785** 040057

[5] Samoylenko V V Polyakov I A Golkovski M G Bataev I A Ruktuev A A 2014 Poverhnostnoe legirovanie titana tantalom i cirkoniem metodom vnevakuumnoj ehlektronno-luchevoj naplavki poroshkovyh materialov *Proceedings of the XXIV International Conference Radiation Solid State Physics* 345-51 [in Russian]

[6] Samoylenko V V Lozhkina E A Polyakov I A Lenivtseva O G Ivanchik I S and Matts O E 2016 The study of the modes of Ta-Zr powder mixture non-vacuum electron-beam cladding on the surface of the cp-titanium plates *IOP Conf. Series: Materials Science and Engineering* **156** 012024

[7] Bataev I A Zhuravina T V Ruktuev A A Lenivtseva O G Romashova Y N 2012 Structural investigation of "titanium-tantalum" coatings obtained by non-vacuum electron beam cladding *Obrabotka metallov* **56** 56-59 [in Russian]

IOP Conf. Series: Materials Science and Engineering **286** (2017) 012028 doi:10.1088/1757-899X/286/1/012028

Structural features of the coatings of the Ni-Al system obtained by air-plasma spraying

E E Kornienko[1], O A Rubtsova[1], I P Gulyaev[2] and V I Kuz'min[2]

[1]Department of Material Science in Engineering, Novosibirsk State Technical University, Prospekt K. Marksa 20, Novosibirsk, 630073, Russia
[2]Khristianovich Institute of Theoretical and Applied Mechanics SB RAS, Institutskaya str. 4/1, Novosibirsk, 630090, Russia

E-mail: oksana.rubtsova@yandex.ru

Abstract. This article deals with the structural features of nickel aluminides coatings obtained by air-plasma spraying with the unit of annular injection of powder. The subsonic (plasma velocity is 1200 m/s) and supersonic (plasma velocity is 1900 m/s) regimes were used for coatings deposition. Ni_3Al powder (Ni is base; 13-15 wt. % Al) was applied to the pipes of low-carbon steel 20 (0.2 % C). The structure and phase composition of the coatings were investigated by optical microscopy, scanning electron microscopy, transmission electron microscopy, and X-ray phase analysis. It is established that the porosity of the coatings obtained at supersonic regimes is reduced by a factor of 2. Ni_3Al and Ni_5Al_3 are shown to be the main phases of coatings. Ni_5Al_3 phase volume fraction of the coatings obtained at supersonic regimes is revealed to be lower which reduces the microhardness of these coatings. It is shown that Ni_3Al phase has subgrain structure and Ni_5Al_3 phase has lamellar structure.

1. Introduction

Plasma spraying is a widely used method of powder materials coatings deposition [1-5]. This technology is known to allow spraying of any materials (metals, ceramics, intermetallics) to any substrate [1, 2]. Plasma coatings are usually used for corrosion protection, for electrical insulation, for enhancing wear resistance and heat resistance etc. [2, 5, 6-9]. During the spraying, the powder particles are heated to molten or plasticized state, accelerated by the gas flow, and then deformed and crystallized after hitting the substrate surface thus forming laminated coating [1].

Currently the main method of powder injection into a plasma jet is cross-point injection through the powder supply tube [10-12]. This method of powder injection leads to raising the heterogeneity of temperature fields and plasma jet velocity which reduce the quality of the formed coatings. In this research the air-plasma spraying with the unit of the annular injection of powder was used for the coatings formation. Previously we showed that this design of the plasmatorch allows obtaining high quality dense coatings from the materials of the system NiCrSiB [9] and B_4C-Ni [13]. The maximum porosity of the obtained coatings does not exceed 1…2 %. Unlike the listed materials, the porosity of the coatings of Ni_3Al powder obtained by this technology reached 6 % [8]. In addition, unmelted particles are presented at the obtained coatings. It is known that one of the ways to raise the density of ceramic and intermetallic coatings is the increasethe sprayed particles velocity in the plasma jet [14,

15]. This article presents results of structural investigations of Ni_3Al coatings obtained by air-plasma spraying at subsonic and supersonic regimes.

2. Materials

Ni_3Al powder 40…100 μm diameter was used as material for coating forming. The chemical composition of the powder was: Ni - base; 13…15 wt. % Al.

The air-plasma spraying was carried out at the Institute of Theoretical and Applied Mechanics SB RAS using the plasmatorch PNK-50 with the unit of the annular injection of powder. The cylindrical nozzle and the Laval's nozzle [16] were used to provide the subsonic (plasma velocity is 1200 m/s) and the supersonic (plasma velocity is 1900 m/s) regimes respectively. Pipes of steel 20 (0.2 % C) with 60 mm internal diameter and 3 mm wall thickness were used as the substrate. Previously, we determined the spraying optimal regimes for Ni_3Al powder: the arc current was 200 A, the voltage was 223 V [8]. The air was used as a transporting, focusing and orifice gas, the blend of the air and the propan-butan was used as a protective gas. The spraying distance was 170 mm. To reduce the level of residual stresses after spraying, we carried out the annealing of steel pipes with coatings at temperature 300 °C for 5 hours.

The velocity and temperature measurement of Ni_3Al particles during spraying at the subsonic regime was carried out using an optical measuring complex [1, 17]. It is established that the average values of the particles velocity at the spraying distance at various regimes were within the range of 200…230 m/s, and temperatures within the range of 2290…2410 K. The velocity and temperature measurement of Ni_3Al particles in the supersonic regime were not carried out. However, authors in the paper [16] showed that at spraying of 20…63 μm diameter NiCrSiB powder on the same equipment the transition from the subsonic to the supersonic regime leads to increase of average velocity of particles from 300 to 500 m/s and to slight rice of their temperature.

We investigated the coatings structure using optical microscope Carl Zeiss Axio Observer Alm, scanning electron microscope Carl Zeiss EVO50 XVP and transmission electron microscope FEI Tecnai G2 20 TWIN. The transversal cross-sections were objects for structural investigations using methods of optical and scanning electronic microscopy. The samples were pressed into the polymer matrix and prepared by standard method including mechanical grinding and polishing. For transmission electron microscopy we cut 3 mm diameter workpieces from the coatings. The workpieces were mechanically thinned to the thickness of 90...100 μm. Then spherical holes were ground using abrasive paste and ion thinning. The phase composition of the coatings was studied using the X-ray diffractometer ARL X'TRA in CuKα radiation. The coatings porosity was evaluated with the help Axio Vision Multiphase (Carl Zeiss) software. The coatings microhardness was estimated using microhardness tester Wolpert Group 402MVD at 100 N load.

3. Experimental details

Figure 1a and b shows the cross-sections of the compounds "coating-substrate" obtained at the subsonic and the supersonic regimes, respectively. Such defects as cracks or delaminating of coatings are not observed. The coatings material consists of particles melted in a plasma jet and plastically deformed particles. It can be noticed that unmelted (round) particles are also presented in the coatings obtained at subsonic regime (figure 1a, marked by circles). This type of particles are not observed in the coatings obtained at the supersonic regime. The porosity of the coatings obtained at the supersonic regime is 2 times lower (~ 2-3 %).

Figure 1c shows that two various phases (1 and 2) as well as regions with a mixed two-phase structure (3) are observed in the coatings obtained at the supersonic regime. Only mixed two-phase areas are presented in the coatings obtained at the subsonic regime [8]. Figure 1d shows a more detailed image of the two-phase area. There are dark gray grains with streaked structure. The plates direction in individual grains has different orientation. A light-gray phase is distributed on the grain boundaries.

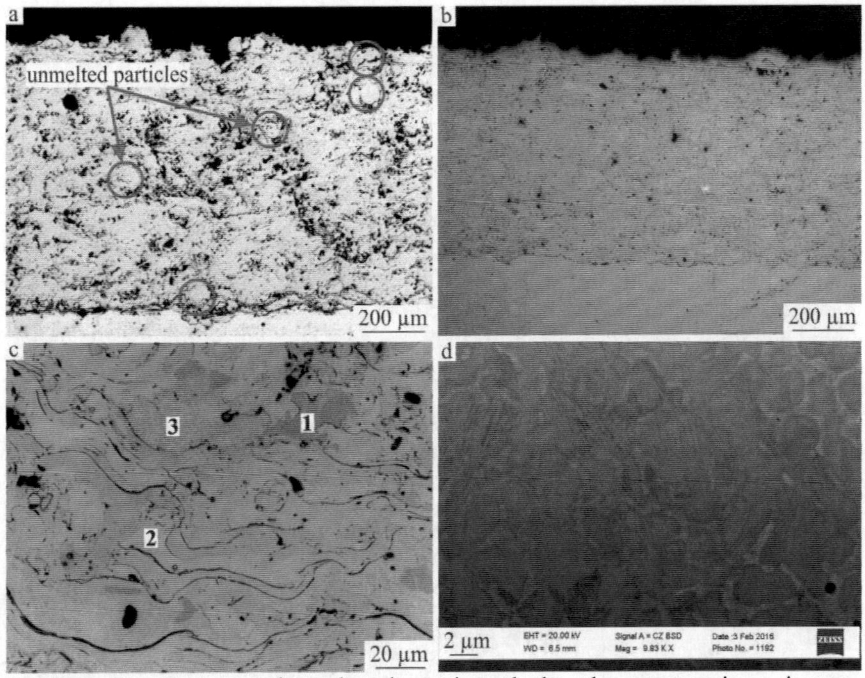

Figure 1. Coatings structure: a – the subsonic regime; b-d – the supersonic regime. a-c – optical microscopy; d – scanning electron microscopy in back-scattered electrons.
1, 2 – areas with various phases; 3 –two-phase area.

X-ray diffraction analysis identified that Ni_3Al and Ni_5Al_3 are the main phases of plasma coatings (Figure 2). The volume fraction of Ni_5Al_3 phase in the coatings obtained at the subsonic regime is higher. It is also confirmed by microhardness measurements. The average microhardness for the coatings obtained at the subsonic regime is 630 HV, while only 430 HV for coatings obtained at the supersonic regime.

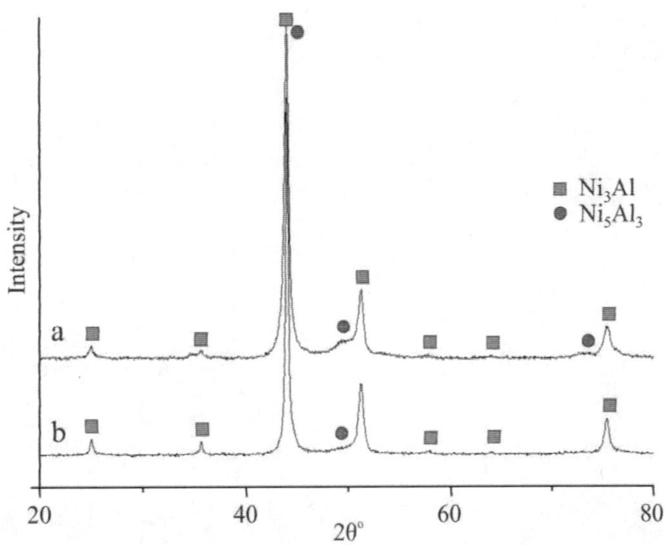

Figure 2. X-ray diffraction pattern of coatings: a – the subsonic regime; b – the supersonic regime.

The method of transmission electron microscopy was used for studying the features of the fine structure of nickel aluminides coatings. High speeds of mixing and cooling of the coating material lead to the formation of various phases. Electron diffraction patterns analysis indicates the presence of intermetallic phases Ni_3Al and Ni_5Al_3.

Figure 3a,b shows the image of the two-phase area of the coating. The grains with streaked structure (1 in Figure 3a) and the phase formed at the grains boundaries (2 in Figure 3a) are clearly visible. The dark-field image and the electron diffraction pattern confirm that the area 1 in Figure 3a is Ni_3Al phase (Figure 3b). Figure 3c shows the subgrain structure of the Ni_3Al phase. It can be seen that subgrains size does not exceed 500 nm. Figure 3d shows typical image of Ni_5Al_3 phase. This phase is represented by the packets of internally microtwinned plates that are equally oriented with every other plate. The plates thickness does not exceed 100 nm and their length is limited by the size of NiAl phase grains. It can be seen that the phase plates consist of thin plates that are twinned to each other in orientation.

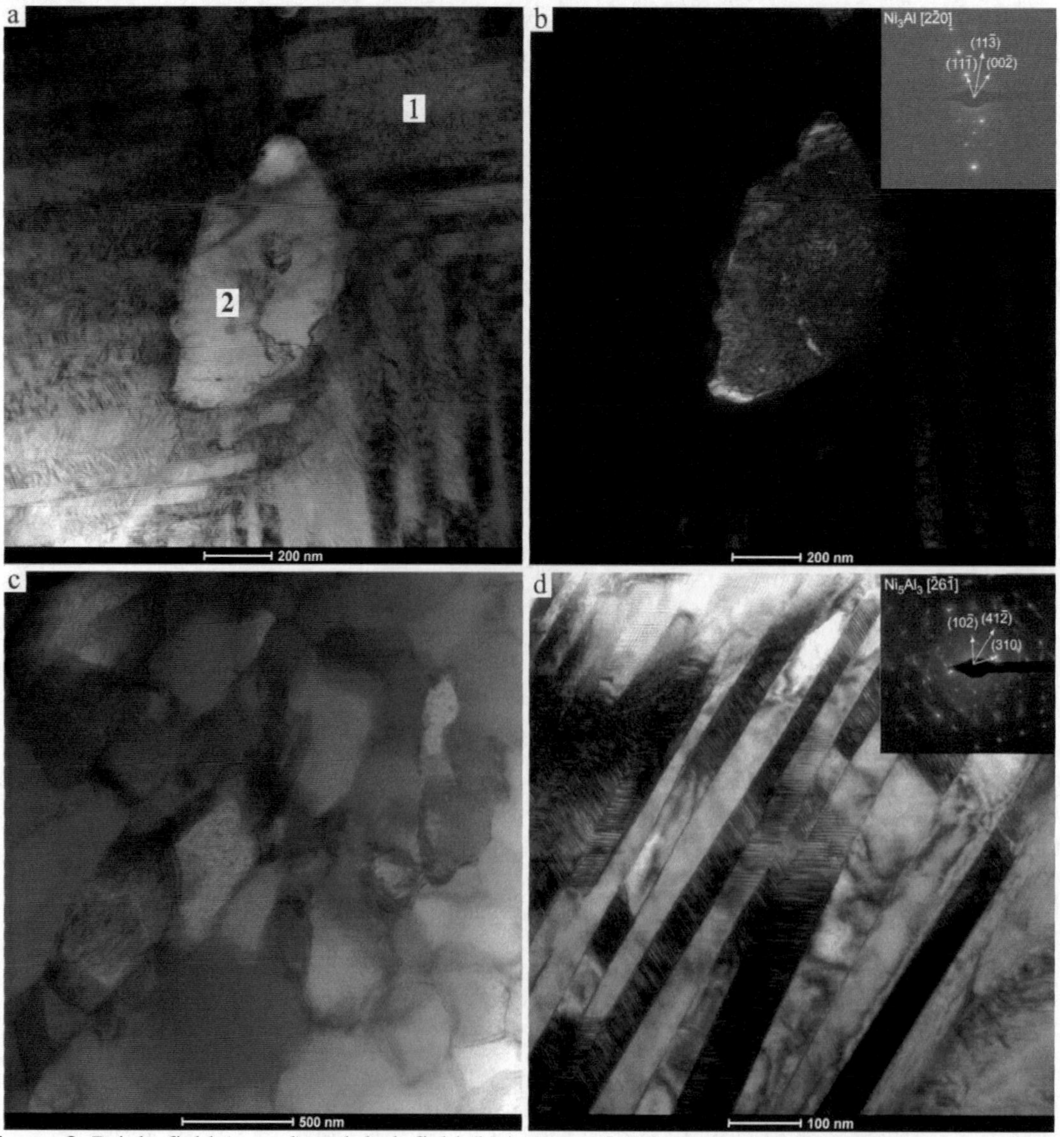

Figure 3. Bright field (a, c, d) and dark field (b) images of plasma coating structure. a, b – two-phase area of the coating; c – Ni_3Al phase; d – Ni_5Al_3 phase.

IOP Conf. Series: Materials Science and Engineering **286** (2017) 012028 doi:10.1088/1757-899X/286/1/012028

4. Conclusions

1. The spraying of coatings from Ni_3Al powder at the supersonic regime allows to reduce porosity by a factor of 2 (up to 2-3 %).
2. X-ray diffraction analysis identified that Ni_3Al and Ni_5Al_3 are the main phases of plasma coatings. The volume fraction of Ni_5Al_3 phase in the coatings obtained at the supersonic regime is lower.
3. The structural investigation identified that the coatings formed at the supersonic regime consist of two various phases as well as two-phase areas. The coatings obtained at the subsonic regimes are characterized mainly by two-phase structure. The transmission electron microscopy showed that the Ni_3Al phase has subgrain structure, Ni_5Al_3 phase has lamellar structure.
4. The microhardness measurements established that the coatings obtained at the supersonic regime have lower hardness.

References

[1] Hasui A and Morigaki O 1985 *Surfacing and spraying* (Moscow: Mashinostroenie)
[2] Hocking M G, Vasantasree V and Sidky P S 2000 *Metallic and Ceramic Coating. Production, High Temperature Properties and Applications* (London: Longman Scientific & Technical).
[3] Tushinski L I, Plohov A V, Tokarev A O, etc. 2004 *Methods of Materials Research: Structure, Properties and Processes of Applying of Inorganic Coatings* (Moscow: Mir, Russia)
[4] Puzryakov A F 2003 *The Theory of Plasma Spraying* (Moscow: MGTU im. N.E. Baumana)
[5] Simunovic K, Saric T and Simunovic G 2014 *Tribology Transactions* **57** 955-979
[6] Hsiao W T, Sua C Y, Huang T S and Liao 2011 W H *Journal of Alloys and Compounds* **509** 8239-45
[7] Hearley J A, Little J A and Sturgeon A J 2000 *Surface and Coatings Technology* **123** 210-218
[8] Kornienko E E, Mul' D O, Rubtsova O A, Vaschenko S P, Kuzmin V I, Gulyaev I P and Sergachev D V 2016 *Thermophysics and Aeromechanics* **23** 919-927
[9] Kornienko E E, Nikulina A A, Bannov A G, Kuz'min V I, Mildebrath M, Bezrukova V A and Zhoidik A A 2016 *Obrabotka metallov (metal working and material science)* **4(73)** 52-62
[10] Fauchais P, and Vardelle M *J.* 2010 *Therm. Spray Technol.* **19** 668-694
[11] An L, Gao Y, and Zhang T J. 2007 *Therm. Spray Technol.* **16** 967-973
[12] Mauer G, Vassen R, Stover D and etc. 2011. *J. Therm. Spray Technol.* **20** 3-11.
[13] Bezrukova V A, Kornienko E E and Kuz'min *V I* 2016 *V sbornike: AKTUAL'NYE PROBLEMY SOVREMENNOJ NAUKI materialy V regional'noj nauchno-prakticheskoj konferencii s mezhdunarodnym uchastiem* 7-11 (in Russian)
[14] Deevi S C, Sikka V K, Swindeman C J, and Seals R D 1997 *Journal of Thermal Spray Technology* **6** 335–344
[15] Bacciochini A, Radulescu M I, Charron-Tousignant Y, Van Dyke J, Nganbe M, Yandouzi M, Lee J J and Jodoin B 2012 *Surface & Coatings Technology* **206** 4343–48
[16] Kuz'min V I, Gulyaev I P, Kornienko E E, Sergachev D V, Dolmatov A V and Tyutyun'kova M K 2016 Sbornik dokladov mezhdunarodnoj nauchno-tekhnicheskoj konferencii. «Klimovskie chteniya - 2016: perspektivnye napravleniya razvitiya aviadvigatelestroeniya» 309-316 (in Russian)
[17] Dolmatov A V, Gulyaev I P, Kuzmin V I. Lyskov E A and Ermakov K A 2017 *Thermophysics and Aeromechanics* **24 (1)** 83-94

IOP Conf. Series: Materials Science and Engineering **286** (2017) 012029 doi:10.1088/1757-899X/286/1/012029

Numerical prediction of the elastic characteristics of spatially reinforced composite materials

A N Anoshkin, P V Pisarev, D A Ermakov and K A Maksimova

Perm National Research Polytechnic University, 29 Komsomolski Ave., Perm, Russian Federation

E-mail: pisarev85@live.ru

Abstract. In this paper, the effective elastic characteristics of spatially reinforced composite materials is calculated. The problem is solved by the finite element method in the Ansys Workbench software using the periodicity cell of the material. When constructing the model, were considered the weaving scheme, the volume fraction of the reinforcing skeleton, and the geometric shape of the reinforcing filaments. The effective elastic modulesare determined by the volume averaging method. To obtain nine independent characteristics of the orthotropic material, six calculation modes were used: tensionalong the axis and ashear in the planes. The stress-strain state of reinforcing filaments is analyzed.

1. Introduction

Currently, more attention is being paid to spatially reinforced composite materials (SRCM). These materials differ from traditional laminates by the reinforcement in third direction. In layered composite materials, the layers are joined together by a binder. In the SRCM manufacturing, additional bonds are created between the layers by introducing reinforcement in directions that differ from the direction of layer reinforcement. The use of such manufacturing method increases the specifications of composite materials in shear strength and tear resistance. Thus, these materials can be used in aircraft and rocketry to produce radar fairings (with glass fiber reinforcement), brake discs for aerospaceindustry (carbon-carbon materials) [1]. SRCM are used for the manufacturing of turbine blades, nozzle blocks, nose cones of rocket engines, sealing elements for hydraulic and pneumatic systems and many other structural elements that are operated under conditions of intense thermal (or cryogenic) and mechanic impact.

When designing structures from SRCM, it becomes necessary to develop methods for predicting the physico-mechanical properties of the materials. Existing experimental methods allow us to determine with sufficient accuracy only six elastic constants: elastic moduli E_1, E_2, Poisson's ratio μ_{12}, shear modulus in G_{12} plane and interlayer shear moduli G_{13} and G_{23} [2]. Therefore, the development of a technique for numerical prediction of the elastic characteristics of spatially reinforced composite materials is an urgent task.

2. Problem formulation

The subject of this study is the SRCM periodicity cell (Figure 1). The structure of the periodicity cell is a system of weft and sewing threads where parts of the reinforcing skeleton are connected by matrix. For carrying out a number of computational experiments, a geometric model was constructed using the TexGen software package. The initial parameters (the cross-sectional area of the filaments and the distance between their centers), necessary for constructing the geometric model of the unit cell of the fabric, were determined on the basis of the specified parameters – the linear density of the filaments and the density for the base and weft. During the geometric model design, the following assumptions are used: solid state threads, the interaction of the filament fibers is ignored, there is no interfacial layer between the matrix and the filaments, technological dispersion in the geometry of the filaments is not taken into account. The deformation of threads during production process is also neglected.

Taking into account the given technological parameters of weaving, the dimensions of the unit cell of this fabric are 4.76 mm along the X axis, 3.68 mm along the Y axis and 2.57 mm along the Z axis. Assuming an ideal packing of fibers in the complex filament, its conditional cross-sectional area can be determined by the formula:

$$S = \frac{T}{pk} \tag{1}$$

where T is the linear density of the complex filament; p is the density of the yarn material, kg/m^3; k is the degree of volume filling of the filament.

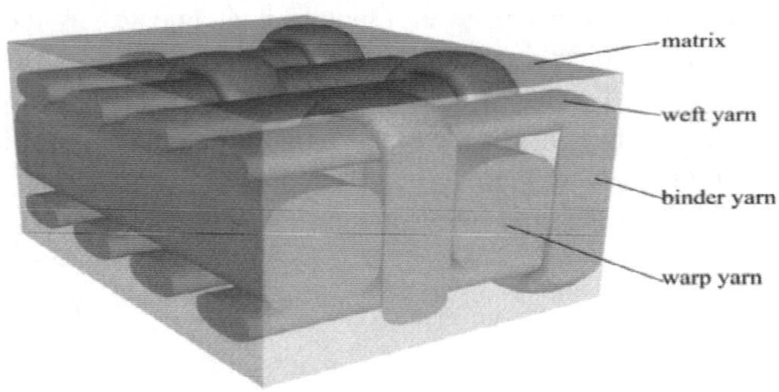

Figure 1. General view of the geometric model of the periodicity cell of a spatially reinforced composite material.

The mathematical formulation of the problem corresponds to the theory of elasticity of an anisotropic body. In the variational formulation this problem consists of finding the minimum of the Lagrange functional with additional conditions in the form of geometric Cauchy relations [3]. The variation of the functional in the absence of mass forces has the form:

$$\delta J_u = \int_V \varepsilon_{ij} C_{ijkl} \delta\varepsilon_{kl} dV - \int_{S_t} F_i \delta u_i dS \tag{2}$$

where ε_{ij} and $\delta\varepsilon_{ij}$ are the tensor and the variation of the strain tensor correspondingly, C_{ijkl} is the tensor of elastic modules, δu_i is the variation of the displacement vector, F_i is the vector of external force.

Additional conditions for the functional (2) are geometric equations:

$$\varepsilon_{ij} = \frac{1}{2}(u_{i,j} + u_{j,i}). \tag{3}$$

Elastic constants of transversal-isotropic yarns based on carbon fibers and epoxy binder for the numerical simulation were used from papers [4-7].

The SRCM sample considered is orthotropic and is characterized by nine independent elastic constants:Young's modules E_X, E_Y, E_Z, Poisson's ratiosν_{XY}, ν_{YZ}, ν_{XZ}, and shear modules G_{XY}, G_{YZ},

IOP Conf. Series: Materials Science and Engineering **286** (2017) 012029 doi:10.1088/1757-899X/286/1/012029

G_{XZ}. Where X is the direction of the base, Y is the direction of the weft. Between the interacting threads of the model the ideal contact interaction type was used.For the periodicity cell, six types of experiments are realized: stretching along the X, Y, Z axes to determine the effective Young's modules and Poisson's ratios and a shear in the planes XY, XZ, YZ in order to determine the shear modules [8]. Effective modules are obtained by averaging over the volume [9] using the following dependencies:

$$E_X^* = \frac{\langle\sigma_X\rangle}{\varepsilon_X}, E_Y^* = \frac{\langle\sigma_Y\rangle}{\varepsilon_Y}, E_Z^* = \frac{\langle\sigma_Z\rangle}{\varepsilon_Z},$$
$$G_{XY}^* = \frac{\langle\tau_{XY}\rangle}{\gamma_{XY}}, G_{YZ}^* = \frac{\langle\tau_{YZ}\rangle}{\gamma_{YZ}}, G_{XZ}^* = \frac{\langle\tau_{XZ}\rangle}{\gamma_{XZ}}, \quad (4)$$
$$\nu_{XY} = \left|\frac{\varepsilon_Y}{\varepsilon_X}\right|, \nu_{YZ} = \left|\frac{\varepsilon_Z}{\varepsilon_Y}\right|, \nu_{XZ} = \left|\frac{\varepsilon_Z}{\varepsilon_X}\right|,$$

where deformations ε_X, ε_Y, ε_Z and γ_{ij} were predetermined in numerical experiments.

The average stresses were calculated from the following relationships:

$$\langle\sigma_{ij}\rangle = \frac{\sum_{k=1}^{n}\sigma_{ij}^{k}V^{k}}{V}; \mathrm{i,j} = \mathrm{x,y,z}, \quad (5)$$

where n is thenumber of finite elements in a representative volume, σ_{ij}^k is the stress in the final element, V^k is the volume of the final element, $V$is the total volume of the model.

The discretization of the representative volume was carried out using the spatial finite elements SOLID186. The minimum size of the final element was 0.01 mm, the total number of finite elements was 3 million. It was taken into account that the numerical experiment is considered to converge, in the case when the result of numerical simulation does not change with further refinement of the finite element grid.

3. Results

Inhomogeneous fields of equivalent, normal and shear stresses for the matrix and filaments in the local coordinate system were obtained from the results of numerical simulation of the periodicity cell stress-strain state (Figure 2-7). The inhomogeneity was caused by the three-dimensional weaving of SRCM. The macroscopic stresses and deformations were determined in the cell under study, and the effective elastic characteristics of SRCM were found at the ratio 4.5 (Table 1).

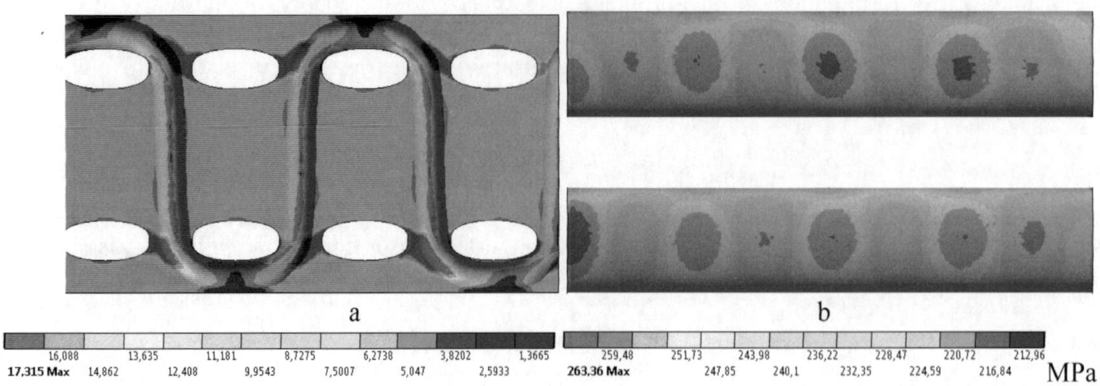

Figure 2. Distribution of stress fields when stretching along the X axis: a – equivalent von Mises stresses in matrix; b – normal σ_{11} in warp threads.

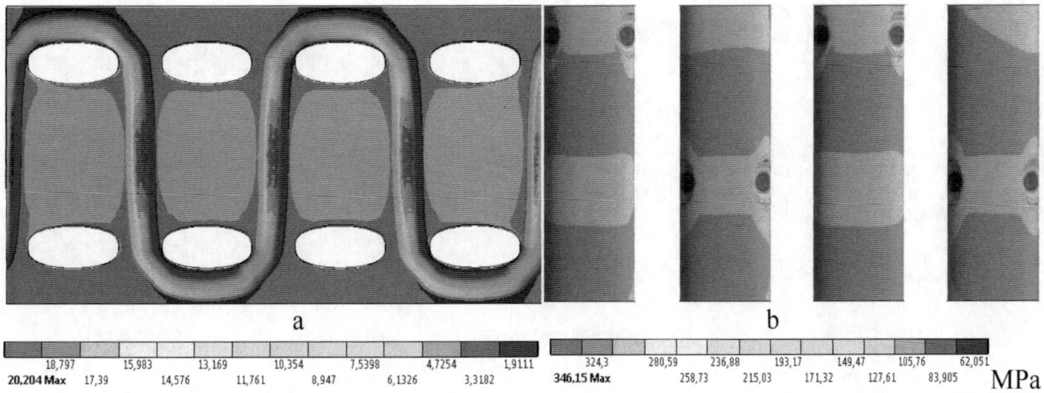

Figure 3. Distribution of stress fields when stretching along the Y axis: a – equivalent von Mises stresses in matrix; b – normal σ_{22} in weft threads.

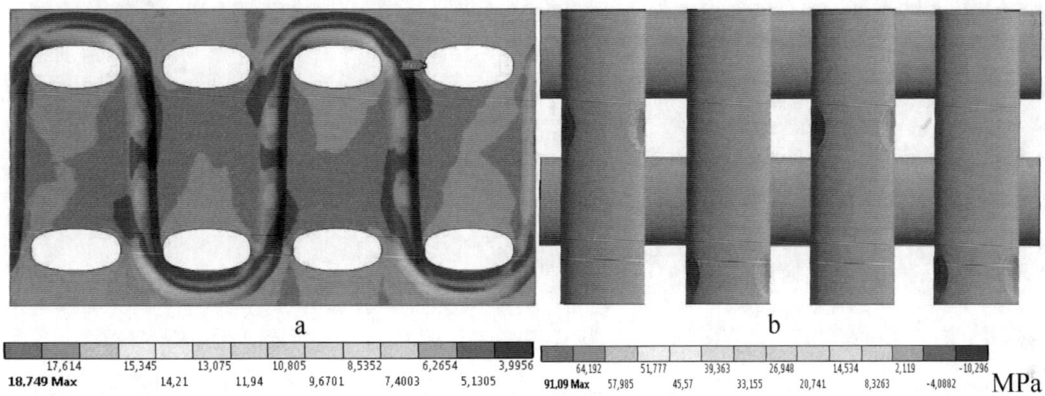

Figure 4. Distribution of stress fields when stretching along the Z axis: a – equivalent von Mises stresses in matrix; b – normal σ_{33} in the warp and weft yarns.

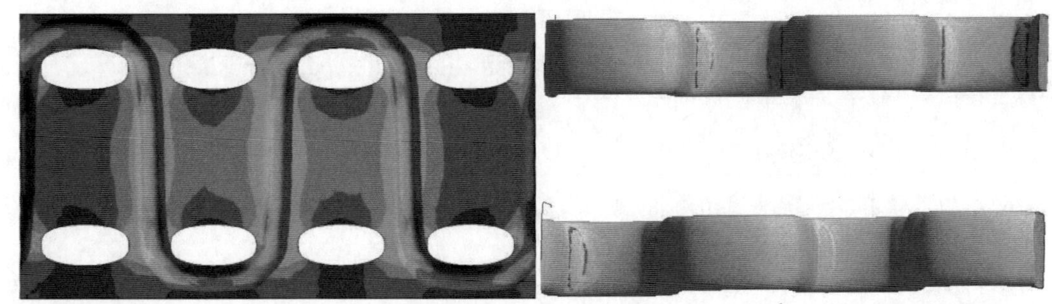

Figure 5. Distribution of stress fields with a shear in the XY plane: a – equivalent von Mises stresses in matrix; b – tangents τ_{12} in the sewing threads.

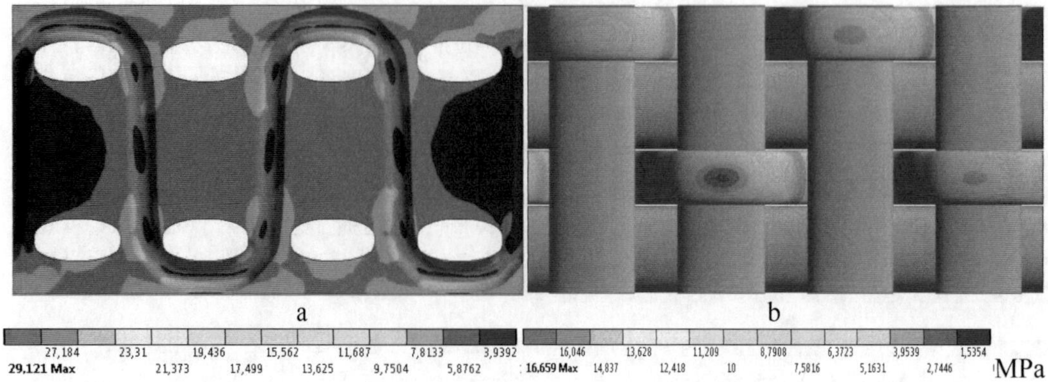

Figure 6. Distribution of stress fields with shear in the plane XZ: a – equivalent von Mises stresses in matrix; b – the tangent τ_{13} in the threads of the weft and the base.

Figure 7. Distribution of equivalent stress fields in the matrix by the Mises criterion in the longitudinal sections along the center: a – the warp threads; b – the sewing thread, withshear in the YZ plane.

Table 1. Effective elastic properties of a textile-reinforced composite.

E_x (GPa)	E_y (GPa)	E_z (GPa)	ν_{xy}	ν_{yz}	ν_{xz}	G_{xy} (GPa)	G_{yz} (GPa)	G_{xz} (GPa)
62.4	42.3	7	0.08	0.3	0.28	4.52	2.55	2.49

Analysis of equivalent stress fields of the matrix revealed that the maximum values 29 MPa are observed with a shear in the XZ plane, near bend of the sewing threads, and equalto approximately 29 MPa. These stresses will be the most dangerous, determining the safety margin of a structure made using this type of weaving. The places of localizations of maximum stresses in the framework were also revealed. When stretching along the X axis, the maximum normal stresses σ_{11} were detected in the warp yarns, in the regions between the base threads and weft threads and amounted to approximately 263 MPa. The greatest normal stresses σ_{22} under tension along the Y axis are observed in the contact areas of the weft yarns and bonding yarns and are 346 MPa. The analysis of stresses along the Z axis shows that the maximum stresses σ_{33} occur in the binding zone of the bonding yarns and equal 91 MPa. With transversal shear in the XY plane, the maximum shear stresses τ_{12} are localized at the bends of the connecting yarns at the contact points with weft threads and are about 12.5 MPa. With shear in the XZ and YZ planes, the greatest shear stresses are observed at the upper inflection point of the bonding yarns and equal to 16.6 MPa and 16.3 MPa for τ_{12}andτ_{23}.correspondingly

In the framework of verification of the developed numerical model, the results obtained were compared with the results of mechanical tests. Comparison showed the following differences from natural tests: 2.2%when stretching along the base (E_x), 15% when a shearing across the main threads (G_{xy}), 12% for the Poisson's ratio (ν_{xy}). The difference can be explained by the spread of the characteristics of the filaments and the binder and by the presence of technological defects in the full-scalespecimens.

4. Conclusion

Thus, a numerical model and a technique for calculating the effective properties of SRCM have been developed the model takes into account the properties of the material components, their volume content, and the spatial framework architecture, which is determined by the technological parameters of the weaving (number of basic and filling layers, stacking density, interlock machine type, etc.). Effective elastic characteristics for a spatially reinforced composite material are obtained. Comparison of the results of numerical simulation with mechanical tests is performed, good matching of the experimental and calculated values of the effective properties of SRCM is obtained. Analysis of stresses in the SRCM matrix for six calculated cases revealed that the most dangerous, determining the safety factor of a structure made using this type of weave, are the stresses τ_{13} observed under shear in the XZ plane.

Acknowledgments

The research was carried out at the Perm National Research Polytechnic University with the support of the Russian Science Foundation (project No. 9.7955.2017/9.10).

References

[1] Bogomolov P I, Kozlov I A and Biryulya M A 2017 *Technico-tehnologicheskie problemy servisa* **1(39)** 22-27

[2] Pazmino J, Carvelli V and Lomov S V 2014 *Composites Part B* **65** 147-157

[3] Grinev M A, Anoshkin A N, Pisarev P V, Zuiko V Yu and Shipunov G S 2015 *PNRPU Mechanics Bulletin* **4** 293-307

[4] Drach A, Drach B, Tsukrov I, Bayraktar H and Goering J 2013 *Proc. of the 19th Int. Conf. on composite materials (Canada)*

[5] Moussaid M E, Wahl J C, Perry N and Bois C 2014 *Proc. of the 16th European Conf. on composite materials (Spain)*

[6] Anoshkin A N, Zuiko V Yu, Osokin V M and Pisarev P V 2016 *Proc. Int. Conf. Youth Scientific School on "Materials and Technologies of New Generations in Modern Materials Science"* vol 156 (IOP Conf. Series: Materials Science and Engineering) p 42

[7] Drach A, Drach B, Tsukrov I, Penverne M and Lapusta Y 2014 *Proc. of the 29th Technical Conf. on composite materials (America)*

[8] Soloviev A N, Ziborov E N and Shevtsov S N 2016 *Science in the South of Russia* **12(2)** 3-10

[9] Anoshkin A N, Zuiko V Yu, Tchugaynova A V and Shustova E N 2016 *Solid State Phenomena* 10.4028/www.scientific.net/SSP.243.1

IOP Conf. Series: Materials Science and Engineering **286** (2017) 012030 doi:10.1088/1757-899X/286/1/012030

Design-experimental analysis of strength of composite flange with a delamination defect

P V Pisarev, A N Anoshkin, V M Osokin and S V Turyshev

Perm National Research Polytechnic University, 29 Komsomolski Ave., Perm, Russian Federation

E-mail: pisarev85@live.ru

Abstract. This paper presents the results of a computational and experimental study of the strength and rigidity of a composite flange with a defect in form of delamination. The locations of interlayer cracks at different loading levels are determined. It is shown that the L-shaped structurally similar element is destroyed by the delamination in the corner zone, without breaking the fibers of the layers with partial preservation of the bearing capacity. Based on the results of numerical simulation, stress and strain fields were obtained in and the strength of the structure was evaluated by the criterion of maximum stresses in layers. Analysis of stress fields showed that when testing flange samples, the most dangerous, determining the fracture initiation of the structure, are regular tensile interlayer stresses. The verification of the developed numerical model is carried out.

1. Introduction

Successful realization of the potential capabilities of polymer composite materials (PCM) in the creation of composite parts largely depends on the design and selection of the reinforcement scheme. The scheme of reinforcement and material laying of composite parts is chosen on the basis of strength calculations in accordance with the standards of strength and rigidity taking into account the features of the manufacturing technology. To calculate parameters of the composite flange units of aircraft engine parts, mathematical models should be used taking into account the inhomogeneity of the part, the anisotropy of the materials in layers and the presence of technological defects. It should be noted that a complex stress state arises in the nodes of the flange joints, high strength characteristics of the composite material in the reinforcement direction are not fully realized, the destruction of the composite flanges begins by delamination with loss of stiffness and bending of the structure [1, 2].

At the same time, the values obtained by the calculation require experimental confirmation. Carrying out experimental studies for such structures is individual and rather difficult task. So, if there are generally accepted standards for testing materials (GOST, ASTM, etc.), then for designs it is necessary to find techniques that must take into account all the features of the mechanical behavior of the object under study and ensure the possibility to use them in the experiment. In addition, it is also necessary to confirm physic-mechanical properties of the composite material used in the part [3, 4]. It should be noted that the composite material and the part are created in one technological process, while the properties of the material on standard samples may differ from the properties in the part, in addition, technological defects, including delamination, may arise during manufacturing [5, 6]. Therefore, the development of a methodology for computational and experimental research of stress-strain state and evaluation of the strength of a flange with a delamination defect is an actual task.

The object of this study is the carbon-fiber segment of the flange of the casing of an aircraft engine with a delamination defect. In this paper, the results of a design-experimental analysis of strength of a composite flange with a delamination defect are presented.

2. Mechanical tests

At the first stage of the study, the technique of mechanical testing of flange segments with a delamination defect and without it was tested. For that purpose, various dimensions of the object under study, different fixing methods were examined and specialized equipment was developed. The method for fixing the samples with the equipment on the testing machine is shown in figure 1 (a, b).

a b

Figure 1. Methods for fixing samples on the testing machine: a - rigid fixing of the lower part of the segment and application of a vertical tensile load to the upper shelf; b - rigid fixing of the lower part of the segment and application of a vertical tensile load to the upper rigidly fixed shelf.

Among the two options for fixing, the option was chosen with rigid fastening of the bottom of the segment and the upper shelf (Figure 1 b), because when using the first method of fixing (Figure 1 a), the upper cross head slipped along the inner surface of the frame shelf.

All mechanical tests of the samples were carried out at room temperature + 22°C and with a constant loading rate of 5 mm/min. During the tests, the loading force and movement of the grippers of the testing machine were recorded. In total, 5 samples without defects and 5 with a delamination defect were tested. For all samples, deformation diagrams were obtained in the "force-displacement" axle.

During the loading, after reaching a load of ~ 16 kN, acoustic signals were received - "clicks", after which there was a decrease in load. This indicates the appearance of interlayer cracks in the part. The appearance of the interlayer crack at the end surface of the sample was detected visually. Just as for a non-defective specimen, cracks are found in the upper corner zone of the inner surface segment near the layer #6. When the load increases, further cracking of the sample occurs. The tearing up of layers and fiber rupture does not occur, only the rigidity and bearing capacity of the damaged structure decreased. The sample breaks down, at a load of ~ 19 kN.

The tests of the flange segments made it possible to evaluate their strength and stiffness an indirect evaluation of the mechanical properties of the composite material was made in the presence of a delamination defect.

In all tested samples, a mechanism of loss of the load-carrying capacity has been determined, locations and indicative sizes of interlayer cracks occurring at different load levels have been identified. In all samples, cracks were found in the upper corner zone of the inner surface segment in the 5th-7th layer region. Analysis of the results obtained from the samples with a defect revealed that they are

IOP Conf. Series: Materials Science and Engineering **286** (2017) 012030 doi:10.1088/1757-899X/286/1/012030

destroyed by delamination in the corner zone, without breaking the fibers of the layers while maintaining 86.11% of the load-carrying capacity, for samples without defects this value was about 65%.

3. Numerical simulation

To calculate the stress-strain state of the flange, a three-dimensional model was constructed using the Siemens NX and Fiber SIM CAD systems. In terms of design, in the segment of the flange, a long base can be found, with four holes for the bolted connection and a shelf, which also has three holes equidistant from each other. According to the testing scheme, the lower base is fixed to the steel equipment with four bolts, the shelf is attached to the upper steel traverse of the testing machine. The delamination defect was set in the central part of the bend between layers 10 and 11, in the form of a circle with a radius of 5 mm and a thickness of 0.02 mm.

The mathematical formulation of the problem corresponded to the theory of elasticity of an anisotropic body. In the variational formulation this problem consists of finding the minimum of the Lagrange functional with additional conditions in the form of geometric Cauchy relations [7].

The boundary conditions for the computational model include the prohibition of axial displacements along X and Y for the bolt joint holes of the upper flange and a 6kN tensile load applied to the inner surfaces of the holes. For the holes for the bolted connection of the equipment and the long base of the flange, a restriction of movements in all directions is specified. For the outer surface of the tooling, the movable joint defines a boundary conditions. The contact between the layers of the flange was considered ideal. The considered boundary conditions correspond to the conditions for laboratory mechanical tests of the segment of the flange.

Evaluation of the strength of the flange segment from PCM was carried out by the criterion of maximum stresses. The strength of the orthotropic material corresponds to the satisfaction of the system of inequalities at each point of the part:

$$S_{11}^{-}\leq\sigma_{11}\leq S_{11}^{+},\ S_{22}^{-}\leq\sigma_{22}\leq S_{22}^{+},\ S_{33}^{-}\leq\sigma_{33}\leq S_{33}^{+},\ S_{12}\leq\sigma_{12},\ S_{13}\leq\sigma_{13},\ S_{23}\leq\sigma_{23}, \quad (1)$$

where S_{ii}^{+}, S_{ii}^{-}, S_{ij} – the material steady-state limits for tension, compression and shear, respectively, in the local coordinate system of the layer.

The safety factor of carbon fiber reinforced plastic (CFRP) on different components of the stress state was estimated from the equation:

$$n_{ij} = \min_{r\in V^{(1)}} \left(\frac{S_{ij}}{\sigma_{ij}(r)k}\right), \quad (2)$$

where k is the coefficient of reduction of the static strength of the material due to the influence of various environmental factors.

Discretization of the model was carried out using the spatial tetrahedral ten nodal finite elements SOLID187 with nonlinear approximation of the displacement function The total number of analyzed elements was 5 million.

Based on the results of numerical simulation of mechanical tests of the flange, stress and strain fields were obtained. Analysis of the axial displacement fields of the flange showed that the central part of the flange along the Z axis has maximum deviation from its original condition. The maximum displacement along the Y axis was 0.51 mm. The obtained values of the maximum deflection in the center of the flange shelf are in good agreement with the results obtained during laboratory tests.

Analysis of the distribution of stress fields in layers' reinforcement directions along the base and weft in the local coordinate system shows that they reach maximum value in the vicinity of the zone of layers' bend. The zones of concentration of these stresses are near the bends of the layers, in the central part of the flange. In this case, the maximum values of these stresses arise on the surface of the first and fifth layers. The maximum normal tensile stresses along the base are σ_{11} = 248.86 MPa, along the weft σ_{22} = 217.17 MPa, and the maximum compressive stresses are σ_{11} = -248.08 MPa, σ_{22} = -251.04 MPa (Figure 2 a, b).

The fields of normal interlayer stresses and tangent stresses in the plane σ_{33} and τ_{12} are shown in figure 3 (a, b). The maximum values of the normal interlayer stresses σ_{33} are observed between the first

IOP Conf. Series: Materials Science and Engineering **286** (2017) 012030 doi:10.1088/1757-899X/286/1/012030

and second layers of the central part of the bend of the CFRP layers, as well as between the 5th and 6th layers. The maximum value of the tensile interlayer stresses σ_{33} was 17.40 MPa. The compressive stresses reach an absolute value of -14.05 MPa.

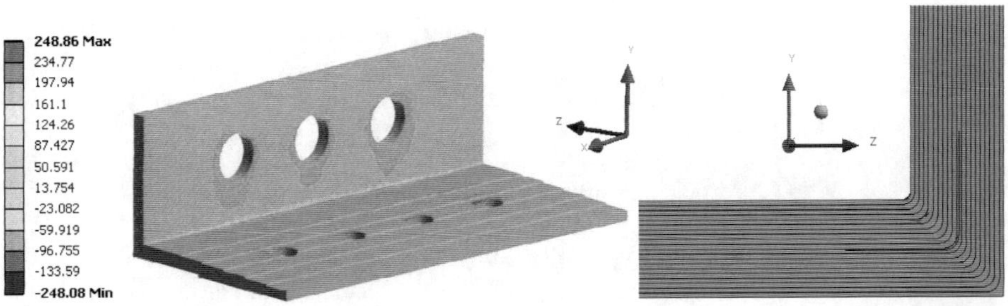

a

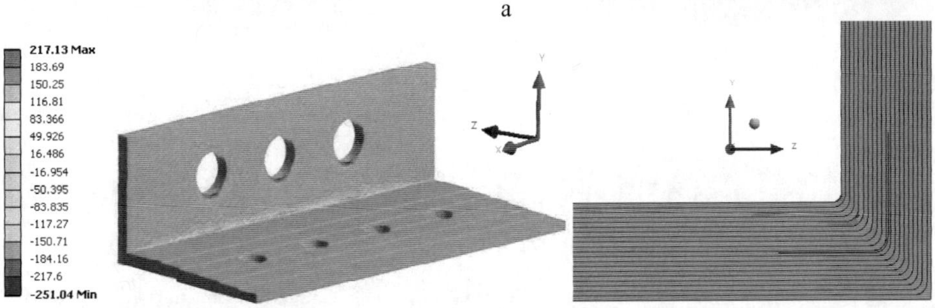

b

Figure 2. Distribution of stress fields along the base σ_{11} (a) and along the weft σ_{22} (b) in the isometry and in longitudinal plane of flange symmetry [MPa].

a

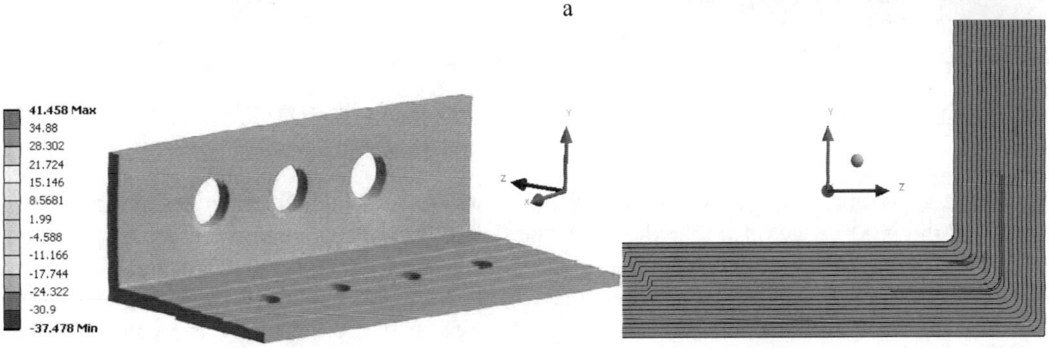

b

Figure 3. The distribution of interlayer normal σ_{33} (a) and tangent τ_{12} (b) stresses in isometric view and longitudinal plane of flange symmetry [MPa].

IOP Conf. Series: Materials Science and Engineering **286** (2017) 012030 doi:10.1088/1757-899X/286/1/012030

For tangent stresses τ_{13} and τ_{23}, concentrations are observed in the central layers near the bolt holes for flange fixing. The maximum value of the interlayer tangent tensile stresses τ_{13} and τ_{23} is 7.57 and 13.73 MPa, respectively. The maximum value of the interlayer compressive stresses τ_{13} and τ_{23} is 7.57 and 13.73 MPa, respectively (Figure 4 a, b).

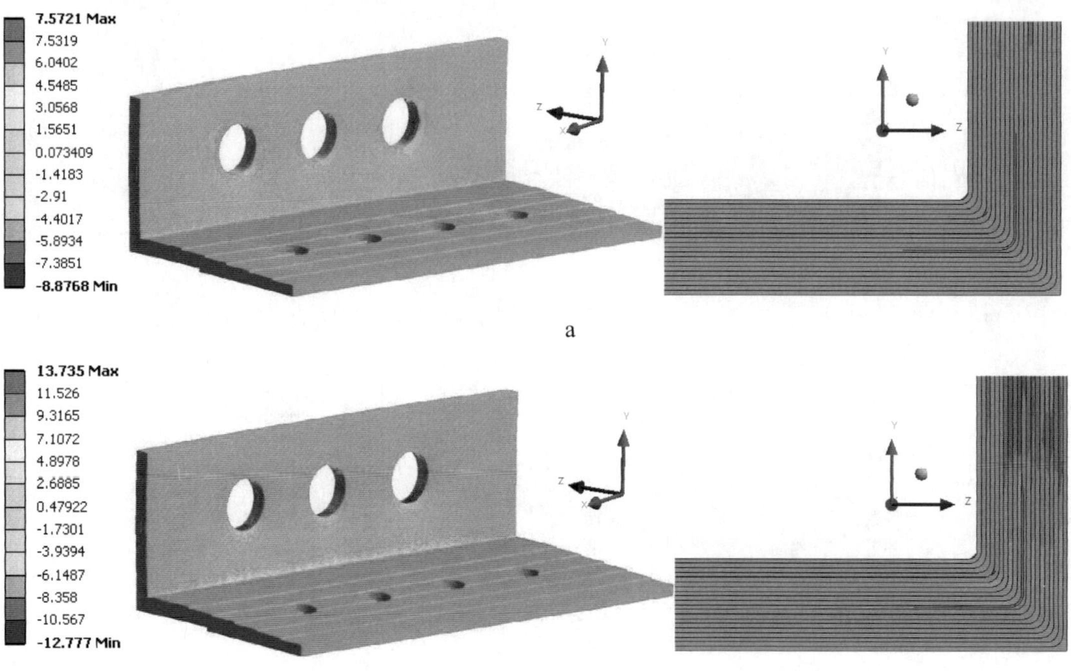

Figure 4. Distribution of interlayer tangent stresses τ_{13} (a) and τ_{23} (b) stresses in isometry and in the longitudinal symmetry plane of the flange [MPa].

The static strength of the flange segment was evaluated using the maximum stress criterion (1). The values of the static strength of CFRP are shown in Table 1. The presented data was obtained experimentally, with the exception of the strength characteristics crosswise of the S_{33} layer, the values of which were taken from [8]. The coefficient k in equation (2) taking into account the decrease in the static strength of the material due to the influence of various environmental factors, was taken to be 1.2, which corresponds to the requirements imposed on structures of this type.

Table 1. CFRP static strength properties.

Material	S_{11}^{+} (MPa)	S_{11}^{-} (MPa)	S_{22}^{+} (MPa)	S_{22}^{-} (MPa)	S_{33}^{+} (MPa)	S_{33}^{-} (MPa)	S_{12} (MPa)	S_{13} (MPa)	S_{23} (MPa)
CFRP	809	804	809	804	56.4	128	150	77	77

The evaluation of the strength of a composite flange with a defect under calculated loads (Table 2) revealed that the most dangerous in the design of the flange from PCM are normal interlayer stresses. The margin of the static strength of the carbon-fiber flange is defined by these stresses and equal to 2.59. It can be expected that the destruction of the structure will begin by the delamination in the regions of the bending in the central layers of the structure in the region where the defect is located. It can be predicted that the damage of the structure will occur at a load of 15.54 kN.

Table 2. Coefficients of static strength of CFRP flange design.

Material	n_{11}	n_{22}	n_{33}	n_{12}	n_{13}	n_{23}
CFRP	3.72	3.26	2.59	3.61	10	5.72

4. Conclusion

Thus, in this work, samples of flanges from PCM without defects, as well as with delamination defects were tested. In total, 10 samples were tested. The locations of interlayer cracks at different loading levels are determined. It is shown that the structurally similar element of the L-shaped profile is damaged by the delamination in the corner zone, without breaking the fibers of the layers with partial preservation of the bearing capacity. For samples with defects in laboratory tests, characteristic features of mechanical behavior were revealed: a low difference between the delamination load and fracture load was found (13.89%), while for samples without defects, the difference was 35%.

Based on the results of numerical simulation of mechanical tests of the flange samples, stress and strain fields in the layers were obtained and the strength of the structure was evaluated by the criterion of maximum stresses in the layers. Analysis of stress fields showed that when testing flange samples, the most dangerous, determining the beginning of the destruction of the structure, are normal tensile interlayer stresses. Calculations on the developed model showed that the destruction of the structure will occur at a load of 15.54 kN. A comparison with the results of the tests showed that the calculated value of the critical load of the beginning of the delamination is 10.41% larger than the observed value in the experiment.

Thus, mathematical modeling using the structural phenomenological model allowed to determine the rigidity of the structure, the critical load and the region of the appearance of the delamination recorded during the laboratory tests. The difference in the calculated and experimental data can be caused both by the indicative value of the strength of the laminate composite for the layer peeling-off used for estimating the strength of the structure, and by the scatter of the characteristics of materials observed in laboratory tests.

Acknowledgments

The research was carried out at the Perm National Research Polytechnic University with the support of the Russian Science Foundation (project No. 15-19-00259).

References

[1] Anoshkin A N, Zuiko V Yu, Osokin V M, Tretyakov A A and Pisarev P V 2015 *PNRPU Mechanics Bulletin* **2** 5-21

[2] Anoshkin A N, Zuiko V Yu, Osokin V M and Pisarev P V 2016 *Proc. Int. Conf. Youth Scientific School on "Materials and Technologies of New Generations in Modern Materials Science"* vol 156 (IOP Conf. Series: Materials Science and Engineering) p 42

[3] Sorrentino L, Esposito L and Bellini C 2017 *Composites Part B* **109** 187-196

[4] Clifford J, Majumdar P, Katiyar P and Wilkes R 2016 *Proc. of the 31st Technical Conf. on composite materials (USA)*

[5] Grunenfelder L K, Dills A, Centea N and Nutt S 2017 *Composites Part A* **93** 88-99

[6] McElroy M, Leone F and Ratcliffe J 2015 *Composites Part A* **79** 192-202

[7] Anoshkin A N, Fedorovtsev D I, Pisarev P V and Osokin V M 2015 *PNRPU Mechanics Bulletin* **43** 116-129

[8] Grinev M A, Anoshkin A N, Pisarev P V, Zuiko V Yu and Shipunov G S 2015 *PNRPU Mechanics Bulletin* **4** 293-307

IOP Conf. Series: Materials Science and Engineering **286** (2017) 012031 doi:10.1088/1757-899X/286/1/012031

Fiber optics method for nanomaterials diagnostics

O O Ovseychook, V I Ivanov, A V Myagotin and G D Ivanova

Department of Physics and Theoretical Mechanics, Far Eastern State Transport University, Khabarovsk, Serysheva street, 47, Russia

E-mail: tmeh@festu.khv.ru

Abstract. This paper analyzed the light induced lens response at the fiber optics scheme in transparent nanosuspension with electrostrictive nonlinearity. The theoretical analysis of the light-induced mass transfer in the nanosuspension was carried out for large intensities of the Gaussian beam radiation, when the concentration change is comparable to the primary. The nonlinear lens in this mode is the exponential function of the incident light intensity. The results are relevant to the study of the radiation self-action in the nanosuspension and optical diagnostics of such materials.

1. Introduction

Thermal lens technique is widely used for the optical diagnostics of materials [1-3]. The light-induced thermal lens in a homogeneous fluid is formed as a result of thermal expansion of a medium. In two-component fluid the heat flow also can cause concentration stream arising from occurrence of thermodiffusion (Soret effect [2]). Another mechanism of optical nonlinearity of the medium is due to the forces operating on the particles of the dispersed phase in gradient light field. Gradient electrostrictive forces may cause concentration streams in nanosuspension [4]. Depending on the sign of polarizability nanoparticles drift from areas with higher intensity of the electromagnetic wave (if their refractive index is more than one for liquid) [5]. This optical nonlinearity was studied experimentally and theoretically in nanosuspensions [5] and microemulsions [6] (artificial Kerr media). The nonlinear response corresponds to cubic nonlinearity for small intensities of radiation, because changing of the particle concentration (and effective refractive index) is proportional to radiation intensity [7].

The aim of this work is to analyze the light induced lens response at the fiber optics scheme in transparent nanosuspension at high intensities of radiation.

2. Light induced lens response

Let us take a single beam light lens fiber scheme [8]. In our scheme, two co-axis fibers are inserted to the optical cell with nanosuspension. The first fiber generates the Gaussian beam and the second one is connected with the photodiode to detect the passed beam power.

For a Gaussian beam we have the next distribution in a plane perpendicular to the optical axis z:

$$I = I_0[1+(z/l_0)^2]^{-1} \exp\{-r^2(z)/r_1^2(z)\}, \tag{1}$$

$$r_1(z) = r_0[1+(z/l_0)^2]^{0.5}, \tag{2}$$

where I_0 is the intensity of radiation on the axis in the plane of the beam waist, λ is radiation wavelength, r is the distance from the axis of the beam, r_0 is the radius of the beam waist, $r_1(z)$ is the radius of the beam at a distance z from the beam waist, $l_0 = (\pi r_0^2 / \lambda)$ is the confocal parameter.

Thermal lens signal $\Theta(t)$ shows the change of the beam intensity $I(t)$ on the optical axis behind the nonlinear layer [7]:

$$\Theta(t) = \frac{I(t) - I(0)}{I(0)}. \tag{3}$$

The light lens signal is defined by the lens transparency of the cell [9]:

$$\Theta(t) = \frac{2(z_1 / l_0) F_{nl}(0,t)}{(1 + z_1^2 / l_0^2)(1 + 3z_1^2 / l_0^2)}, \tag{4}$$

where the distance between the two fiber ends is $2z_1$, $F_{nl}(0,t)$ is nonlinear phase of the optical path on the beam axis.

3. Light induced mass transport task

We will consider the transparent nanosuspension under the influence of laser radiation with Gaussian intensity profile [9].

Balanced equation describing the dynamics of concentration of nanoparticles considering diffusion and electrostrictive flows can be written as [4]:

$$\frac{\partial C}{\partial t} = D\Delta C - div(\gamma C \nabla I), \tag{5}$$

where $C(r,t)$ is mass concentration of dispersed nanoparticles, D is the diffusion coefficient, $\gamma = 4\pi\beta D(\bar{c} n k_B T)^{-1}$, β is the particle polarizability, k_B is the Boltzmann constant, n is effective refractive index, $\bar{c}$ is the speed of light in vacuum, $I(r)$ is the intensity of radiation.

The source in (4) is due to the gradient force F_∇ by the electric field of the light wave [5]:

$$F_\nabla = 4\pi\beta \nabla I / \bar{c} n. \tag{6}$$

In the stationary mode equation (5) takes the next form:

$$0 = D\Delta C - div(\gamma C \nabla I). \tag{7}$$

The general solution to the equation (6) is looking for in the form of:

$$C = B\exp\{I / I_s\}, \tag{8}$$

where $I_S = \gamma D^{-1}$ is the “saturation” intensity, B is the normalization constant.

Let us introduce dimensionless parameter of light intensity $\alpha = I_0 I_S^{-1}$. For small intensities of radiation the change of the particle concentration is proportional to the radiation intensity ($\alpha << 1$).

The constant of normalization $B(\alpha)$ is given from conservation of particle number:

IOP Conf. Series: Materials Science and Engineering **286** (2017) 012031 doi:10.1088/1757-899X/286/1/012031

$$\int_0^R C2\pi r dr = \pi R^2 C_0, \tag{9}$$

where R is the radius of the cylindrical cell, C_0 is the initial concentration of dispersed particles.

The concentration dependence versus the distance from the axis of the beam $\rho = (r/r_0)$ was calculated for $R/r_0 = 5$.

Figure 1 shows that the concentration of the nanoparticles $C_{st} = C/C_0$ on the axis of the optical beam depends exponentially versus the beam intensity (as opposed to the usual cubic nonlinearity).

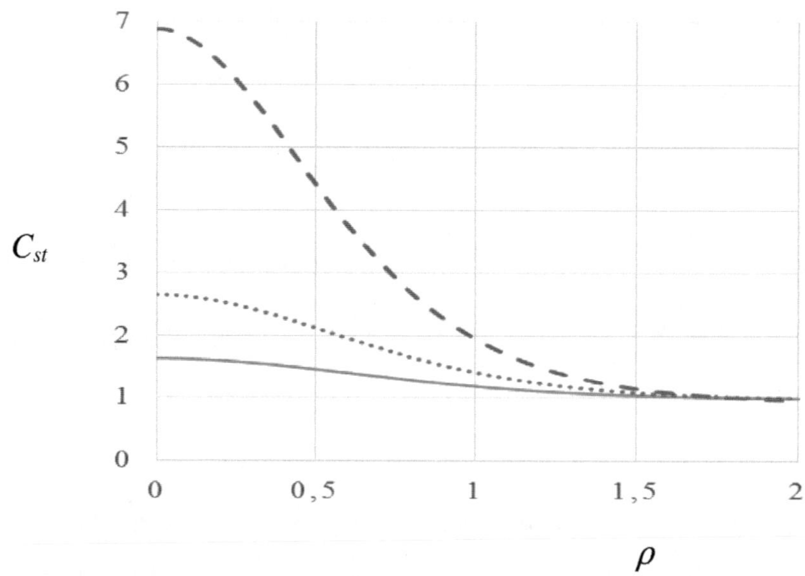

Figure 1. The concentration of nanoparticles $C_{st} = C/C_0$ (a.u.) versus distance ρ (a.u.) from the axis of the beam for different values of the intensity of the light: $\alpha_1 = 0{,}5$ (___); $\alpha_2 = 1$ (.......); $\alpha_3 = 2$ (_ _).

The nonlinear phase $F_{nl}(0)$ is proportional to the change of the concentration ΔC_{st} on the axis of the beam (for the stationary regime):

$$F_{nl} = 2d(\partial n/\partial C)C_0 \Delta C_{st}, \tag{10}$$

$$\Delta C_{st} = B(\exp\alpha - 1). \tag{11}$$

where $\Delta C_{st} = (C_{st} - 1)$.

Figure 2 plots the change of the concentration ΔC_{st} on the axis of the beam versus normalized intensity α.

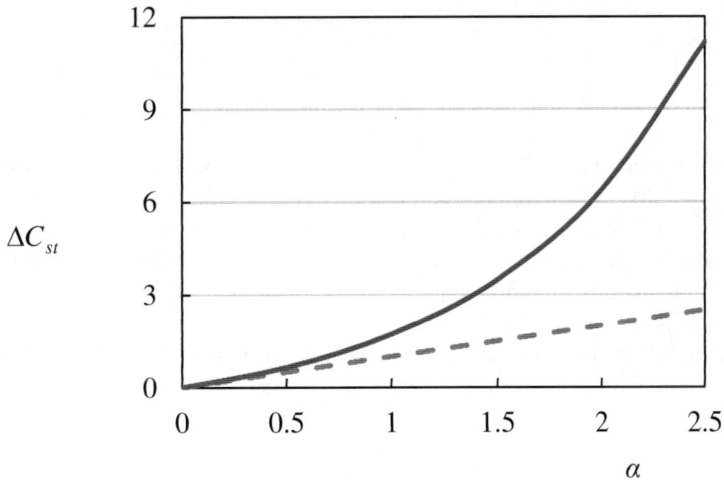

Figure 2. The change of the concentration ΔC_{st} on the axis of the beam versus normalized intensity α (——). The linear dependence $\Delta C_{st} = \alpha$ is shown for comparison (- - -).

In nanosuspension the particle radius is much smaller than the radiation wavelength λ, therefore the refractive index of the medium is proportional to the concentration of particles [8]:

$$n = n_1(1 + f\delta), \tag{12}$$

where $\delta = (n_2 - n_1)/n_1$; n_1 and n_2 are the refractive indices of the substance and the dispersion medium of the dispersed phase, respectively; $f = v_0 C$ is the volume fraction of the dispersed phase, r is the radius of the nanoparticle, $v_0 = (4/3)\pi r^3$ is the volume of the one nanoparticle, C is the concentration of nanoparticles.

$$F_{nl}(0) = 2dn_1 v_0 \delta C_0 \Delta C_{st}, \tag{13}$$

Thus, the expression for stationary response was achieved for the light induced lens response in nanosuspension. As the lens grows exponentially, it is fair to call it "super lens".

4. Conclusions

The light induced lens response is analyzed in the fiber optics scheme. We have analyzed the two-dimensional diffusion in the nanosuspension with electrostrictive nonlinearity in a Gaussian beam radiation field. As a result the exact expressions were achieved for the light induced lens response in nanosuspension.

The analysis gives you the ability to determine not only concentration of nanoparticles, but also the transport coefficients of nanoparticles [10-12]. The results are relevant to optical diagnostics of dispersed liquid nanomaterials, including the thermal lens spectroscopy [13-15].

References

[1] Cabrera H, Sira E, Rahn K and García-Sucre M 2009 *Appl. Phys. Let.* **94** 051103
[2] Ivanov V I and Ivanova G D 2016 *Journal of Physics: Conference Series* **735** 012037
[3] Vicary L 2002 *Philos. Mag.* B. **82** 447-452
[4] Ivanov V I, Ivanova G D and Khe V K 2017 *Proc. SPIE* **10176** 1017607

IOP Conf. Series: Materials Science and Engineering **286** (2017) 012031 doi:10.1088/1757-899X/286/1/012031

[5] Freysz E, Afifi M, Ducasse A, Pouligny B and Lalanne J R 1984 *J. Opt. Soc. Amer. B* **1** 433
[6] Smith P W, Maloney P J and Ashkin A 1982 *Opt. Lett.* **7**(8) 347.
[7] Ivanov V I , Ivanova G D, Okishev K N and Khe V K 2016 *Proc. SPIE* **10035** 100354Y
[8] Lee W, El-Ganainy R, Christodoulides D, Dholakia K and Wright E 2009 *Optics Express* **17** 10277
[9] Ivanov V, Ivanova G, Okishev K and Khe V 2017 *IOP Conf. Ser.: Mater. Sci. Eng.* **168** 012045
[10] Leppla C and Wiegand S 2003 *Philosoph. Mag.* **83** 1989
[11] Giglio M and Vendramini A 1974 *Appl. Phys. Lett.* **25** 555
[12] Chintamani P, Shalini M, Agnel P, Meera V, Tejas I H and Radha S 2014 *International Journal of Chemical and Physical Sciences* **3**(5) 44
[13] Sheldon S J, Knight L V and Thorne J M 1982 *Appl. Opt.* **12** 1663
[14] El-Ganainy R, Christodoulides D, Rotschild C and Segev M 2007 *Opt. Express* **15** 10207
[15] Smith P W, Ashkin A and Tomlinson W J 1981 *Opt. Lett.* **6** 284

IOP Conf. Series: Materials Science and Engineering **286** (2017) 012032 doi:10.1088/1757-899X/286/1/012032

On the symmetry identification for the multi-level models of polycrystalline materials

K V Ostapovich and P V Trusov

Department of Mathematical Modeling of Systems and Processes, Perm National Research Polytechnic University, 29 Komsomolsky prospekt, Perm, 614990, Russian Federation

E-mail: ostkirvad@gmail.com

Abstract. An approach to identify the elastic symmetry of anisotropic materials is considered and applied together with the multi-level constitutive models of polycrystals. By using the two-level model of elasto-visco-plasticity, the changes in the symmetry of the elasticity tensor for the representative volume element of a polycrystalline copper during inelastic deformation are studied. The two-link strain trajectories of the "simple shear – simple shear" type are analyzed and the problem of the compatibility of the properties under investigation with any symmetry classes is formulated and solved at various deformation stages.

1. Introduction

Physic-mechanical properties of materials are dependent on their microstructure so the control of its formation during different technological processes makes it possible to obtain components and structures with unique exploitation characteristics. During the large inelastic deformation of polycrystalline metals and alloys, the evolution of inner structure is especially influenced by the mechanisms related to the rotation of crystallites. The essentially inhomogeneous distribution of crystal lattice orientations (i.e. crystallographic texture) in such materials may results in the significant anisotropy of their macroscopic properties. Herewith, the establishing of the orthogonal transformations, with respect to which the constitutive equations are invariant plays an important role in different technical applications.

In this paper, with the help of a multi-level approach and crystal elasto-visco-plasticity, changes in the elastic properties of the representative volume element (RVE) of a polycrystalline copper during inelastic deformation along two-link trajectories are numerically investigated. The incompatibility of these properties with some special classes is analyzed at the various stages of the loading processes in question. The advantage of the designated approach is the possibility of the direct consideration of the material's evolving microstructure, which allows the explicit determination of all macroscopic characteristics.

2. Elastic symmetry identification problem

Let us consider an anisotropic material, the elastic deformation of which is governed by the generalized Hooke's law:

$$\boldsymbol{\sigma} = \boldsymbol{\Pi} : \boldsymbol{\varepsilon}, \tag{1}$$

where $\boldsymbol{\sigma}$ and $\boldsymbol{\varepsilon}$ are the Cauchy's stress and the small strain (2-rank) tensors, respectively; $\boldsymbol{\Pi}$ is the elasticity (4-rank) tensor. Depending on the symmetry group of $\boldsymbol{\Pi}$, the material under examination can be assigned to some of the eight *elastic symmetry classes* [1]. It is of a certain interest to solve the *symmetry identification problem (SIP)* [2, 3], which is to form the hierarchy of the classes related to this tensor with a determined error.

The structure of a symmetry class allows us to represent its tensor, $\boldsymbol{\Pi}^{(s)}$, where s is the significator of the class in the framework of the accepted classification in the following form (hereinafter we adopt the summation convention for repeated indexes, which are not in the parentheses):

$$\boldsymbol{\Pi}^{(s)} = \Pi_\alpha \mathbf{O} * \mathbf{K}_\alpha^{(s)} . \tag{2}$$

Here Π_α are arbitrary independent constants, $\mathbf{O}$ is an orthogonal tensor, $\mathbf{K}_\alpha^{(s)}$ are the linearly independent 4-rank tensors defined in the terms of some orthonormal basis, $\{\mathbf{l}_i\}$ (hereinafter we associate this basis with the laboratory coordinate system) and the «$*$» operation is the Rayleigh product [4]. Equation (2) implies that the multi-dimensional matrix of $\boldsymbol{\Pi}^{(s)}$ has the special form in the orthonormal basis, $\{\mathbf{k}_i\}$, where $\mathbf{k}_i = \mathbf{O} \cdot \mathbf{l}_i$, which is characteristic of the tensors from the same class. In this regard the specified basis can be termed *canonical.*

Let $\|\cdot\|_r$ be the Frobenius norm on the space of r-rank tensors. We call a quantity, $ME\left[\boldsymbol{\Pi}^{(s)}, \boldsymbol{\Pi}\right]$, such that $\left\|\boldsymbol{\Pi}^{(s)} : \boldsymbol{\varepsilon} - \boldsymbol{\Pi} : \boldsymbol{\varepsilon}\right\|_2 \le ME\left[\boldsymbol{\Pi}^{(s)}, \boldsymbol{\Pi}\right]\|\boldsymbol{\varepsilon}\|_2$ holds for all $\boldsymbol{\varepsilon}$ the *mismatch estimate (ME)* for the approximation, $\boldsymbol{\Pi}^{(s)}$, with respect to the tensor, $\boldsymbol{\Pi}$. This estimate can be associated with the level of the maximal absolute error, which occurs in stresses determined by (1) when approximating the elasticity tensor by its symmetric analogue. Thus, it is possible to reduce the SIP for a given elasticity tensor to finding its approximation with the acceptable small value of the ME [3].

One should note that the "exact" ME is defined by the so-called operator norm, $\|\cdot\|$: $\left\|\boldsymbol{\Pi}^{(s)} - \boldsymbol{\Pi}\right\| = \inf\limits_{\|\boldsymbol{\varepsilon}\|_2 = 1} \left\|\boldsymbol{\Pi}^{(s)} : \boldsymbol{\varepsilon} - \boldsymbol{\Pi} : \boldsymbol{\varepsilon}\right\|_2$, the minimization of which over the admissible sets of the tensors of the form (2) for different classes gives the solution of the SIP. Nevertheless, from the computational point of view, its upper bound, $\left\|\boldsymbol{\Pi}^{(s)} - \boldsymbol{\Pi}\right\|_4$ (i.e. the quantity defined by the Frobenius norm), is more convenient to use it as an object function. In the case when $\mathbf{K}_\alpha^{(s)}$ are biorthonormal with respect to the total scalar product, the corresponding optimization problem can be formulated as follows:

$$\textit{Find } \breve{\mathbf{O}} \textit{ such that } \sum_\alpha \left(\left(\mathbf{O}^{\mathrm{T}} * \boldsymbol{\Pi}\right) \circ \mathbf{K}_\alpha^{(s)}\right)^2 \to \max_{\mathbf{O}} . \tag{3}$$

In the general case, it is sufficient to consider the special orthogonal group of 2-rank tensors as the admissible set for the values of $\mathbf{O}$. The solution, $\breve{\mathbf{O}}$, of the stated problem determines the orientation of the canonical basis of the tensor, $\breve{\boldsymbol{\Pi}}^{(s)} = \left(\boldsymbol{\Pi} \circ \left(\breve{\mathbf{O}} * \mathbf{K}_\alpha^{(s)}\right)\right)\left(\breve{\mathbf{O}} * \mathbf{K}_\alpha^{(s)}\right)$, which is the closest to $\boldsymbol{\Pi}$ with respect to the Frobenius norm. As it can be seen, $\breve{\boldsymbol{\Pi}}^{(s)}$ is the orthogonal projection of $\boldsymbol{\Pi}$ on the span of $\breve{\mathbf{O}} * \mathbf{K}_\alpha^{(s)}$. It is worthy to remark that such projection preserves positive definiteness [5].

A canonical basis for a tensor from a symmetry class is not unique. In particular, the basis obtained from the canonical one by a transformation from the symmetry group of the given tensor is also canonical. Such fact allows us to reduce the admissible set of the problem (3) to tensors, with respect to which the special orthogonal group is defined as the union of the left cosets [6] in the symmetry group. Thus, in the framework of the proposed approach, optimization is not required to find the isotropic approximation and it is sufficient to consider only the orientation of the isotropy axis for the

IOP Conf. Series: Materials Science and Engineering **286** (2017) 012032 doi:10.1088/1757-899X/286/1/012032

transversal isotropic one. To describe this orientation, we shall use the angles, (θ, φ), of the spherical coordinate system, which parameterize the component matrix of **O** in the problem (3) as follows:

$$\{O_{ij}\} = \begin{pmatrix} \cos\theta\cos^2\varphi + \sin^2\varphi & (\cos\theta - 1)\cos\theta\sin\varphi & \sin\theta\cos\varphi \\ (\cos\theta - 1)\cos\varphi\sin\varphi & \cos\theta\sin^2\varphi + \cos^2\varphi & \sin\theta\sin\varphi \\ -\sin\theta\cos\varphi & -\sin\theta\sin\varphi & \cos\theta \end{pmatrix}. \quad (4)$$

As for other cases, the orientation of the canonical basis can be expressed e.g. by the Euler angles, (α, β, γ), in the form:

$$\{O_{ij}\} = \begin{pmatrix} \cos\alpha\cos\beta\cos\gamma - \sin\alpha\sin\gamma & -\sin\alpha\cos\gamma - \cos\alpha\cos\beta\sin\gamma & \cos\alpha\sin\beta \\ \sin\alpha\cos\beta\cos\gamma + \cos\alpha\sin\gamma & \cos\alpha\cos\gamma - \sin\alpha\cos\beta\sin\gamma & \sin\alpha\sin\beta \\ -\sin\beta\cos\gamma & \sin\beta\sin\gamma & \cos\beta \end{pmatrix}. \quad (5)$$

A symmetric approximation, $\bar{\mathbf{\Pi}}^{(s)}$, obtained from the numerical solution of the problem (3) causes an absolute error with the exact upper estimate defined by $\left\|\bar{\mathbf{\Pi}}^{(s)} - \mathbf{\Pi}\right\|$. To characterize the relative error of $\bar{\mathbf{\Pi}}^{(s)}$, we introduce the so-called *relative mismatch estimate (RME)* as $\overline{RME}^{(s)} = \left\|\mathbf{\Pi}'^{-1}\right\|\left\|\bar{\mathbf{\Pi}}^{(s)} - \mathbf{\Pi}\right\|$, where $\mathbf{\Pi}'^{-1}$ is the generalized inverse [2] tensor of $\mathbf{\Pi}$. It is simply to show that the introduced quantity satisfies $\left\|\bar{\mathbf{\Pi}}^{(s)} : \boldsymbol{\varepsilon} - \mathbf{\Pi} : \boldsymbol{\varepsilon}\right\|_2 \le \overline{RME}^{(s)} \left\|\mathbf{\Pi} : \boldsymbol{\varepsilon}\right\|_2$ for all $\boldsymbol{\varepsilon}$.

3. Numerical experiment

By using the two-level constitutive model of elasto-visco-plasticity [7], a numerical investigation on elastic symmetry changes during inelastic deformation was performed for the RVE of a polycrystalline copper. A single realization of the aggregate of 1000 crystallites with initially uniform orientation distribution was examined. The analysis was focused on the kinematic two-link strain trajectories of the type "simple shear – simple shear" with the constant velocity gradients of the following forms.

Stage I: $\nabla\mathbf{v}_I = 5\times10^{-4}\left(\mathbf{l}_1\mathbf{l}_1 - \mathbf{l}_2\mathbf{l}_2 + \mathbf{l}_1\mathbf{l}_2 - \mathbf{l}_2\mathbf{l}_1\right)$.

Stage II:

$$\nabla\mathbf{v}_{II} = 5\times10^{-4}\left(\mathbf{l}_1\mathbf{l}_1 - \cos^2\psi\mathbf{l}_2\mathbf{l}_2 - \sin^2\psi\mathbf{l}_3\mathbf{l}_3\cos\psi\left(\mathbf{l}_1\mathbf{l}_2 - \mathbf{l}_2\mathbf{l}_1\right) + \sin\psi\left(\mathbf{l}_1\mathbf{l}_3 - \mathbf{l}_3\mathbf{l}_1\right) - \cos\psi\sin\psi\left(\mathbf{l}_2\mathbf{l}_3 + \mathbf{l}_3\mathbf{l}_2\right)\right).$$

The value of the parameter, ψ, was varied from 0° to 180°. Each stage was followed by an elastic unloading down to zero macroscopic stress. The time durations of the active loadings are $T_{I,II}$ = 2000 s. Let us note that the process according to the provided scheme corresponds to the two consecutive stages of equal channel angular pressing with the channels intersection angle of 90° and the rotation of the sample around its axis by the angle, ψ, before the second route.

4. Results

The direct pole figures (DPFs) for the polycrystalline aggregate in different configurations are shown in figure 1 (hereinafter the value of ε is equal to the accumulated effective strain). The considered cases show the acquirement of the significant inhomogeneity by the crystallites' orientation distribution.

IOP Conf. Series: Materials Science and Engineering **286** (2017) 012032 doi:10.1088/1757-899X/286/1/012032

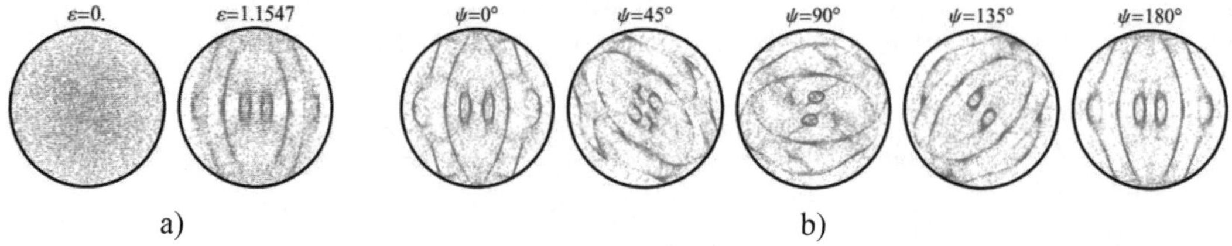

Figure 1. DPFs for <111> directions (projecting along $\mathbf{l}_3$):
a) initial and final configurations of the Stage I; b) final configurations of the variants of the Stage II.

The plot presented in figure 2a describes the changes in $\overline{RME}^{(s)}$ of the aggregate elastic properties with respect to some symmetry classes during the Stage I. In figure 2b, the same quantity is plotted for the final configurations of the Stage II in dependence on ψ. All outcomes of the process demonstrate the amplification of the elastic properties non-conformance with the classes under consideration compared to the initial configuration.

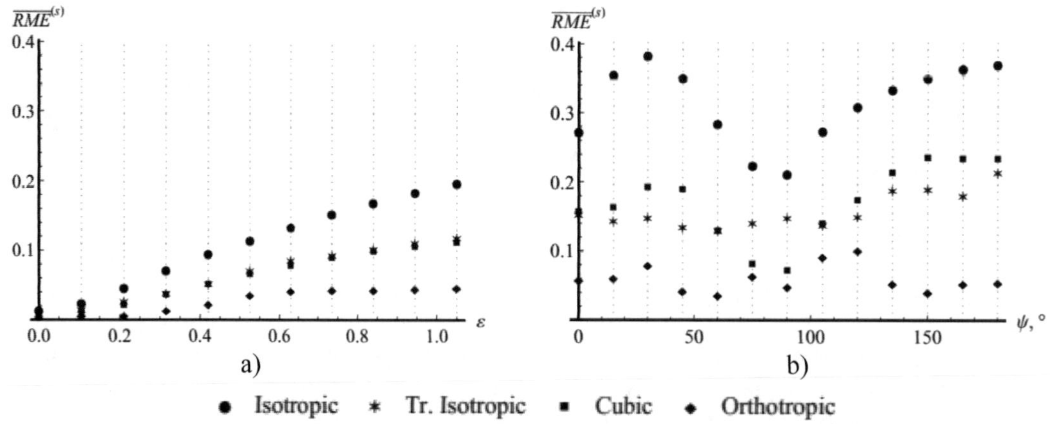

Figure 2. Changes in RMEs of the aggregate elastic properties in some symmetry classes:
a) during the Stage I; b) in the result of the variants of the Stage II.

One should note that, in some cases, the RMEs are not strictly increasing during the Stage II as shown e.g. in figure 3a, where the values of $\overline{RME}^{(iso)}$ (in the isotropy class) are provided for different ψ. The curves interpolating the dependence of the same quantity on this angle at various moments of the Stage 2 are plotted in figure 3b. As it is easy to see, the curves have pronounced minima at $\psi = 90°$, near which the changes in $\overline{RME}^{(iso)}$ are not monotonic.

Figures 4-6 show the orientations of the canonical bases of the optimal symmetric approximations obtained for the aggregate in some configurations. At each plot, the family of the points refers to parameters of the orientation tensors, (4) or (5), differing from each other by a transformation from the related symmetry group. In the case of orthotropic approximations, the provided orientations are associated with the longitudinal elastic moduli in the non-increasing order.

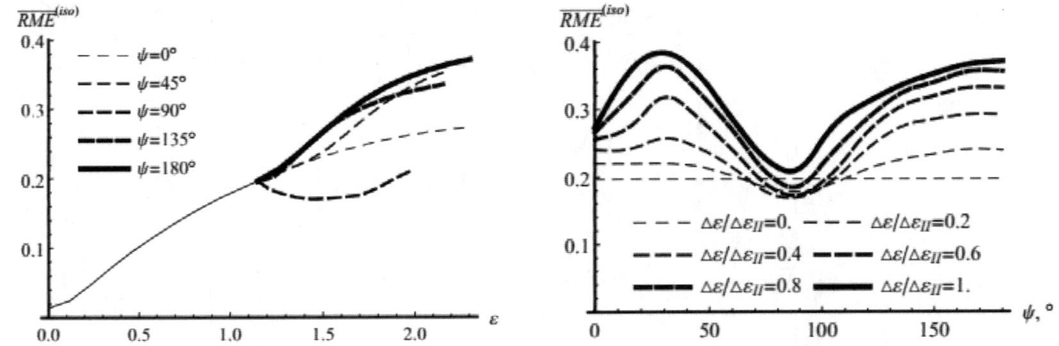

a) b)

Figure 3. RMEs of the aggregate elastic properties in the isotropy class:
a) different variants of the Stage II; b) various moments of the Stage II.

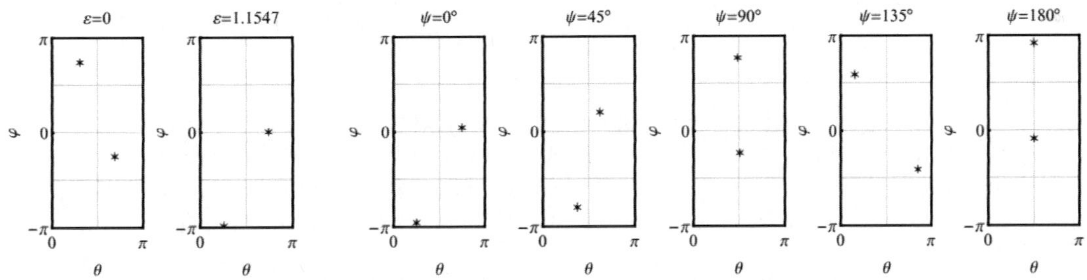

a) b)

Figure 4. Orientations of the canonical bases
for the optimal transversal isotropic approximations of the aggregate elasticity tensor:
a) initial and final configurations of the Stage I; b) final configurations of the variants of the Stage II.

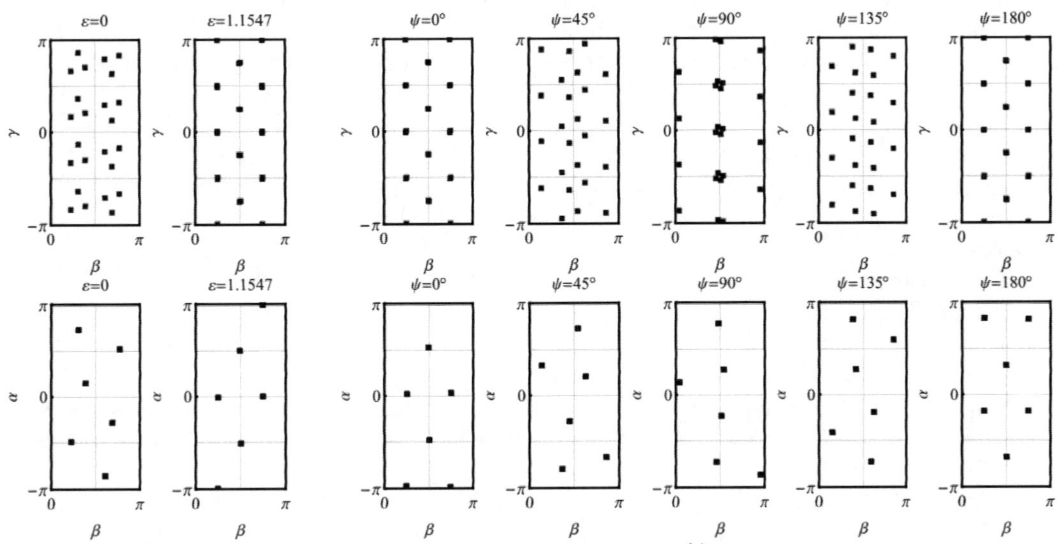

a) b)

Figure 5. Orientations of the canonical bases
for the optimal cubic approximations of the aggregate elasticity tensor:
a) initial and final configurations of the Stage I; b) final configurations of the variants of the Stage II.

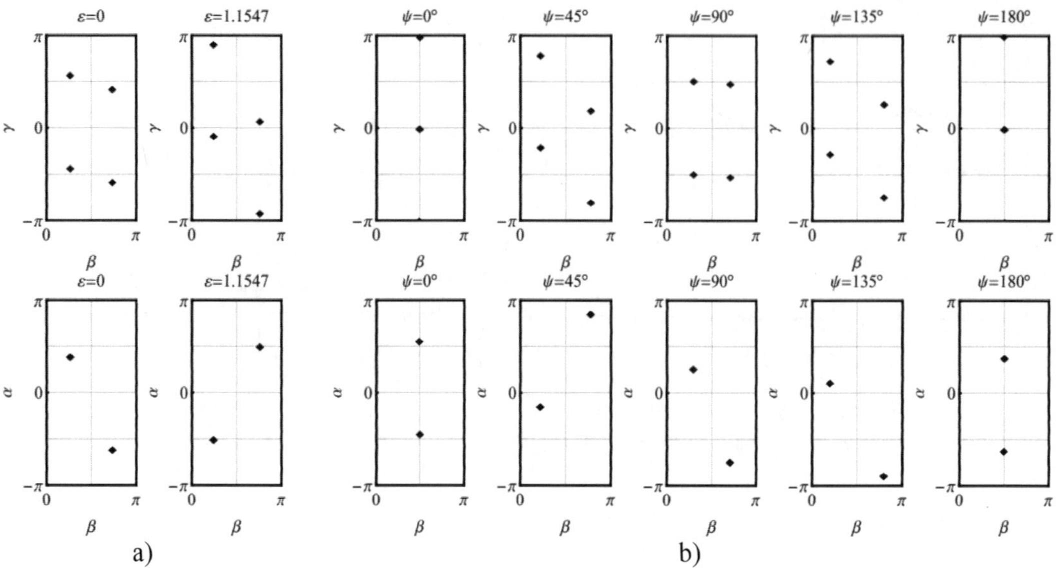

Figure 6. Orientations of the canonical bases for the optimal orthotropic approximations of the aggregate elasticity tensor: a) initial and final configurations of the Stage I; b) final configurations of the variants of the Stage II.

5. Conclusion

In this work, an approach to solve the SIP for anisotropic materials is considered in application to multi-level-modeling. An investigation on changes in the elastic symmetry of the RVE of a polycrystalline copper during inelastic deformation is performed with the help of a two-level model of crystal elasto-visco-plasticity. Kinematic two-link strain trajectories of a special form are analyzed. The used approach is based on the approximation of the macroscopic elasticity tensor by the tensors of various symmetry types, which minimize some upper estimate of the residual norm in the generalized Hooke's law. As a result, it is possible to form the hierarchy of the symmetry classes, which the material in given configurations can be related to (by its elastic properties) with a known error.

Acknowledgments

The work was supported by the Russian Foundation for Basic Research (grants No. 17-41-590694-p_a, No. 17-01-00379-a).

References

[1] Bóna A, Bucataru I, Slawinski M A 2004 Material symmetries of elasticity tensors *Q. J. Mech. Appl. Math.* **57** 583–598
[2] Ostapovich K V, Trusov P V 2016 On elastic anisotropy: symmetry identification *Mekhanika Kompositsionnykh Mater. i Konstr.* **22** 69–84
[3] Ostapovich K V, Trusov P V 2017 On elastic symmetry identification of polycrystalline materials *Non-Equilibrium Processes in Continuous Media* (Perm: Perm State University) **2** 122–124
[4] Bertram A 2012 *Elasticity and Plasticity of Large Deformations* (Berlin: Springer-Verlag)
[5] Gazis D C, Tadjbakhsh I, Toupin R A 1963 The elasticity tensor of a given symmetry nearest to an anisotropic elastic tensor *ActaCryst.* **16** 917–922
[6] Baumslag B, Chandler B 1968 *Theory and problems of group theory* (New York: McGraw-Hill)
[7] Shveikin A I, Trusov P V 2016 Correlation between geometrically nonlinear elasto-visco-plastic constitutive relations formulated in terms of the actual and unloaded configurations for crystallites *Phys. Mesomech.* **5** 48–57

IOP Conf. Series: Materials Science and Engineering **286** (2017) 012033 doi:10.1088/1757-899X/286/1/012033

Analysis of structure of hyperfine poly(3-hydroxybutyrate) fibers (PHB) for controlled drug delivery

A A Olkhov[1,3], R Yu Kosenko[3], V S Markin[3], A K Zykova[1,2], P V Pantyukhov[1,2], S G Karpova[2] and A L Iordanskii[3]

[1]Department of Chemistry and Physics, Plekhanov Russian University of Economics, 36 Stremyanny Lane, Moscow, 117997, Russia
[2]Emanuel Institute of Biochemical Physics Russian Academy of Sciences, 4 Kosygina Street, Moscow, 119334, Russia
[3]Semenov Institute of Chemical Physics Russian Academy of Sciences, 4 Kosygina Street, Moscow, 119334, Russia

E-mail: zykovaak@yandex.ru, aolkhov72@yandex.ru

Abstract. Hyperfine fibers based on biodegradable poly (3-hydroxybutyrate) with encapsulated drug substance (dipyridamol) were obtained by using electrospinning method. Addition of dipyridamol has a significant effect on geometrical shape and structure of microfibers as well as total porosity of fibrous material. Observation of fibers using scanning electron microscopy (SEM) method showed that without or at lower dipyridamol content (<3%) fibers consisted of interleaved ellipsoid and cylindrical fragments. At higher dipyridamol content (3-5%) anomalous ellipsoid structures did not practically form, and fiber's shape became cylindrical. The totality of morphological and structural characteristics determined the rate of dipyridamol diffusive transports. The simplified model of drug desorption from fibrous matrix was presented. In current work it was showed that the rate-limiting stage of transport was the diffusion of dipyridamol in the bulk of cylindrical fibers.

1. Introduction

The development of polymer matrices for controlled and targeted drug delivery is the topical issue of modern therapy of many illnesses and traumas [1]. Generally for the development of such therapeutic systems the poly(3-hydroxybutyrate) (PHB) is used as a polymer matrix, which is characterized by high biocompatibility and optimal terms of biodegradation [2]. High physical and mechanical properties of PHB films and fibers allow consider this biopolymer as one of the most promising biopolymers in medicine [3].

In recent research the physical, chemical and dynamic properties of fibrous materials on PHB and solid nanoparticles basis were considered. The nanoparticles were as follows, such as modified titanium dioxide, nanocrystalline silicium in titanium dioxide cover and chitosan. It was concluded that physical, mechanical and diffusive properties of fibrous materials depend on degree of crystallinity and amorphous phase of PHB. It was found that addition of nanoparticles at low concentration (from 0.1 to 0.5%) into forming PHB solution causes the significant changes of polymer structure at supramolecular level. These data are in agreement with works [4-6], in which the gelatin, calcspar and L-lactide-ε-caprolactone copolymer were used as modifying additives. Reciprocal action of crystalline and intercrystalline regions in biodegradable polymers and its blends still has been a challenging and poorly known topic of modern polymer material science.

2. Materials and Methods

In this work, polyhydroxybutyrate of the 16F series (BIOMER®, Germany) with a molecular weight *Mw* = 2.06×105 g/mol) was used to produce the investigated fibers. As a model drug substance for controlled release was used dipyridamol (2,2',2'',2'''-(4,8-di(piperidin-1-yl)pyrimido[5,4-d]pyrimidine-2,6-diyl)bis(azanetriyl)tetraethanol). Pharmacological class of dipyridamol (DPD) is «antiplatelet, angioprotectors and correctors of microcirculation, adenosyn synergistic agent». Hyperfine PHB fibers were obtained by the method of electrospinning from a solution. PHB and PHB with 1%, 3% and 5% DPD were soluted in chloroform.

The electron paramagnetic resonance spectrum (X-band) was recorded on an EPR-V automated spectrometer (Semenov Institute of Chemical Physics, RAS, Moscow) at microwave power of 7 mW and modulation amplitude of 0.5 gauss. As a probe, the stable nitroxyl radical 2,2,6,6-tetramethylpiperidin-1-oxyl (TEMPO) was used. The radical was incorporated into fibers from vapor at a temperature of 50°C to a concentration not exceeding 3–10mol/l. The differential scanning calorimetry (DSC) measurements were performed on a Netzsch DSC 204 F1 calorimeter in argon atmosphere at a heating rate of 10 K/min. The morphology of the fibers was studied via electron microscopy on Hitachi TM-1000 scanning electron microscope (Japan).

3. Results and Discussion

Analysis of morphology of obtained fibers (Figure 1) revealed that quantity of thickenings decreased with increase of DPD content, and thickenings practically were not observed at highest DPD content (5%). The emergence of thickenings may be explained by either insufficient conductivity of electrospinning solution or its nonoptimal viscosity.

Study of thermophysical characteristics using DSC method showed that specific melting enthalpy of PHB fibers increased with addition of DPD. With increasing DPD content in PHB fibers the formation of equilibrium and more superior structure was determined.

Obtained results should be taken into consideration while studying the segmental mobility and diffusive kinetics. These two fundamental processes determine rate and mechanism of controlled drug release from polymeric therapeutic systems.

Structure and molecular dynamics of intercrystalline parts of samples were examined by using EPR method. Experimental EPR spectrum is presented in Figure 2.

It was found that as degree of crystallinity in intercrystalline parts of samples increased, the radical rotation speed decreased, which can be explained by delaying of molecular mobility of segments. With increasing DPD content, the correlation time rose in following sequence: 6.6 10^{-9} sec. (0% DPD) > 7.1 10^{-9} sec. (1% DPD) > 8.8 10^{-9} sec. (3% DPD) > 9.0 10^{-9} sec. (5% DPD), which can be explained by increase of PHB degree of crystallinity in fibers.

Curves of DPD release from fibers are presented in Figure 3. All curves were characterized by the presence of initial non-linear and last linear fragments. At the linear fragment of the curves the concentration of desorpted DPD increased linearly.

Non-linear fragment of kinetic curves of DPD controlled release is determined by the diffusion process. Rapid change of values of diffusion coefficient was correlated with structural transition from spindle shaped fibers to cylindrical. Moreover, the rate of degradation process decreased exponentially with DPD increase, which was confirmed by the previous data.

Non-linear fragment of kinetic curves of DPD controlled release is determined by the diffusion process. Rapid change of values of diffusion coefficient was correlated with structural transition from spindle shaped fibers to cylindrical. Moreover, the rate of degradation process decreased exponentially with DPD increase, which was confirmed by the previous data. Exhibition of two release processes (diffusive and kinetic) allowed suppose that DPD existed in PHB fibers in two forms. The first form was presented by DPD in a free state, and what was highly important, only this form was capable of desorption from PHB fibers according to diffusive mechanism. Another form of DPD existed in immobilized state and was characterized by extremely low diffusive mobility. Although, the release of such DPD form from the polymer matrix was explained by partial degradation of PHB (weight loss of PHB fibers including immobilized form of DPD) according to the zeroth-order equation.

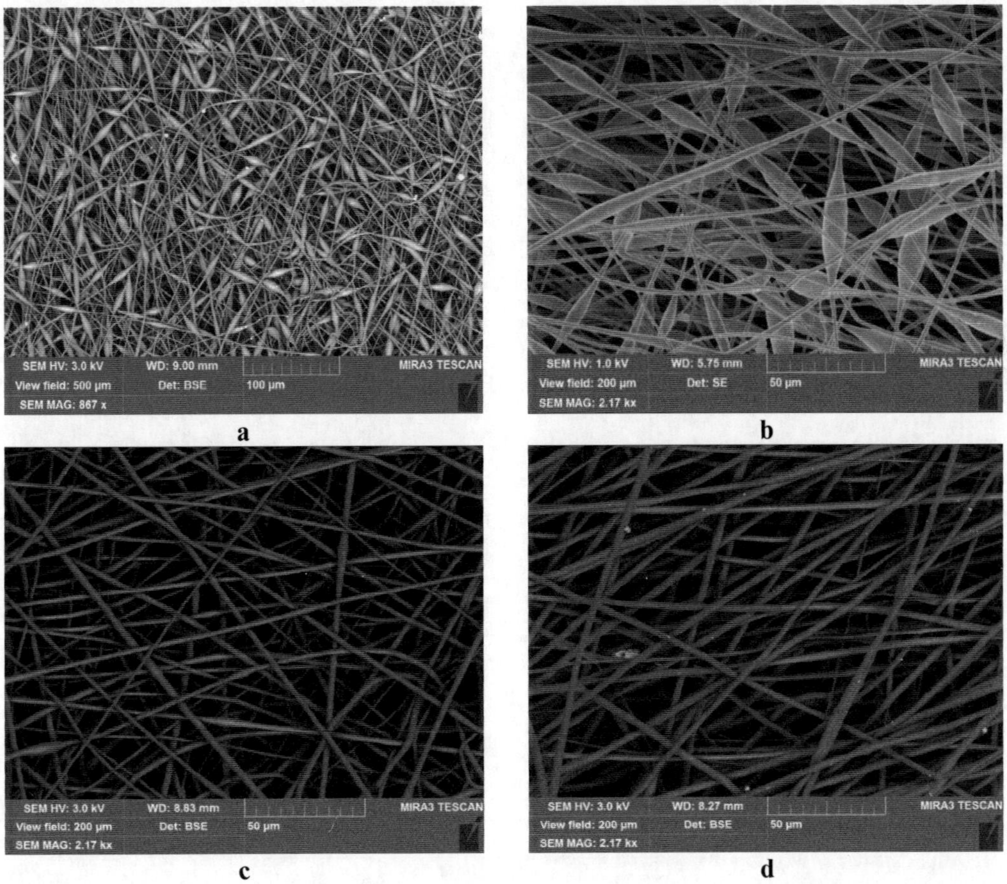

Figure 1. SEM photographs of fibrous materials obtained by electrospinning method from polymeric solutions: initial PHB (a), PHB +1% DPD (b), PHB+3% DPD (c), PHB+5% DPD (d).

4. Conclusion

Encapsulation of DPD within hyperfined PHB fibers had an influence on PHB geometrical shape, degree of crystallinity, and packing density in non-woven matrix. The totality of those characteristics determined the kinetics of diffusive release of DPD. Structural and thermophysical characteristics of fibrous materials and kinetic curves of DPD release allowed suppose that rate-limiting stage of general release process was the first stage i.e. DPD diffusion in the bulk of fibers.

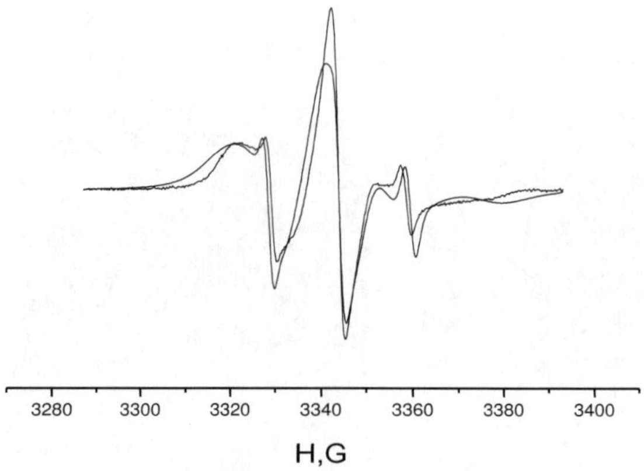

Figure 2. EPR spectrum of nitroxyl radical TEMPO in PHB fibers with 1% of DPD.

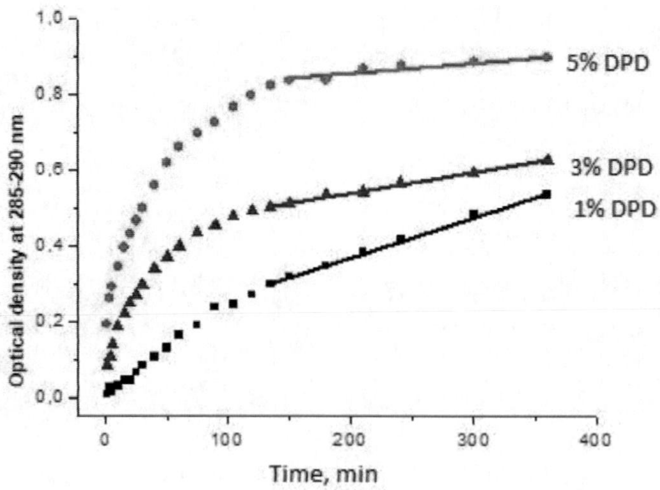

Figure 3. Kinetic curves of DPD controlled release from hyperfine PHB fibers.

Acknowledgements
This study was financed by a grant from the Plekhanov Russian University of Economics.

The authors express an acknowledgement to the common use centre «New materials and technologies» at the Emanuel Institute of Biochemical Physics of Russian Academy of Sciences.

This work was supported by the Russian Foundation for Basic Research (project no. 15-59-32401 RT-omi and 15-29-04862 ofi-m) and the Program of Basic Research of the Division of Chemistry and Materials Sciences, Russian Academy of Sciences, «Development of Macromolecular Structures of a New Generation» (OKhNM-1-2014).

We are grateful to Prof. U.J. Haenggi (BIOMER®, Germany) for providing us with poly(3-hydroxybutyrate).

References

[1] Martinho N, Damgé C, Reis C P 2011 *J. Biomater. Nanobiotechnol.* **2** 510-526

[2] Karpova S G, Olkhov A A, Iordanskii A L, Lomakin S M, Shilkina N S, Popov A A, Gumargalieva K Z, Berlin A A 2016 *Rus. J. Phys. Chem. B* **10** 687-698
[3] Filatov Yu N 1997 *Electrospinning method of fibrous materials manufacturing* (Russia: Neft I Gas) p 298
[4] Nuge T, Hoque M E, Yeow T K, Nordin N, Chowdhury M 2013 *Regenerative Res.* **2** 39-42
[5] Guan D H, Huang C P, Liu J, Tian K, Niu L, Lin Y H, Qu Y L, Wang P, Ye Y M, Chen Z Q 2007 *Key Eng. Mater.* **330-332** 695-698
[6] Daranarong D, Chan R T H *Biomed. Res. Int.* 2014 **2014** 741408

IOP Conf. Series: Materials Science and Engineering **286** (2017) 012034 doi:10.1088/1757-899X/286/1/012034

Spark plasma sintering of ceramic matrix composite based on alumina, reinforced by carbon nanotubes

A A Leonov, A O Khasanov, V A Danchenko and O L Khasanov

Department of nanomaterials and nanotechnologies, National Research Tomsk Polytechnic University, 30 Lenin Ave., Tomsk, 634050, Russia

Email: laa91@tpu.ru

Abstract. Alumina composites reinforced with 3 vol.% multi-walled carbon nanotubes (MWCNTs) were prepared by spark plasma sintering (SPS). The influence of sintering temperature (1400-1600 °C) on the composites microstructure and mechanical properties was investigated. Microstructure observations of the composite shows that some CNTs site along alumina grains boundary, while others embed into the alumina grains and shows that CNTs bonded strongly with the alumina matrix contributing to fracture toughness and microhardness increase. MWCNTs reinforcing mechanisms including CNT pull-out and crack deflection were directly observed by scanning electron microscope (SEM). For Al_2O_3/CNT composite sintered at 1600 °C, fracture toughness and microhardness are 4.93 MPa·m$^{1/2}$ and 23.26 GPa respectively.

1. Introduction

Carbon nanotubes are widely used as reinforcements in polymers, metals, and ceramics to improve their mechanical and functional properties [1]. CNT has been considered as an ideal candidate for reinforcing/functioning elements because of small size, low density, high aspect ratio, and outstanding mechanical, electrical, and thermal properties. The demand for advanced composite materials comprising of high performance characteristics and improved functional properties is always rising [2]. Alumina, along with other ceramics like zirconia, silicon nitride, tungsten carbide, titania, etc., is not only being used as a biomedical material but also for high temperature applications in aerospace and automobile industries. This is due to its high hardness, good wear resistance, thermal/electrical insulation properties and excellent chemical inertness. But the main problem associated with alumina is its low fracture toughness which sometimes hinders its application as a structural material [3, 4]. Various types of particulate, whisker and fiber reinforcements have been used to improve the fracture toughness of alumina in the last few decades. In this regard, CNTs are achieving a lot of attention as a reinforcement candidate due to their good thermal stability up to 1800 °C [5] and excellent mechanical properties after sintering with alumina. CNTs connect the alumina grain boundaries and retard grain growth resulting in the grain refinement during the sintering at elevated temperatures which help in the improvement of fracture toughness, Young's and shear modulii [6]. The toughening mechanism of CNT-alumina composites is mainly attributed to the crack deflection at CNT-alumina interface, crack bridging and CNT pull-out from alumina grains [7]. Most studies have been based on laboratory-scale self-made composite powders, which cannot meet the requirements of wider applications. In the present study, commercial Al_2O_3/CNT composite powders were used as raw materials and sintered by spark plasma sintering.

2. Materials and methods

Commercial α-Al_2O_3/CNT composite powder (Applied Carbon Nanotechnology Co., Ltd., Korea) with MWCNT contents of 3 vol.% were used as the starting powder. The particle morphology of initial composite powder and microstructures of the fracture surfaces were observed by means of a scanning electron microscope (JSM-7500FA, JEOL). A small amount of composite powder was dispersed onto a TEM grid to be characterized by transmission electron microscopy (TEM, JEM-2100F, JEOL). The measurements of particle size distribution of composite powder by means of laser diffraction method were performed with SALD-7101 (Shimadzu). Composites were prepared by spark plasma sintering (SPS-515S, Syntex) in a graphite die with an inner diameter of 15 mm at different sintering temperatures (1400, 1500, and 1600 °C), constant load of 40 MPa, heating rate 100 °C/min, and holding time 10 min. After sintering, the composites were allowed to cool naturally to room temperature. The density of the sintered disk-shaped composites was determined using the geometrical method. The theoretical densities were calculated according to the rule of mixtures, in which densities of 3.99 g/cm^3 and 2.10 g/cm^3 were assumed for the alumina and MWCNTs, respectively [8]. The relative density ρ is the ratio of the measured density to the theoretical density. Measurements of the microhardness H_V were performed with a PMT-3M (LOMO) microhardness tester under ambient conditions. The Al_2O_3/CNT composites were indented using a Vickers diamond pyramid with a load of 4.9 N applied on the surface for 15 s. At least 10 measurements were carried out for each sample. The diagonal lengths of the indents were measured using the attached microscope, converted to Vickers hardness number (HV) and further converted to GPa. A Vickers hardness tester TP-3R-1 was used for fracture toughness measurements using a load of 49 N. The fracture toughness values K_{IC} were calculated using the Anstis method [9]:

$$K_{IC} = 0.016\left(\frac{E}{H_V}\right)^{1/2}\left(\frac{P}{c^{3/2}}\right),$$

where E is the Young's modulus, H_V is the microhardness, c the radial crack length generated by Vickers's indentation and P the load at failure.

3. Results and discussion

Figure 1a shows morphology of the Al_2O_3/CNT raw powder. The alumina particles are irregular in shape with size ranging from 100 nm to 9 μm. Aggregation of alumina particles is not observable.

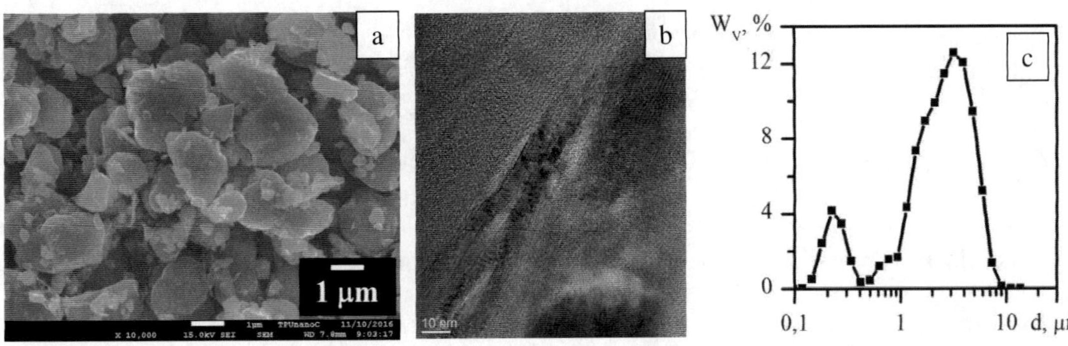

Figure 1. (a) SEM image showing morphology of composite powder, (b) TEM image showing morphology of MWCNT and (c) particle size distribution of composite powder.

Figure 1c shows the particle size distribution of composite powder. The average particle size of powder is approximately 1.7 μm. TEM images shown in Figure 1b display the morphology and surface features of the CNTs used in this research. The surface of the CNTs is not smooth, and hollow core and numerous graphitic layers of the CNTs are clearly visible, although the layers are not concentrical and some compartments exist.

Let us now examine the three SEM micrographs of the sintered products obtained by SPS. First of all, it is seen that all of the samples, consist of 1-2 μm sized grains (Figure 2a, b, c) and there is no evidence of CNTs damage during sintering. Employing the SPS sintering, the densification of samples can be realized in short time without a considerable grain growth process. The CNTs are located mainly in the inter-granular places and in some cases incorporated into grains. As can be seen (Figure 2c), CNTs were distributed inside the alumina grains and strongly bonded with the alumina matrix.

Figure 2. SEM images of the fractured surface of composites sintered at temperatures: (a) – 1400; (b) – 1500; (c) – 1600 °C.

As the sintering temperature rises from 1400 to 1600 °C, the density of the Al_2O_3/CNT composite increases (Table 1). It is evident that the temperature increase intensifies the processes of diffusion in solid and plastic deformation with attainment of creep threshold stress, and, therefore, of the material yield [10]. In this case, the whole free volume in solid is filled with the material, pore disappear, and a compressed (monolith) compound is formed (Figure 2c).

Table 1. Properties of Al_2O_3/CNT composites.

T (°C)	ρ (%)	H_V (GPa)
1400	89.80	14.69
1500	97.84	17.52
1600	98.60	23.26

Microhardness of composite Al_2O_3/CNT is 14.69 GPa after sintering at 1400 °C. With increasing the sintering temperature to 1500 and 1600 °C, microhardness values increase by 19 % and 58 % respectively (Table 1). For comparison, in [11-13] composites Al_2O_3 containing 3, 2.4 and 0.9 vol.% of MWCNTs fabricated by SPS had microhardness values of 18.40, 18.50 and 17.00 GPa, respectively. The composite Al_2O_3 with 2.5 vol.% of MWCNT obtained by hot pressing [14] and Al_2O_3 composite with 0.1 wt.% of MWCNT were densified by pressureless sintering [15] had microhardness values of 18.00 and 16.66 GPa, respectively. It is also worth noting that at a load of 4.9 N, no cracks appeared on the sample surfaces, which indicates an exceptionally high fracture toughness of the samples.

More detailed views about the characteristics of microstructure of composites are shown in Figure 3a, b. MWCNT tend to assemble in bundle of nanotubes as can be seen in Figure 3a which consist of several tubes and aligned along their length in van der Waals bonding with one another. CNT bundle (d ~ 180 nm) is embedded into the matrix and this bundle is divided into three little bundles (Figure 3a, d ~ 20-35 nm). MWCNTs, which seem to have been entrapped in the matrix grains during SPS, are cut near the grain surface (Figure 3b). As shown in Figure 3b, observed the formation of strong interface between CNTs and matrix. Also shown clear evidence of pulled-out

CNTs, which indicates that the CNTs bear significant stress by sharing a portion of the load and, at the same time, toughen the matrix by a bridging effect.

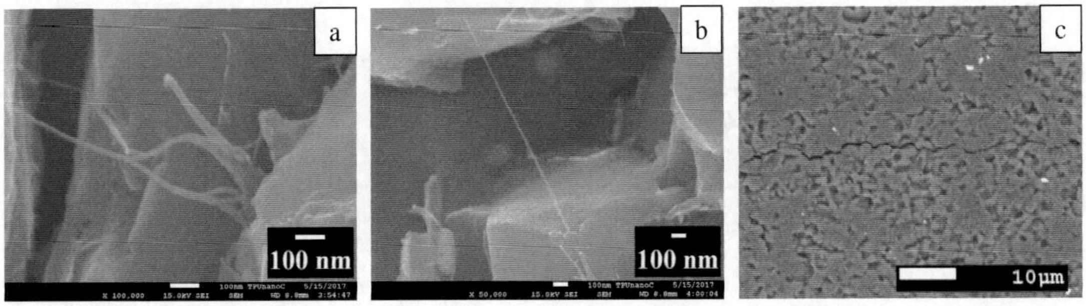

Figure 3. (a, b) SEM images of the fractured surface and (c) SEM images of crack on the surface of composite by indentation.

The fracture toughness of the Al_2O_3/CNT composite after sintering at 1600 °C is measured to be 4.93 MPa·m$^{1/2}$, which is high enough due to crack deflection, crack bridging and pull-out mechanism [16, 17]. The improvement in fracture toughness is also attributed to minimization of CNTs damage during sintering [18]. The crack is characterized by a small number of deviations at large angles and therefore, a very tortuous path occurs explaining the high efficiency of the deflection mechanism and therefore, an increase in fracture toughness (Figure 3c). MWCNT pull-outs were also detected on the fracture surface suggesting a strong bonding between CNT and Al_2O_3 matrix and significant load transfer from the matrix to carbon nanotubes during loading. In intergranular fracture the crack propagates along the grain boundary thus consumes more fracture energy, which is favorable for the increment of fracture toughness. However, our results shows that the toughness of the Al_2O_3/CNT composite sintered at 1600 °C is 4.93 MPa·m$^{1/2}$, which is 105 and 83 % increase in toughness over pure alumina with the same-as grain size (1.66 and 2 µm) 2.41 and 2.70 MPa·m$^{1/2}$ respectively [19, 20].

The SPS method has been useful to give high mechanical properties of the Al2O3/CNT composites. The SPS technique is a pressure-assisted fast sintering method based on high-temperature plasma momentarily generated in the gaps between powder materials by electrical discharge during on-off direct current pulsing. It has been suggested that the direct current pulse could generate several effects such as spark plasma, spark impact, Joule heating, and an electrical field diffusion [21]. Through these effects, the formation of a comparatively stronger bond between MWCNTs is expected for the SPS samples.

4. Conclusions

The investigated composites produced by SPS of the Al_2O_3/CNT composite powder presented strongly bonded CNTs with the alumina matrix. Microstructure investigation of the samples showed that structure of the composite sintered at 1600 °C is comprised of densely sintered grains of alumina surrounded by carbon nanotubes and some CNTs embedded into the alumina grains. The Al_2O_3/CNT composites thus fabricated showed effectively enhanced hardness and toughness compared to monolithic materials, which is based on the mechanisms CNT pull-out, CNT bridging and crack deflection. Multi-wall carbon nanotubes are attractive materials for reinforcement (strength and toughness) of ceramics and strong bonding of CNTs with matrix play an important role for reinforcement.

Acknowledgements

The authors thank M S Petyukevich of Department of nanomaterials and nanotechnologies, National Research Tomsk Polytechnic University, for the SEM analysis.

IOP Conf. Series: Materials Science and Engineering **286** (2017) 012034 doi:10.1088/1757-899X/286/1/012034

References

[1] Thostenson E T, Ren Z and Chou T W 2001 Advances in the science and technology of carbon nanotubes and their composites: a review *Composites Science and Technology* **61** 1899–1912

[2] Cho J, Boccaccini A and Shaffer M 2009 Ceramic matrix composites containing carbon nanotubes *Journal of Materials Science* **44** 1934–1951

[3] Ahmad I, Cao H, Chen H, Zhao H, Kennedy A and Zhu Y Q 2010 Carbon nanotube toughened aluminium oxide nanocomposite *Journal of the European Ceramic Society* **30** 865–873

[4] Mukhopadhyay A and Basu B 2007 Consolidation-microstructure-property relationships in bulk nanoceramics and ceramic nanocomposites: a review *International Materials Reviews* **52** 257–288

[5] Zhang H L, Li J F, Yao K F and Chen L D 2005 Spark plasma sintering and thermal conductivity of carbon nanotube bulk materials *Journal of Applied Physics* **97** 114310

[6] Zhan G D, Kuntz J D, Wan J and Mukherjee A K 2003 Single-wall carbon nanotubes as attractive toughening agents in alumina-based nanocomposites *Nature Materials* **2** 38–42

[7] Peigney A, Garcia F L, Estournès C, Weibel A and Laurent C 2010 Toughening and hardening in double-walled carbon nanotube/nanostructured magnesia composites *Carbon* **48** 1952–1960

[8] Yamamoto G, Shirasu K, Nozaka Y, Wang W and Hashida T 2014 Microstructure-property relationships in pressureless-sintered carbon nanotube/alumina composites *Materials Science and Engineering: A* **617** 179–186

[9] Anstis G R, Chantikul P, Lawn B N and Marshall D B 1981 A critical evaluation of indentation techniques for measuring fracture toughness: I, direct crack measurements *Journal of the American Ceramic Society* **64** 533–538

[10] Mehrer H 2007 *Diffusion in Solids: Fundamentals, Methods, Materials, Diffusion-Controlled Processes* (Berlin: Springer) p 654

[11] Kasperski A, Weibel A, Estournès C, Laurent C and Peigney A 2014 Multi-walled carbon nanotube-Al_2O_3 composites: Covalent or non-covalent functionalization for mechanical reinforcement *Scripta Materialia* **75** 46–49

[12] Sikder P, Sarkar S, Biswas K G, Das S, Basu S and Das P K 2016 Improved densification and mechanical properties of spark plasma sintered carbon nanotube reinforced alumina ceramics *Materials Chemistry and Physics* **170** 99–107

[13] Yamamoto G, Shirasu K, Hashida T, Takagi T, Suk J W, An J, Piner R D and Ruoff R S 2011 Nanotube fracture during the failure of carbon nanotube/alumina composites *Carbon* **49** 3709–3716

[14] Hanzel O, Sedlacek J, Sajgalik P 2014 New approach for distribution of carbon nanotubes in alumina matrix *Journal of the European Ceramic Society* **34** 1845–1851

[15] Aguilar-Elguézabal A and Bocanegra-Bernal M H 2014 Fracture behaviour of α-Al_2O_3 ceramics reinforced with a mixture of single-wall and multi-wall carbon nanotubes *Composites Part B: Engineering* **60** 463–470

[16] Zhang S C, Fahrenholtz W G, Hilmas G E and Yadlowsky E J 2010 Pressureless sintering of carbon nanotube-Al_2O_3 composites *Journal of the European Ceramic Society* **30** 1373–1380

[17] Zhang T, Kumari L, Du ., Li W, Wang Q, Balani K and Agarwal A 2009 Mechanical properties of carbon nanotube-alumina nanocomposites synthesized by chemical vapor deposition and spark plasma sintering *Composites Part A* **40** 86–93

[18] Zhan G D and Mukherjee A K 2004 Carbon nanotube reinforced alumina-based ceramics with novel mechanical, electrical, and thermal properties *International Journal of Applied Ceramic Technology* **1** 161–171

[19] Maensiri S, Laokul P, Klinkaewnarong J and Amornkitbamrung V 2007 Carbon nanofiber-reinforced alumina nanocomposites: Fabrication and mechanical properties *Materials Science and Engineering: A* **477** 44–50

[20] Fedosova N A, Faikov P P, Popova N A, Kol'tsova É M and Zharikov E V 2015 Ceramic composite material obtained by spark plasma sintering technology with carbon nanotubes *Glass and Ceramics* **72** 13–16

[21] Omori M 2000 Sintering, consolidation, reaction and crystal growth by the spark plasma system (SPS) *Materials Science and Engineering: A* **287** 183–188

IOP Conf. Series: Materials Science and Engineering **286** (2017) 012035 doi:10.1088/1757-899X/286/1/012035

Analysis of type IIb synthetic diamond using FTIR spectrometry

I V Klepikov[1,2], A V Koliadin[2] and E A Vasilev[3]

[1]A.P.Karpinsky Russian Geological Research Institute, 74 Sredniy prospect, Saint Petersburg, 199106 Russia
[2]LLC “New Diamond Technology”, 2 Voskova st, Sestroretsk, Saint Petersburg
[3]Saint-Petersburg mining university, 2 21st line, Saint Petersburg, 199106 Russia

E-mail: Klepikov_Igor@mail.ru

Abstract. Analysis of internal structure in large IIb-type high pressure-high temperature (HPHT) synthetic single-crystal diamond are presented. The concentration of boron impurity in different growth sectors varies from 0.02 to 10.3 ppm. It is shown that in the manufacturing of synthetic diamond plates, internal inhomogeneities of the diamond should be taken into account; plates with different characteristics can be cut from one diamond, each of which can be used for its own purpose.

1. Introduction

Nowadays, methods of synthesis of single-crystal diamonds came to a high level, thus diamonds of different physical types are grown commercially in laboratories. Diamond plates have extreme hardness and high thermal conductivity, that is why they are in demand for medicine, military and space industries, science applications. They are used as UV and IR windows, components of laser optics, lenses for gyrotrons, monochromators and X-ray equipment, sensors and detectors of particles. There are two main competing with each other methods of diamond growth: high temperature – high pressure (HPHT) and chemical vapor deposition (CVD). In this paper, we present data on internal inhomogeneities of 10.87 ct IIb type synthetic single-crystal diamond grown by high pressure-high temperature method.

Synthetic HPHT single-crystal diamonds have sectorial structure (figure 1c). Presence and predominance of different crystallographic forms in synthetic diamonds depends on metals catalysts and an arrangement of seeds in a growth cell (figure 1 a,b). Using of an alloy catalyst based on Fe with incorporation of Aluminum is the cause of {111} and {100} facets predominance, while incorporation of Co, Ti, Sr leads to {113}, {110} and {115} facets development [2].

2. Materials and methods

11.46x11.04x0.51 mm single-crystal diamond plate was cut from a synthetic diamond of IIb type (crystal weight is 10.87 ct) perpendicularly to [111]. This diamond was grown on (111) seed. Except predominant {111} facets, there were different facets of {100}, {110}, {113} and {115} orientations in the diamond. This crystal was dark blue due to strong doping of boron to initial fusion mixture. Studied single-crystal diamond plate had blue color zonality (figure 2) clearly visible by naked eye.

IOP Conf. Series: Materials Science and Engineering **286** (2017) 012035 doi:10.1088/1757-899X/286/1/012035

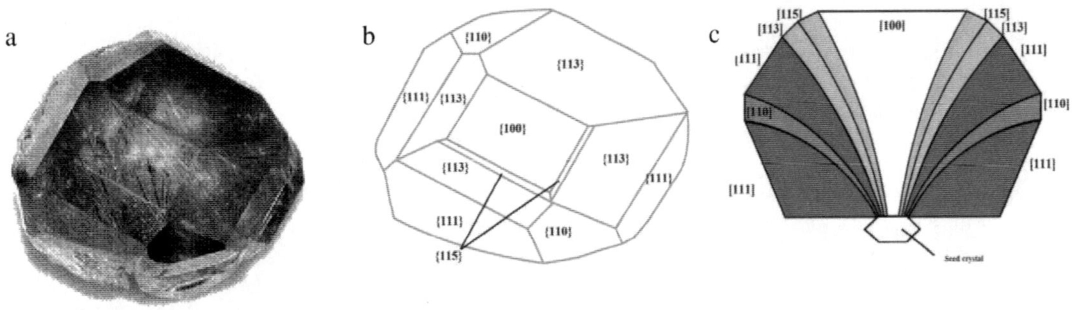

Figure 1. Synthetic HPHT 17.50 ct diamond from "New Diamond Technology" company: a-photo of the crystal, b-scheme of dominant facets, c-scheme of growth sectors.

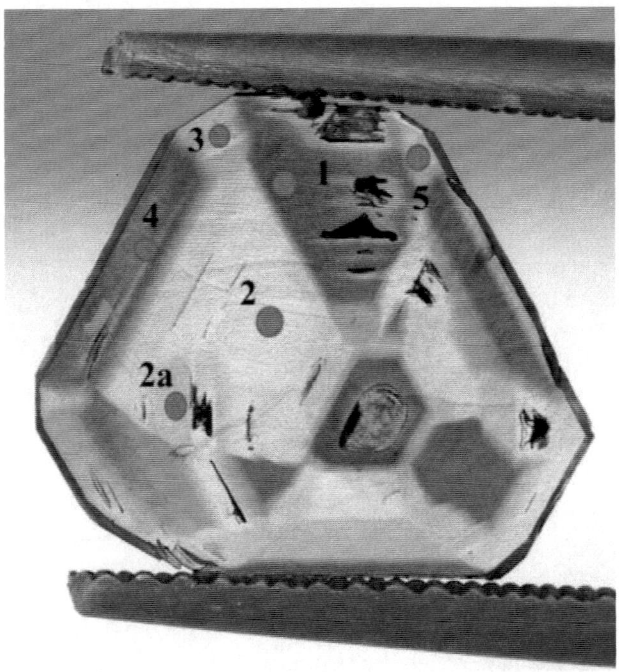

Figure 2. The plate from the synthetic single-crystal 10.87 ct diamond of IIb type.1, 2, 2a, 3, 4, 5 – IR spectrum points.

FTIR spectra were recorded on Vertex 70 spectrometer with Hyperion1000 microscope with the resolution of 4 cm^{-1} (32 scans averaging). FTIR spectra were registered in 6 points corresponding to main growth sectors with different blue color saturation. Presence of boron in the diamond was determined by FTIR spectra by a number of bands: 1290, 2460, 2802, 2936 cm^{-1} [3]. Generally, concentration of boron was calculated by 2802 cm^{-1} band (figure 4), but if the concentration of boron was extremely high, then it was calculated by 1290 cm^{-1} band (figure 3), because it is impossible to find distinct peaks in three-phonon area in such situation due to strong absorption of boron.

IOP Conf. Series: Materials Science and Engineering **286** (2017) 012035 doi:10.1088/1757-899X/286/1/012035

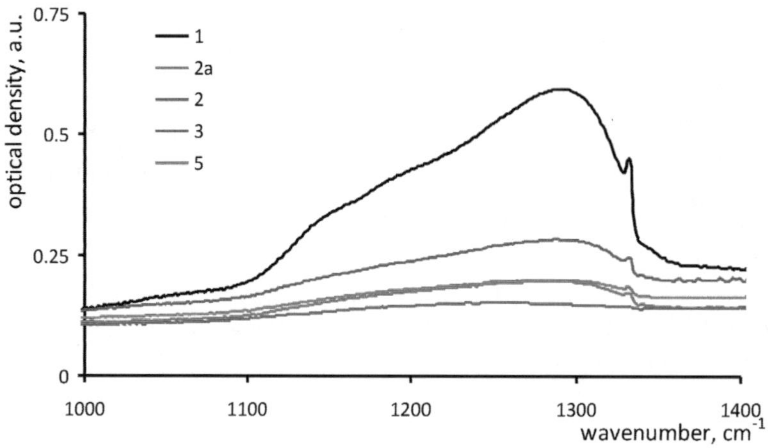

Figure 3. FTIR spectra for tested points (number refers to figure 1) in the range 1000-1400 cm^{-1}.

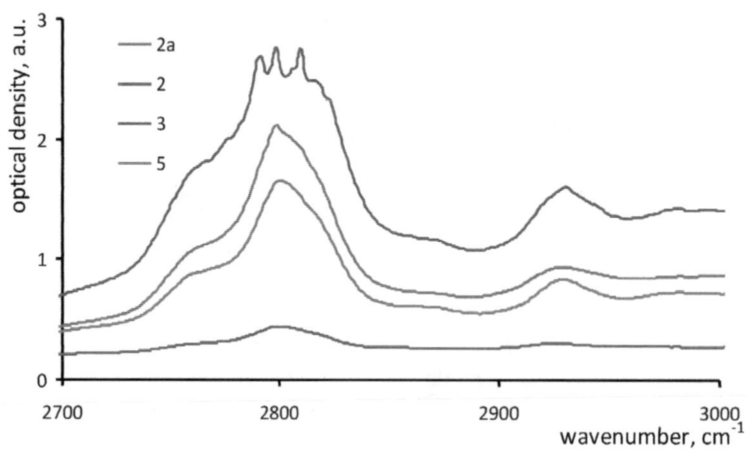

Figure 4. FTIR spectra for points 2, 2a, 5 in the range 2700-3000 cm^{-1}.

3. Results and discussion

Calculations were made by the following equations:

$$[B](ppm)=(5.53\times10^{-4})\times I_{2802}(cm^{-2}) \quad (1)$$

$$[B](cm^{-1})=(2.1\times10^{17})\times\alpha_{1290}(cm^{-1}) \quad (2)$$

where α = A/ t – is an absorption coefficient (A – optical density, t – sheet gage); I – is integral absorbance of peak [1,3-6].

Values of boron concentration from 0.02 to 10.29 ppm (table 1) were obtained. They show that concentrations of boron impurity in a diamond can be varied by 500 times. The most saturated was [111] sector (points 1 and 4). By contrast, the minimum value was registered in [100] sector. Inconspicuous change of intensity of light blue color within light zones of the plate was observed. In fact the value of boron concentrations was 50 times different (points 2 and 2a). This inconspicuous color zonality was explained by the fact that except[100] sector, there are other growth sectors in the diamond, e.g. [110], [113] and [115], which have their own growth nature and are characterized by low concentrations of boron impurity. For example, the point 3 corresponds to the growth sector [110], and here the value of boron concentration is higher, than in sector [100]. In the exit zone to the lateral surface of {113} and {115} facets, the IR spectrum of a point 5 was made.

Table. 1. Calculated concentration of boron for 6 points of a plate.

Point No	Concentration of boron (ppm)	Growth sector
1	10.29	[111]
2a	0.99	[113], [115]
2	0.02	[100]
3	1.95	[110]
4	9.36	[111]
5	0.79	[113], [115]

We can produce monosectorial homogeneous plates of different orientations 5x2, 3x3, 2x2 and 4x3 mm size from the investigated plate. Plates with the concentration edges of boron impurity (from ppb to ppm) can be prodused. Generally, we can produce square monosectorial plates 2x2, 3x3, 4x4, 5x5 mm size and different thickness (0.1-3 mm) with the concentration of boron from level ppb to 10-15 ppm. The extended monosectorial plates can reach 7x5, 9x7 in size. Multisectorial square plates [100] and [111] with boron are able to reach sizes 8x8 up to 10x10 mm.

It should be noticed that inclusions of metals carbides [2] in different growth sectors also create inhomogeneity in boron impurity distribution –white areas with lower concentration of boron around inclusions were observed.

4. Conclusions

- IIb synthetic HPHT diamonds have strong inhomogeneity of boron impurity distribution: concentration in different growth sectors can be 500 times different.
- The sector[111] is the most active for boron impurities acceptance, the [100] sector - the least; [110], [113] and [115] sectors also participate in creation of the composite picture of zonality and are characterized by low values of boron concentration.
- Inclusions of metals carbides in synthetic diamonds also bring inhomogeneity in distribution of boron impurity on a crystal/plate;
- We can produce monosectorial and multisectorial plates of different sizes (up to 10x10 mm) and of different orientations ([111],[100],[110],[113] and [115]) with the concentration of boron from ppb to ppm level.

References

[1] Collins AT and Williams A W S 1971 *J. Phys. C: Solid St. Phys.* **4** 1789-800
[2] D'Haenens-Johanson U, Katrusha A, Moe K, Johnson P, Wang W 2015 *Gems&Gemology* L **51** **3** 260-79
[3] Dishler B 2012 *Handbook of Spectral Lines in Diamond* (Springer) p 467
[4] Fisher D, Sibley S and Kelly C 2009 *J. Phys.: Condens. Matter.* **21** 364213
[5] Karna S K, Martyshkin D V, Vohra Y K and Weir S T 2013 Synthesis and Characterization of Boron-Doped Single Crystal Diamond *Materials Research Society Proceedings.* **1519**
[6] Walker P L and Thrower P A 1977 *Chemistry and Physics of Carbon 13* (New York: Marcel Dekker inc) p 34

IOP Conf. Series: Materials Science and Engineering **286** (2017) 012036 doi:10.1088/1757-899X/286/1/012036

Structure and properties of steel case-hardened by non-vacuum electron-beam cladding of carbon fibers

A A Losinskaya, E A Lozhkina and A I Bardin

Novosibirsk State Technical University, 20, K. Marksa Prospekt, Novosibirsk, 630073, Russia

E-mail: alexeixxx13@yandex.ru

Abstract. At the present time, the actual problem of materials science is the increase in the steels performance characteristics. In the paper some mechanical properties of the case-hardened materials received by non-vacuum electron-beam cladding of carbon fibers are determined. The depth of the hardened layers varies from 1.5 to 3 mm. The impact strength of the samples exceeds 50 J/cm^2. The wear resistance of the coatings obtained exceeds the properties of steel 20 after cementation and quenching with low tempering. The results of a study of the microhardness of the resulting layers and the microstructure are also given. The hardness of the surface layers exceeds 5700 MPa.

1. Introduction

Year by year, the requirements of the industry for utilizable materials are becoming higher that requires a constant search for the ways to improve materials performance characteristics. To improve the mechanical properties of the surface layers of products, surface hardening technological processes are used. It is possible to change the physical and mechanical properties, as well as the chemical composition of the surface layers. The most economical and widespread method of case-hardening of steel items is the cementation process. The main disadvantage of this process is its long duration, due to the need for a long diffusion saturation of the surfaces of items with carbon.

Alternative solutions to harden the steel surfaces are technologies based on the use of high-energy methods of action. Such methods make it possible to accelerate significantly the hardening process. In the modern literature there are data on the formation of hardened layers using laser, ion, electron, and plasma treatment [1-3]. For example, a plasma-electrolytic treatment process was used to carburize iron to a depth of 20–160 μm and formed a structure with a mixture of martensite, austenite, and cementite [4]. At the moment, the physics of high-energy treatment is well studied [5]. These methods are widely used in industry, providing coatings with high values of hardness, corrosion resistance, wear resistance and heat resistance. The technology of non-vacuum electron-beam processing is of great interest in the problem of creating high-strength coatings [6].

In previous studies, it was found that non-vacuum electron-beam treatment is an effective method to create high-carbon coatings on steels using carbon-containing powders [6]. In this paper, carbon fibers are used as a carbon source.

2. Methods

In this study, using non-vacuum electron-beam treatment of carbon fiber GG 210-P, high-carbon layers were formed on low-carbon structural steel 20 using an industrial electron accelerator ELV-6 (property of the Institute of Nuclear Physics of the SB RAS). The merit of the non-vacuum electron-beam impact is that the kinetic energy of the electron beam formed in a vacuum is discharged into the air atmosphere and converted into a thermal energy in the processing zone.

For an even distribution of carbon in the base material, a wetting agent (an extra-pure grade iron powder) was used. The average powder particle size was 64 μm. To protect the melt bath and powder materials from the air exposure, the flux MgF_2 was used. The glue was used to prevent the blowing of the powders from the surface of the material by the shock wave of the electron beam. A mixture of a binder with iron and flux powders was applied in two ways. The first method consisted in applying the mixture over two layers of fiber, the second method was supplemented by applying a mixture between the fiber layers. Composition of the cladded material was the following: 13 wt.% of carbon, 37 wt.% of iron and 50 wt.% of flux, 100 wt.% of glue. Before cladding, the surface was cleaned of oxides and degreased. After surfacing, the workpieces were dried in a furnace at 40 °C till the glue complete drying out.

The top surface of each workpiece during electron-beam treatment was at a distance of 90 mm from the outlet. The energy of the electron beam was 1.4 MeV, the beam current was 8 and 10 mA. The electron beam was extracted into the atmosphere through a diaphragm of 1 mm diameter. Under these conditions, the Gaussian diameter of the electron beam on the specimen surface was 12 mm. The treatment took place in a track mode, with a speed of 10 and 25 mm per second.

The structural, mechanical, and tribological properties of the obtained materials were studied. Chemical composition of the cladded layers was determined on an ARL 3460 optical emission spectrometer. Studies of the structure of the samples were performed using a Carl Zeiss EVO50 XVP electron-scan microscope in the magnification range from x50 to x30 000. The specimen wee etched using a 3% solution of nitric acid in ethanol. The microhardness of the coatings to be tested was evaluated using a WolpertGroup 402MVD semiautomatic micro microhardness tester in accordance with GOST 9450-76. Impact toughness testing of materials with cladded coatings was carried out on the impact pendulum-type testing machine Metrocom in accordance with GOST 9454-78. The maximum energy of the pendulum was 300 J. The tests on wear resistance in condition of friction on fixed abrasive particles were carried out in accordance with GOST 17367.

3. Results and discussion

As a result of cladding, layers with the thickness up to 3 mm with a carbon content of 2.2% were obtained. The study of the samples structure revealed the following structure: the upper high-carbon layer, the heat affected zone, the zone of the base unchanged metal. Perlite, secondary cementite and ledeburite are observed in the structure of the upper layer (Figure 1). The phase of secondary cementite has a Widmanstatten morphology. The heat affected zone consists of perlite and ferrite of the Widmanstatten type. Macroscopic defects (cracks, pores) were not found in the processed samples.

The maximum microhardness of the coatings obtained reaches 5700 MPa. The obtained values of the microhardness of the cladded layers are lower than that of the steel 20 cemented and hardened by the traditional method. The hardness of the cladded layers can be increased by using a subsequent heat treatment, which can also be carried out on an industrial electron accelerator.

The samples of steel 20 with a ferrite-pearlite structure are characterized by the highest impact strength (125 J/cm^2). After electron-beam cladding, the impact toughness of steel 20 with high-carbon layers is 2.5 times lower compared to the initial unhardened state. It has also been experimentally established that the impact strength of the materials being studied depends to a greater extent not on the carbon content but on its thickness of the coatings.

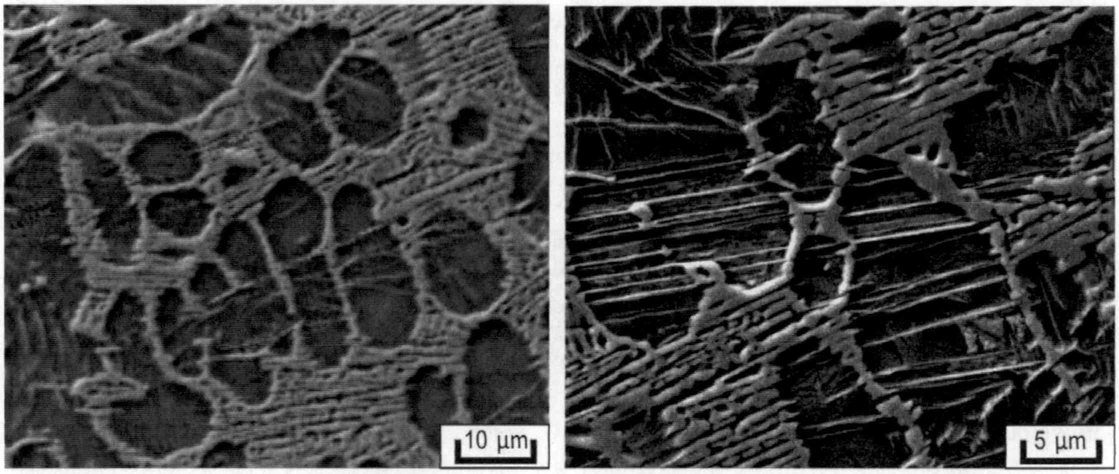

Figure 1. The structure of cladded layer.

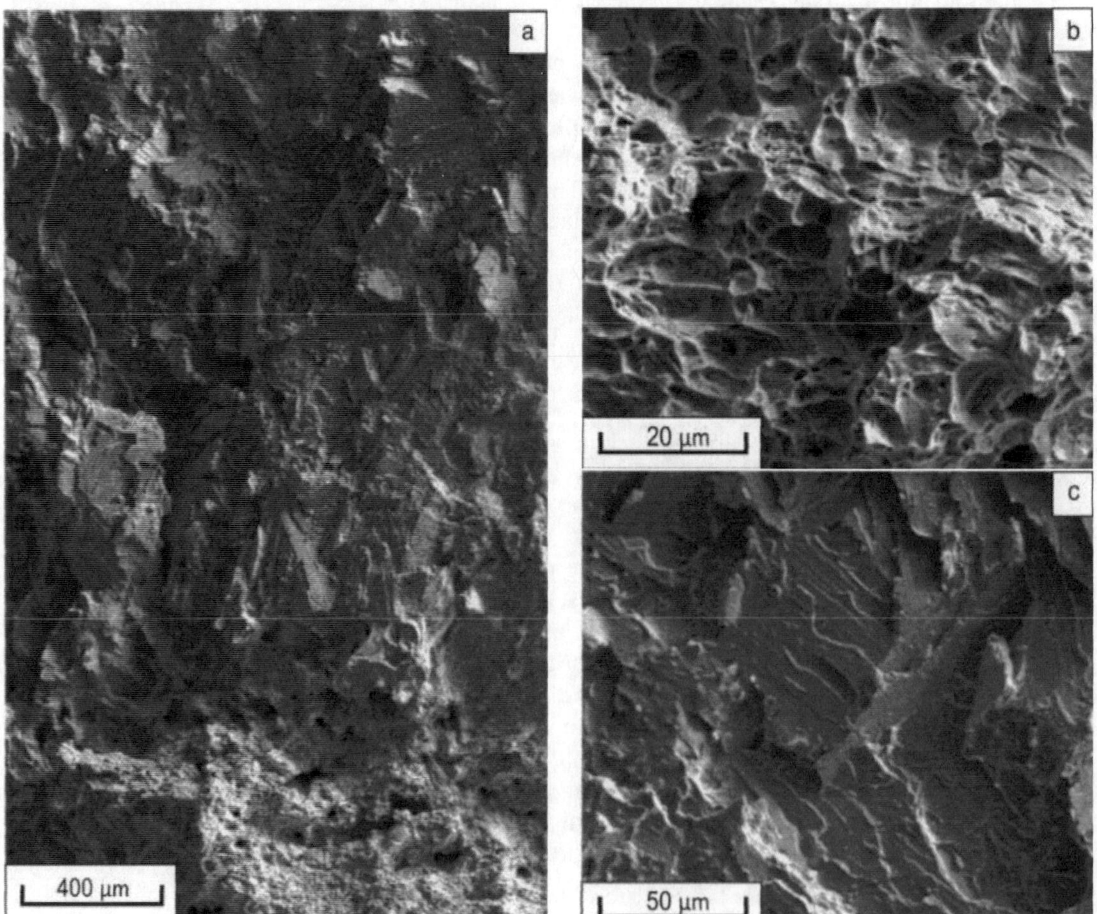

Figure 2. Fracture pattern, obtained in the impact toughness testing: a) general view; b) the fracture of the main material; c) fracture of the cladded layer.

Fractografic studies of the samples after impact toughness testing showed that the destruction of the base material corresponds to the viscous nature of the fracture (Figure 2a, b). The upper part of the fracture is the result of the brittle fracture; the lower part is ductile fracture by means of the micropores fusion. In Figure 2b, c, it is seen that the pits of ductile fracture of the base material are oriented in the direction of shear. Quantitative analysis showed that the size of the pits is in a wide range of sizes (from 1 μm to 20 μm).

The study of coatings obtained with non-vacuum electron-beam cladding of a powdered carbon-containing mixture showed that in most cases coatings are fractured by a brittle mechanism (Figure 2c).

The ability to operate under abrasive wear is one of the most important qualities of the case-hardened materials. Friction tests on fixed abrasive particles simulate the conditions that arise when the monolithic rock interacts with machine parts and structural elements. This type of wear is one of the most severe types of surface loading.

Tribotechnical testing showed that when exposed to fixed abrasive particles, the wear resistance of the cladded layers corresponds to the properties of steel 20 after cementation with quenching and low tempering (Figure 3).

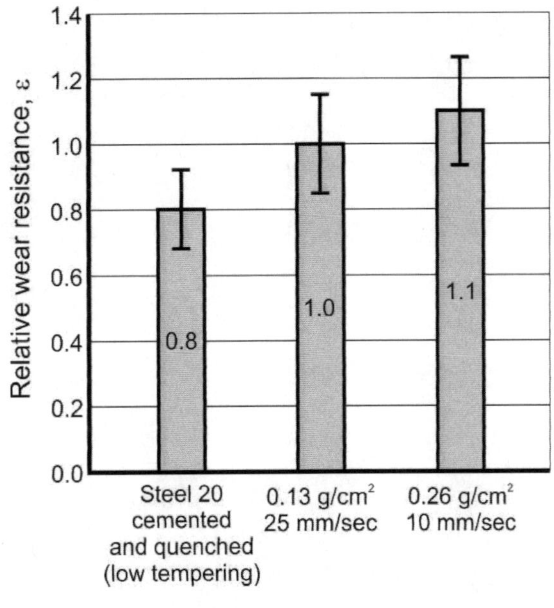

Figure 3. The results of the wear testing.

The hardness of the cladded layers corresponds to the results of Ref. [1] and others in the cases where laser or plasma coatings had hypoeutectic or eutectic structure. Analysis of the microhardness distribution across the thickness of the hardened layer shows that the width of the transition zones within which the microhardness decreases to that of the substrate is ~200…250 μm. Thus, the most significant structural changes, associated with a decrease in microhardness, occur in a layer with a thickness of about 10…15% of the total thickness of the hardened layer.

4. Conclusion

The results of the executed work show that electron-beam cladding of carbon fibers in the air atmosphere is an effective way to obtain hardened high-carbon layers on steels. This technology ensures the formation of high-carbon layers up to 3 mm thick, which have high hardness and wear resistance. High-carbon layers are inextricably linked to the substrate and exert an embrittlement on the starting material, which is revealed during dynamic impact toughness testing.

IOP Conf. Series: Materials Science and Engineering **286** (2017) 012036 doi:10.1088/1757-899X/286/1/012036

Acknowledgments

The research is carried out with the financial support of the Russian Foundation for Basic Research in the framework of a scientific project N 16-33-00141 mol_a.

References

[1] Abboud J H Benyounis K Y Olabi A G Hashmi M S J 2007 *Journal of Materials Processing Technology* **182** pp 427-431

[2] Yao J Zhang Q Gao M Zhang W 2008 *Applied Surface Science* **254** pp 7092-7097

[3] Tushinsky L I Kovensky I M Plokhov A V Sindeyev V I Reshedko P V 2002 *Coated Metal. Structure and Properties of Metal-Coating Compositions* (Berlin; Heidelberg; New-York: Springer) p 457

[4] Cavuslu F Usta M 2011 *Applied Surface Science* **257** 4014-4020

[5] Rahimjanov H M Solonenko O P Alhimov A P Marusin V V 2000 *High-energy treatment of materials* (in Russian) (Novosibirsk) p 425

[6] Golkovskij M G Tushinskij L I Baris N M 2002 *New high-energy technologies of hardening of metals* (in Russian) pp 14-27

[7] Losinskaya A A Drobjaz E A Bataev V A Plotnikova N V Golkovskij M G 2013 *Obrabotka Metallov / Metal Working and Material Science* **4 (61)** 5-11

Microarc synthesis of nanostructured radiation-absorbing coatings on aluminum and titanium surfaces

T A Baranova, A K Chubenko, A E Ryabikov, A I Mamaev, V A Mamaeva and E Yu Beletskaya

National Research Tomsk State University, 36 Lenina ave., Tomsk, 634050, Russia

E-mail: konstantinova.ta9@gmail.com

Abstract. The article is devoted to the investigation of radiation-absorbing properties of the coatings formed on aluminum and titanium surfaces. Microarc oxidation in aqueous solutions was considered as a method of the coating formation. It was shown that the coatings obtained in iron-containing solutions were homogeneous, porous and had high absorption factor of electromagnetic radiation within the terahertz range (up to 70 %), which made this material extremely promising in terms of intensive development of terahertz technologies as well as stealth technologies. Investigations of elemental composition of the coatings were carried out by the X-ray fluorescent method, and the morphology of the surface was represented by the results of raster scanning electron microscopy. Absorption factors were measured using a TDS-21 terahertz spectrometer by «KDP» OJSC.

1. Introduction

The devices that use or produce concomitant electromagnetic radiation (ER) within the microwave range are widely used in modern practice. For example, various electronic devices, communication equipment and radar stations work in this way. The presence of such radiation requires solving the emerging accompanying problems, such as environmental protection of biological objects from its harmful effects, suppression of communication equipment interference, procuring of electromagnetic compatibility of radio electronic equipment, prevention of information leakage through the radio channel, etc. Protection of the equipment and strategic facilities from the radar detection is topical in military activity. Modern advances in electronics production allow the new ranges of radiation with high energies to be used. One of the most promising ranges is the so-called terahertz range. In general, it still remains poorly understood as well as protection from this radiation.

One way of protection from this kind of radiation is using the radiation-absorbing coatings and materials (RAM). The existing materials contain magnetoactive components. Fe, Co and Ni in their metallic forms are widely known and used as well as their mixed oxides – ferrites and organic nanostructural polymeric materials also containing the above-mentioned magnetoactive particles in their dispersed form. These components are either applied in different ways as films to the surface to be protected or integrated by alloyage.

Most modern developed RAM have a sufficient level of ER absorption, but other parameters do not always satisfy the operating conditions. Thus, most of these coatings have low adhesion to the substrate that reduces the operating time. Most of these coatings are also superimposed with several layers on each other to increase the absorption efficiency (sometimes this condition is necessary), so the coatings are quite thick (1-10 mm). Such properties as wear resistance, heat resistance, dielectric strength and others do not always meet the requirements of the operation properties.

In the present paper the production of RAM on aluminum and titanium alloys was carried out in the microplasma mode in aqueous solutions of electrolytes.

2. Materials and methods

In the RAM synthesis the composition and homogeneity of the coatings are of particular importance since the coatings of different composition have different functional properties including the ability to absorb ER. The formation of the coatings in the microplasma mode provides wide opportunities of controlling and changing the composition of the material. Selecting the modes of nanostructured materials formation ensures suitable absorption of ER. The formation of coatings in the microarc mode provides a complex of functional properties in addition to radiation-absorbing ones (wear resistance, corrosion resistance, heat resistance, etc.) due to the variation of the electrolyte compositions and the mode of electrochemical processing. The distinctive features of the obtained coatings are the gradient structure (uniform variation of the concentration of different phases along the thickness), the controlled porosity, the presence of magnetoactive phases in the nanostructured form and the controlled coating thickness (from 10-100 μm).

The formation of the radiation-absorbing coatings was carried out in a homogeneous alkaline electrolyte containing complex compounds of Fe and Co. The homogeneity of such electrolytes was ensured by the ability of these metals to form stable complex compounds with a series of anions (CN^-, $EDTA^{4-}$) even in an alkaline medium without undergoing hydrolysis. Such soluble anion complexes approach the metal surface (which is an anode) during the microplasma process and are inserted into the coating as nanoparticles of the metal while participating in chemical and plasmachemical reactions or form crystal structures of spinels.

The process of microplasma oxidation of the samples was carried out in one step with polarizing voltage of 310 V and 20 minutes of processing time. The samples served as anodes, and plates of stainless steel served as cathodes. The electrochemical cell was continuously cooled during the processing.

The ability of the process to form high-temperature oxide modifications is confirmed either by a number of published sources [1–8] or in the present study.

Based on all of the above, the main purpose of the study is obtaining the coatings that absorb the terahertz radiation range in homogeneous alkaline electrolytes based on anion complexes of iron and cobalt.

3. Results and discussion

3.1. Structural investigations of the coatings

Topographic and qualitative phase analysis of the surface of A1 and A2 samples was carried out using the SEM 515 scanning electron microscope (Philips, Netherlands), and semi-quantitative elemental analysis was performed as well. Secondary or reflected back-scattered electrons were used to investigate the surface structure (Figures 1 and 2).

The presented coatings were dark gray, homogeneous and had developed porous structures with clear outlines of pore boundaries in both cases. However, the pores were small, and this fact was typical for undercoating layers. The phenomenon was indicated also by small thickness of the coatings (15-30 μm). Such result could be explained by complex equilibrium processes established in the electrolyte only over a period of time (a freshly prepared electrolyte was used to form these coatings). Due to the lack of complexation the deposition rate of the components from the electrolyte to the film decreased and this was reflected both in the color of the coating (a notable gray tint appeared with the increase of the deposition rate which indicated a small amount of iron), and in the pore size reducing.

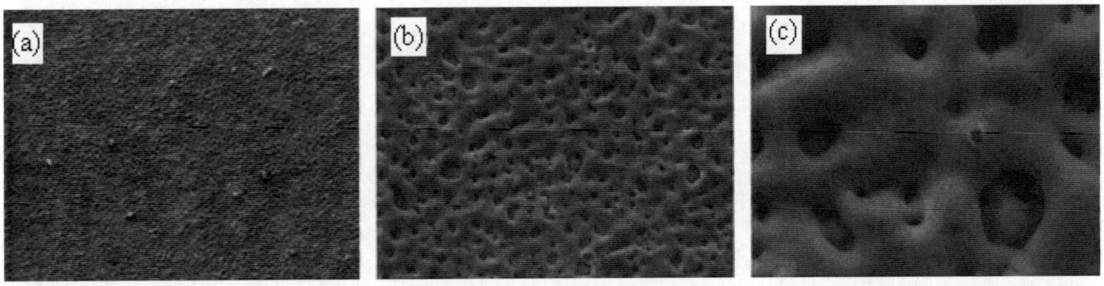

Figure 1. Approximations of the coating surface on titanium using the SEM. (a) – approximation x100, (b) - approximation x500, (c) - approximation x2000.

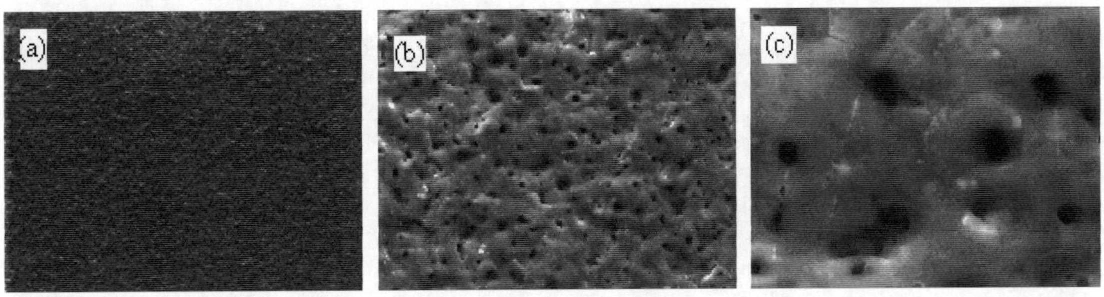

Figure 2. Approximations of the coating surface on aluminum using the SEM. (a) – approximation x100, (b) - approximation x500, (c) - approximation x2000.

It should be noted that the amount and size of pores on titanium substrate was larger than on aluminum one, which was connected with the different behavior of the substrate metal during the microplasma process, different process modes and compositions of the electrolytes. The presence of silicates in the solution for aluminum oxidation reduced the porosity of the coating and made it more smooth and homogeneous.

The presence of magnetoactive elements (Fe and Co) in the form of metallic nanosized particles and in crystalline mixed oxide forms (Fe_3O_4 and $CoFe_2O_4$) was discovered during the X-ray phase analysis. The formation of such phases was possible due to the specificity of the microarc oxidation process. Microplasma discharges occurred during the disruption of the anodic oxide film, and these discharges were characterized by high temperature values (2000-8000 K). The expected mechanism of the process was based on high-temperature dissociation of the anion complexes of magnetoactive particles while approaching the anode surface:

$$[CoEDTA]^{2-} \rightarrow Co^{2+} + EDTA^{2-};$$

$$[Fe(CN)_6]^{3-} \rightarrow Fe^{3+} + 6CN^{-}.$$

Then the metal ions were deposited on the surface of the coating in the form of oxides by the following reactions:

$$Me^{n+} + nOH^{-} \rightarrow Me(OH)_n;$$

$$Me(OH)_n \rightarrow Me_mO_n + nH_2O;$$

$$Me^{n+} + nO^{-} \rightarrow Me_mO_n.$$

Mixtures of the resulting oxides under the influence of high temperature were able to melt, thereby forming crystalline spinel ferrites:

$$Fe_2O_3 + CoO \rightarrow CoFe_2O_4;$$

$$Fe_2O_3 + FeO \rightarrow Fe_3O_4.$$

Some part of those oxides was also deposited on the surface of the metal due to the disruption of the oxide film. The substrate metal was able to reduce Fe and Co from the oxides during that process under the influence of high temperature with forming metallic phase of Fe and Co:

$$Me_mO_n + mAl \rightarrow Al_2O_3 + mMe^{n+}.$$

3.2. Determination of the elemental composition of the coatings

The X-ray fluorescent method was used to determine the elemental composition of the obtained coatings. The X-ray fluorescent sequential wave-dispersive spectrometer Shimadzu XRF-1800 was used in the Material science collective research center of Tomsk State University.

Investigations of the phase composition were carried out using PDF 4+ databases with POWDER CELL 2.5 full-profile analysis program.

The results of elemental analysis are presented in Table 1.

Table 1. The results of elemental analysis of the coatings on titanium and aluminum substrates.

Substrate	Ti (% wt)	Al (% wt)	Fe (% wt)	Co (% wt)	O (% wt)	Si (% wt)
Al	-	20.1	6.9	0.02	48.6	19.9
Ti	38	0.07	6	2.1	41.7	0.06

One could see from the obtained data that magnetoactive Fe and Co had been inserted into the coatings as a result of the microarc processing. Speaking of Fe, it should be noted that despite the difference in the processed metals, modes and compositions of the electrolyte, the Fe percentage in those two coatings was quite close (the difference was about 0.9 %). This fact probably indicated that the potassium hexacyanoferrate (III) used as a Fe-containing substance acted well in the electrochemical process. This complex at pH 7–10 was really stable in the solution, and the complex salt completely dissociated according to the equation:

$$K_3[Fe(CN)_6] = 3K^+ + Fe(CN)_6^{3-}.$$

The anion Fe complex subsequently participated in the transformations described above.

As for Co, its percentage in the coatings sharply differed depending on the oxidized metal. The reason was, first and foremost, in different compositions of the electrolytes. A miscalculation happened during the development of electrolytes for oxidation, namely, the presence of Co^{2+} and $Fe(CN)_6^{3-}$ in the solution that gave an insoluble deposition was not taken into account:

$$3Co^{2+} + 2Fe(CN)_6^{3-} = Co_3([Fe(CN)_6])_2.$$

The insoluble deposition was discovered after preparation of both electrolytes.

The electrolyte contained Trilon B (Na_2H_2Y) during the microarc oxidation of titanium, and Trilon B partially bound Co^{2+} ions into a stable chelate complex thus increasing the percentage of Co ions soluble form, and the resulting CoY^{2-} complex anion took part in the electrochemical process through its migration to the anode. The electrolyte for the microarc oxidation of aluminum did not contain Trilon B, so the percentage of Co in the solution was apparently rather small, that resulted in the elemental composition of the obtained coating. The percentage of the potassium hexacyanoferrate (III) was either much higher (15 g / l versus 1 g / l cobalt acetate in the electrolyte for the microarc oxidation of aluminum) or equal (5 g / l $[Fe(CN)_6]^{3-}$ and CoY^{2-} in the electrolyte for the microarc oxidation of titanium), so that Fe was inserted into the coatings and was quite close in percentage values.

3.3. Measurements of the reflection factors

Measurements of the reflection factors for samples A1 (coating on the titanium substrate) and A2 (coating on the aluminum substrate) were carried out using a TDS-21 terahertz spectrometer within the frequency range from 428 to 973 GHz. Two oscillators based on backward-wave tubes (BWT) were used as radiation sources.

The absolute value of the reflection factor was calculated as the ratio of the power received by the detector when the sample was in the holder to the power at the detector when the sample was absent in the path, and the reflection emerged from the reference metal mirror.

The dependencies of the reflection factors on frequency were obtained on the basis of the measurement results (Figures 3 and 4).

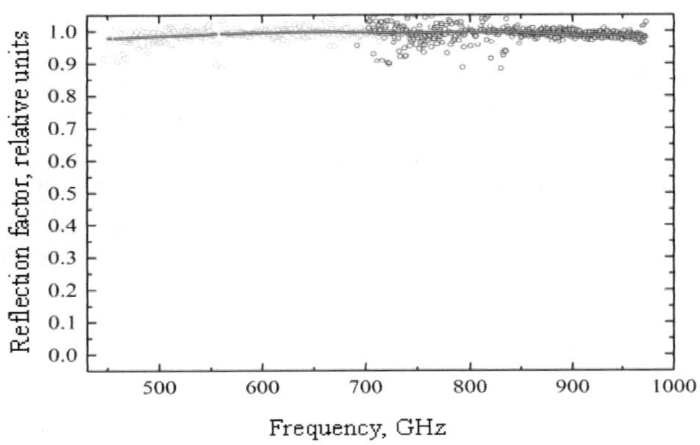

Figure 3. The dependence of the reflection factor on frequency for A1 sample (aluminum).

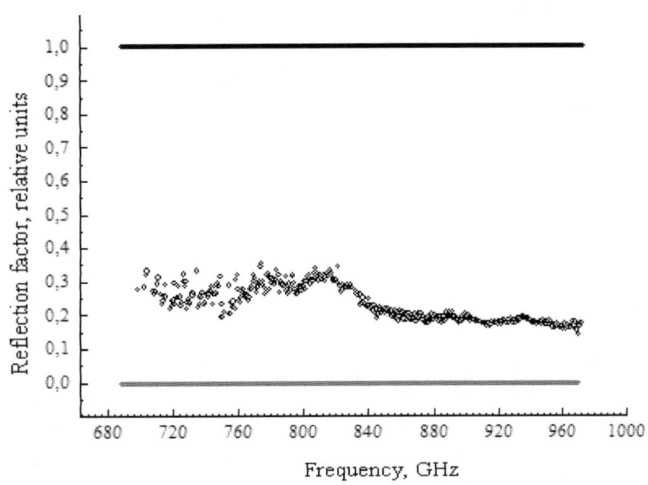

Figure 4. The dependence of the reflection factor on frequency for A2 sample (titanium).

One of the most important tasks in creating RAM was to achieve high absorption factors and low reflection factors simultaneously within a wide range of wavelengths.

The obtained dependencies showed that in the first case the reflection factor decreased rather weakly, however, for titanium this decrease amounted to 70-80 % for the given frequency range. This behavior of the coatings with respect to the given frequency range could be explained by the different percentage of magnetoactive elements, in particular, different percentage of Co and the degree of homogenization. The presence of Co (2 % wt) in combination with Fe gave a wider range of various magnetoactive phases (metallic Co, ferrite $CoFe_2O_4$), which apparently affected the decrease of the reflection factor.

4. Conclusions

Radiation-absorbing within the terahertz frequency range coatings were obtained on aluminum and titanium substrates. Fe and Co were chosen to serve as magnetoactive components ensuring the absorption of ER, and these metals were inserted into the electrolyte in the form of soluble anion complexes ($Fe(CN)_6^{3-}$ and CoY^{2-}). Investigations showed that the magnetoactive particles were included into the coatings during the microplasma process. The obtained percentage of Fe was 6 % wt, the obtained percentage of Co was 2 % wt, and the measured reflection factor within the frequency range 450-1000 GHz was reduced by 15 % for the coating on the titanium substrate. The obtained percentage of Fe was 6.9 % wt, the obtained percentage of Co was 0 % wt for the coating on the aluminum substrate, and the measured reflection factor did not change and was 100 %. Such result could be explained by the complex character of the processes occurring at the stage of the electrolyte formation.

On the one hand, the formation of complex compounds was an equilibrium process characterized by quite long settling time (for Co). On the other hand, some features of the interaction of individual components (Co^{2+} + $Fe(CN)_6^{3-}$) were not taken into account. This phenomenon affected the percentage of the required elements in the solution, and, consequently, the percentage of these elements in the coatings. At the same time, the electrochemical process of the coating formation on aluminum in the presence of EDTA showed the harmful effect of this component on the formed coating during the cycle of experiments on model samples of pure aluminum. Aluminum that was dissolving during the microplasma process, on the one hand, bound into poorly soluble compounds ($Al(OH)_3$, $AlPO_4$, Al_2SiO_3, etc.) and deposited on the surface of the metal, which contributed to the formation of the coating. On the other hand, the released Al^{3+} ions were able to bind into soluble stable compounds and to be ejected from the reaction zone into the solution since their concentration in the said solution was extremely small. This process was exactly the one that caused etching, since there was no solid phase deposited on the surface of the substrate. $Al\text{-}EDTA^-$ complex was characterized by high stability ($\lg\beta = 18$), therefore it could be one of the mechanisms of the negative influence of such complexing agent on the process of the coating formation.

Acknowledgement

The results were obtained in the framework of the state assignment by the Ministry of Education and Science of Russian Federation, project № 16.9621.2017/8.9

References

[1] Mamaev A I and Mamaeva V A 2005 *High-Current Processes in Electrolytic Solutions* (Novosibirsk: The Publishing House of the Siberian Branch of the Russian Academy of Sciences) p 254

[2] Mamaev A I, Dorofeeva T I, Emelyanova E Y, Mamaeva V A and Budnitskay Y Y 2012 *Izv. Vyssh. Uchebn. Zaved.* **55** 107–14

[3] Emelyanova E Y, Dorofeeva T I, Mamaev A I, Mamaeva V A and Budnitskay Y Y 2011 *Izv. Vyssh. Uchebn. Zaved.*, **54** 132–8

[4] Gordienko P S and Rudnev V S 1999 *Elektrohimicheskoe formirovanie pokrytii na aluminii i ego splavah pri potentsialah iskreniya i proboya* (Vladivostok: Dalnauka) p 233
[5] Chernenko V I, Snezhko L A and Papanova I I 1991 *Poluchenie pokrytii anodno-iskrovym elektrolizom* (L.: Himiya) p 128
[6] Kazmina O, Dushkina M, Suslyaev V and Semukhin B 2014 Porous material for protection from electromagnetic radiation. *AIP Conference Proceedings* **1623** DOI:10.1063/1.4898927
[7] Dey A, Mukhopadhyay A K, Gangadharan S, Sinha M K and Basu D 2009 Development of hydroxyapatite coating by microplasma spraing. *Materials and manufacturing processes* **24** 1321–30
[8] Xin S, Jiang Z, Wang F, Wu X and Zhao L 2001 Effect of current density on Al alloy microplasma oxidation *J. Mater. Sci. Technol* **17** 657–60